EAGLE DAYS

Dr Victoria Taylor is an award-winning aviation historian who specialises in the operational capabilities of the Luftwaffe in the period of National Socialism, and the Royal Air Force.
She has contributed to numerous popular history magazines and sits on the Advisory Board for the cross-party Spitfire AA810 restoration project in the House of Lords. She is an assistant editor for the Royal Aeronautical Society's Journal of Aeronautical History.
Eagle Days is her first non-fiction book.

EAGLE DAYS

LIFE AND DEATH *for the* LUFTWAFFE *in the* BATTLE OF BRITAIN

VICTORIA TAYLOR

An Apollo Book

First published in the UK in 2025 by Head of Zeus Ltd,
part of Bloomsbury Publishing Plc

Copyright © Victoria Taylor, 2025

The moral right of Victoria Taylor to be identified
as the author of this work has been asserted in accordance with
the Copyright, Designs and Patents Act of 1988.

All rights reserved. No part of this publication may be: i) reproduced or transmitted in any form, electronic or mechanical, including photocopying, recording or by means of any information storage or retrieval system without prior permission in writing from the publishers; or ii) used or reproduced in any way for the training, development or operation of artificial intelligence (AI) technologies, including generative AI technologies. The rights holders expressly reserve this publication from the text and data mining exception as per Article 4(3) of the Digital Single Market Directive (EU) 2019/790.

9 7 5 3 1 2 4 6 8

A catalogue record for this book is available from the British Library.

ISBN (HB): 9781804549995
ISBN (E): 9781804549971

Maps © Jeff Edwards
Plate section images with permission
of Süddeutsche Zeitung Photo.

Typeset by Siliconchips Services Ltd UK

Printed and bound in Great Britain by
CPI Group (UK) Ltd, Croydon CR0 4YY

Bloomsbury Publishing Plc
50 Bedford Square, London, WC1B 3DP, UK
Bloomsbury Publishing Ireland Limited,
29 Earlsfort Terrace, Dublin 2, D02 AY28, Ireland

Head of Zeus Ltd
5–8 Hardwick Street
London EC1R 4RG

To find out more about our authors and books
visit www.headofzeus.com

For product safety related questions contact productsafety@bloomsbury.com

To the 'Rents' – love from Daught x

Contents

Map . . . . viii–ix
Luftwaffe Ranks and RAF Equivalents x
Introduction 1

PART I – AGITATION

1 Corps of Vengeance 11
2 The Line Was Dead 29
3 How Insane Is This War! 43
4 A Lobster Dinner 59
5 Poor *Poilu*! 73

PART II – CONFRONTATION

6 A Future Without Albion 91
7 Flags Up! Hearts Up! 107
8 The Overture 123
9 A Very Bad Piece of Work 143

PART III – CONSOLIDATION

10 The Hour of Judgement 163
11 Don't Talk so Loud! 183
12 Like a Thunderbolt He Falls 201
13 You *Dummkopf*! 221

PART IV – DIVERSION

14 Great Dark Bloodstains 241
15 Dante's *Inferno* 261
16 Flying, Sleeping, Eating, Waiting 279
17 The Other Faculty 297
18 Better Liars than Flyers 317

Conclusion 335
Acknowledgements 345
Further Notes & References 349
Index 401

Important Luftwaffe airbases in France, Belgium and the Netherlands, 1940

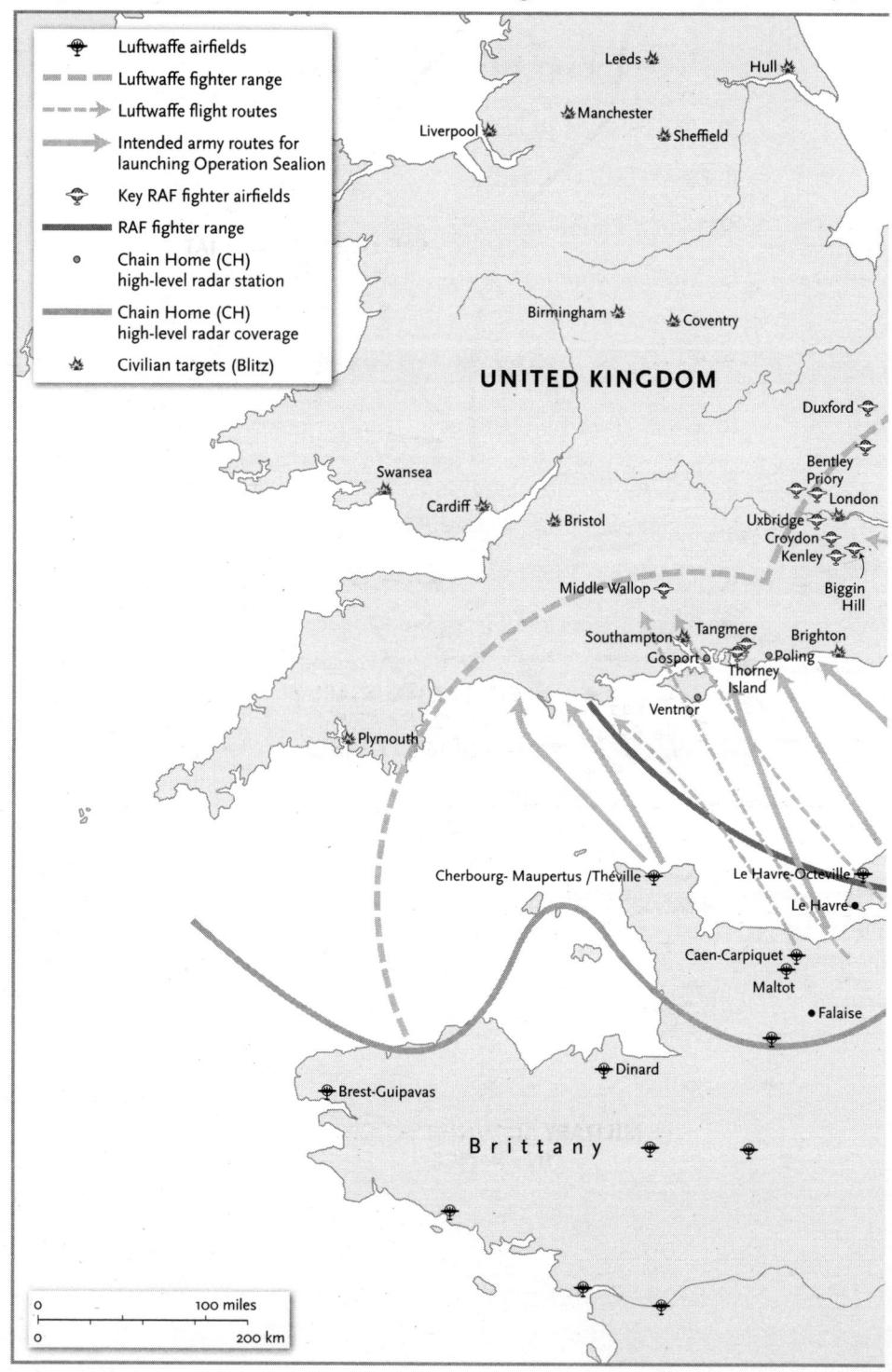

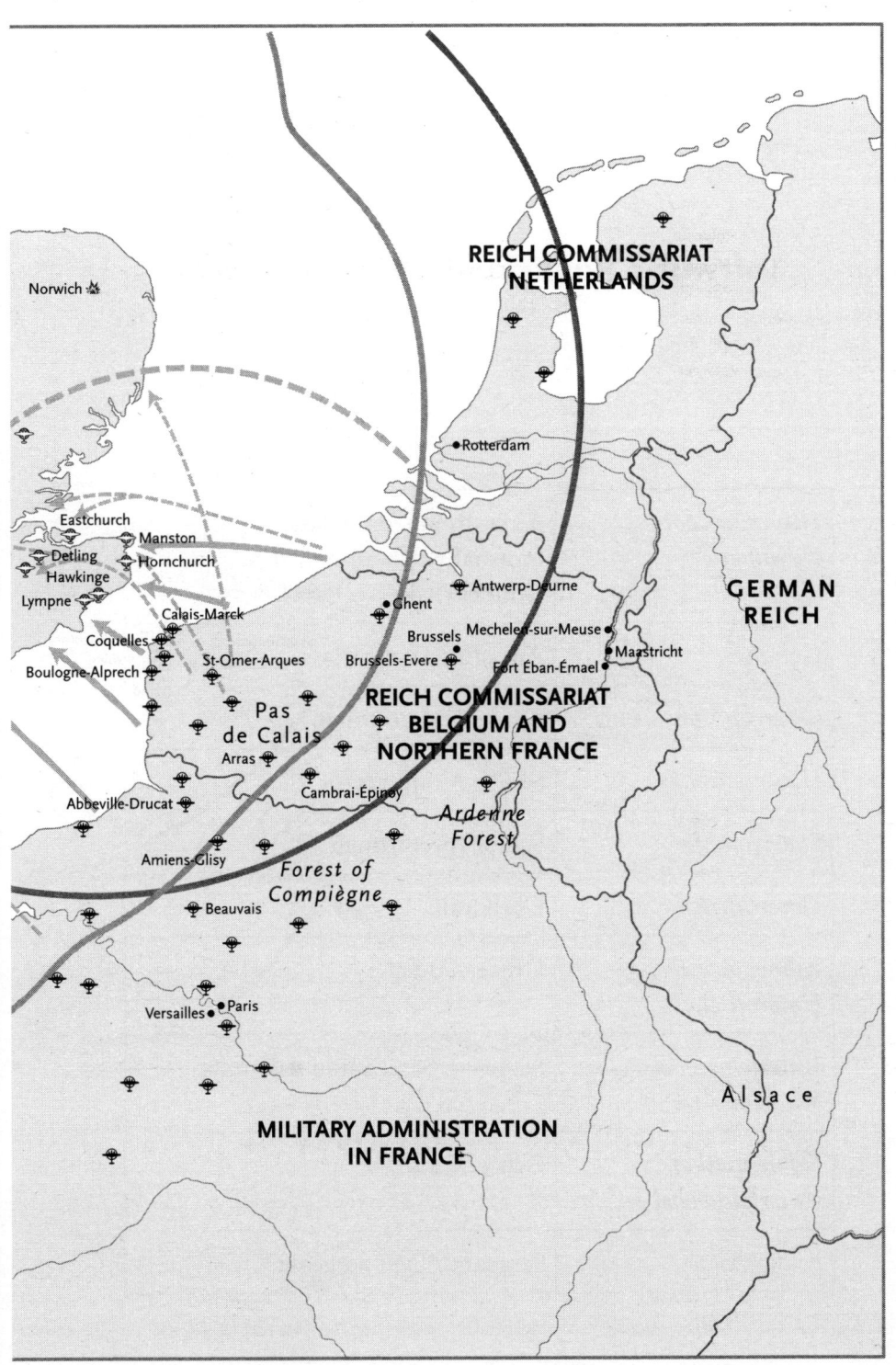

Luftwaffe Ranks and RAF Equivalents

(In Ascending Order)

Flieger; Soldat; Kanonier	Aircraftman 2nd Class; Soldier (normally flak, technical & admin); Gunner (flak)
Gefreiter	Aircraftman 1st Class
Obergefreiter	Leading Aircraftman
Hauptgefreiter	Senior Aircraftman
Stabsgefreiter	Senior Aircraftman
Unteroffizier	Corporal
Fahnenjunker-Unteroffizier	Officer Cadet
Unterfeldwebel; Wachtmeister	Sergeant; Sergeant in flak units
Fahnenjunker-Unterfeldwebel	Officer Cadet
Feldwebel	Sergeant/Flight Sergeant

LUFTWAFFE RANKS AND RAF EQUIVALENTS

Fahnenjunker-Feldwebel	N/A
Oberfeldwebel	Flight Sergeant
Fahnenjunker-Oberfeldwebel	Acting Pilot Officer
Fahnenjunker-Stabsfeldwebel	Between Flight Sergeant and Warrant Officer
Stabsfeldwebel	Warrant Officer
Leutnant	Pilot Officer
Oberleutnant	Flying Officer
Hauptmann	Flight Lieutenant
Major	Squadron Leader
Oberstleutnant	Wing Commander
Oberst	Group Captain
Generalmajor	Air Commodore
Generalleutnant	Air Vice-Marshal
General der Fallschirmtruppe; Flakartillerie; Flieger; Luftnachrichtentruppe; Luftwaffe	Air Marshal (of the Paratroopers; Flak; Aviators; Air Signals Troops; Luftwaffe)

Generaloberst	Air Chief Marshal
Generalfeldmarschall *Reichsmarschall*	Hanfred Schliephake stated that *Generalfeldmarschall* was the equivalent to an Air Chief Marshal – see H. Schliephake, *The Birth of the Luftwaffe* (Chicago: Henry Regnery, 1971), 80. However, as *Reichsmarschall* was an unprecedented rank created for Hermann Göring, *Generalfeldmarschall* is sometimes listed as equivalent to Marshal of the RAF.

Introduction

It was 26 August 1940: *Flieger* Franz Dahm was restless and enervated. Casting a worried eye to the skies, he searched fiercely for the reassuring glints of grey that would signal the homecoming of his comrades. Next to him, the groundcrew of his Luftwaffe bomber *Staffel* ('squadron')* paced back and forth like expectant fathers. 'Where are they?' they kept asking nervously. 'They should have actually been back by now!'[1] 'The industrious guard of our ground staff, whose silent heroic deeds are not spoken of, and on whose tireless work so much depends, are waiting in agony for "their" aircraft,' Dahm mused tensely. 'The *Staffel* is visiting England, but all the hearts and thoughts of the flight mechanics who stayed here flew with them.'

Yet, it was not just news of their safe return that he was eager to obtain: Dahm was a journalist from a Luftwaffe war

---

* This book uses the German term *Staffel* instead of 'squadron', as a Luftwaffe *Staffel* (nine to twelve aircraft) was a slightly different size to an RAF squadron during the Battle of Britain, which had a theoretical strength of sixteen aircraft – twelve of which were required to be operational. *Gruppe* ('group') is also used instead of 'Group' because a *Gruppe* was much smaller than an RAF Group. See J. Holland, *The Battle of Britain: Five Months Which Changed History* (Transworld, 2010) [Kindle Edition], loc. 520–534.

correspondence company. Pencil firmly in hand, and paper clenched in his white-knuckled fist, he had started writing a crowing battle report for the Nazi press. The reporter inside was poised to sniff out stories of heroism, stoicism and – perhaps most desirable of all – progress. Crushing the British Royal Air Force in this new *Luftschlacht um England* ('Air Battle for England') ahead of *Unternehmen 'Seelöwe'* ('Operation Sealion'), a proposed amphibious invasion of Great Britain, was proving to be an irksome anomaly in the Wehrmacht's triumphant track record.

Finally, Dahm spotted the surviving German bombers emerging in the distance. Washed over with relief, the jittering journalist strode across the airfield to get an inside scoop on this intense Luftwaffe raid over RAF Biggin Hill. 'I catch hold of a pilot *Unteroffizier*, who flew with us on the left wing of our formation in this major attack on the pirate island,' Dahm reported breathlessly. The *Unteroffizier* subsequently informed him that 'Tommy has had enough... today, the boys didn't even dare approach us anymore. We completed our combat mission of smashing Biggin Hill airfield as safely and reliably as a practice flight at the home airbase.'

'With that,' as Dahm's report concluded, 'angry at such a shabby opponent, [the *Unteroffizier*] takes off his fur boots, throws the flight suit over his shoulder, and trudges off to the accommodation.' Dahm's account was written for propaganda purposes, and therefore must be taken with a grain of salt, but it certainly encapsulates the Luftwaffe's growing belief that Britain was facing its swan song. As the decorated Luftwaffe fighter ace Hans-Ekkehard Bob confirmed, 'We, the German fighter pilots, really had the impression that the British fighters might have been decimated considerably, since during the late summer of 1940, we had little contact with the enemy, although we flew up to London daily.'[2]

Propaganda reports like Dahm's were hungrily devoured back home: thousands of Third Reich citizens were pinning their hopes on the '*Führer*'s Hammer', as the Nazi press dubbed the Luftwaffe, to end the war against the last official bastion of Allied resistance in Europe. Then, fate – or, more precisely, *Reichsmarschall* Hermann Göring – intervened. Within just two weeks of Dahm's report, Adolf Hitler was forced to postpone indefinitely the amphibious invasion of Britain. Göring, the Luftwaffe's bumptious commander-in-chief, had unexpectedly changed his tactics. Instead of his German air force focusing solely on the RAF, it was now going to pound British cities, industries and civilians – starting the 'Blitz'. Allowing the groggy RAF to recuperate and revitalize itself enough to render Operation Sealion impossible by mid-September 1940, the Luftwaffe's momentum was irretrievably squandered.

The rest, as they say, is history – but it is a history with which Britain remains unshakeably enraptured eighty-five years later.

The Battle of Britain is the most famous aerial campaign in British history, so it is tempting to think we already know the Luftwaffe's role in the story well. We are hardly short of literature that details German aircraft and technology; elite units and fighter aces; tactics, strategy and its fractious leadership during the battle. Nevertheless, this chronic hyperfocus on the campaign's operational minutiae means that the Luftwaffe's story is often relegated to one of three oversimplified narratives within British historiography. The first, and perhaps the most pervasive, is that of the menacing Goliath to the RAF's plucky David.

To Britain, the air campaign was a defining 'backs-against-the-wall' moment where the island proved that it may have been isolated, but it was not down and out. German awareness of this national narrative can be seen with the editor of *Militär & Geschichte* magazine writing in March 2020: 'It is no coincidence that the British, in the Brexit discussions, like to draw parallels to a broad turning point in their more recent national history: to the Second World War, particularly to the years 1940/41, when the "Battle of Britain" had to be fought against the Third Reich and they were largely left to their own devices.'[3]

For the Germans, however, the *Luftschlacht um England* – which, unlike the British timeline of the campaign, also encompassed the Blitz – is from 'a time which cannot be exemplary for us',[4] in the words of former defence minister Ursula von der Leyen. This is not to say that the battle is unknown in Germany, with the subject garnering attention from news media such as *Die Welt*, *Der Spiegel* and *Stern* in the last few years.[5] Yet, despite a handful of commendable forensic studies from historians like Peter Cronauer, Jens Wehner, Falko Bell, Hans Peter Eisenbach and Carolus Dauselt, the Germans understandably lack the appetite to document the Luftwaffe's experiences in the campaign with the same ardour that characterizes much of the British literature on the subject.[6]

The majority of writing on the Luftwaffe in the Battle of Britain, then, falls to the British. But a second trope that British writers are frequently guilty of regurgitating is representing the Luftwaffe's esteemed 'knights of the sky' – its famed *Jagdflieger* (fighter pilots) – as being in an honourable joust with their equals in the RAF's Fighter Command. There are three main issues to this approach. The first is that it excludes parts of the narrative that do not fit in with the swashbuckling dogfights, from long stretches of monotony to the strafing of parachuting airmen and air-sea rescue vessels.

The second issue is that it encourages a hyperfocus on the Luftwaffe's elite – such as decorated generals and illustrious fighter aces – which neglects the experiences of more 'regular' servicemen across all ranks and roles during the *Luftschlacht um England*. The most troubling concern, however, is that the 'knights of the sky' trope often drives the narrative in Britain that Luftwaffe airmen were 'just like us', as Flight Sergeant Ian Walker once wrote of two captured German airmen in his diary on 24 August 1940.[7] This premise often had benign and even benevolent origins, where British airmen forged a mutual understanding with their German counterparts over the sacrifices that both sides were making for their respective nations.

But what do we mean by 'just like us'? We likely mean courageous, scared, honourable, skilful, flighty – *human*. And, in many cases, the Luftwaffe's men showcased some or all of those attributes. They had similar motivations of family and friends; of comrades and duty; of longing to be back at home, no longer sullied by the horrors of war. Many of them wanted to believe, as the famous Luftwaffe fighter ace Johannes Steinhoff once insisted after the war, that 'We came after having conquered Europe, as young men, fighting sportsmanlike.'[8]

Indeed, we should not entirely invalidate the shared experiences between the British and German fighter pilots, with their wartime acts of chivalry and open-minded attitudes towards one another doing much to form the basis of enduring post-war reconciliation in Europe. But to say that they were 'just like us' is to underplay the significant culpability that the Luftwaffe wielded in enabling a genocidal political regime to realize its rapacious territorial ambitions; it also covers up the atrocities committed by a small but significant minority of the Luftwaffe within the Nazi-occupied territories before the Battle of Britain.

Certainly, serving one's country and wanting to fly was a universal motivator among aviators. Yet, the Luftwaffe fighter ace Günther Rall readily admitted that 'We accepted the activities of the National Socialists as ugly side effects of a policy that had so far kept all its lofty promises.'[9] In other words, as a Luftwaffe flight instructor put it after the war, 'They served the devil to conquer heaven.'[10] Willingly becoming Göring's 'corps of vengeance',[11] as the commander-in-chief christened the organization to a group of junior Luftwaffe officers in 1936, meant that the Luftwaffe gained its strength because of political spite, defiance and, eventually, the desire for retribution.

After the First World War, the Treaty of Versailles had completely stripped Germany of its military and naval air forces. This ensured that the *Schwarze Luftwaffe* ('Shadow Luftwaffe'), which secretly emerged shortly afterwards in the Weimar Republic, had an enduring chip on its shoulder that the RAF – forged from the Royal Flying Corps as it edged towards victory in the First World War – did not. Then, under the Nazis, a blazing *lex talionis*, or 'eye for an eye' principle, was imbued within their newly consolidated Luftwaffe that may have occasionally softened upon personal contact with their respected British foes, but nevertheless fuelled every sortie the German air force flew against the island in the Second World War.

To claim its airmen were 'just like us', then, is to misunderstand what Luftwaffe personnel were willing to fly and die for in the Battle of Britain – or at least, what they were expected to do.

Finally, the third reductive presentation of the Luftwaffe is when British historians examine it from a highly clinical perspective. In certain studies of the campaign, the German air force is seen as pieces on a chessboard – a type of aircraft here, a *Staffel* there – where the core of the players' mindset remains largely unconsidered. Each individual move is often mechanically analysed without pausing to consider what the Luftwaffe was truly playing *for*, or without giving them much definition beyond the popular comic book image of a ruthless Nazi *Übermensch*.[12]

Historical attention has frequently been paid to *Reichsmarschall* Göring and his exasperated commanders, but what about the non-commissioned officers and the other ranks, who made up the German air force in far greater numbers? Did *they* regard pressuring the British war cabinet to sue for peace as the overarching goal – or was paving the way for the Sealion invasion at the forefront of their minds? *General der Flieger* Paul Deichmann dismissively claimed after the war that 'The famous Sealion was only an idea.'[13] But, as British air raids intensified over Nazi Germany on 30 August 1940, a low-ranking Luftwaffe signalman of the Air Signals Regiment 35 ruthlessly promised his wife that 'if our house is hit, every house in England that I can lay my hands on will burn'.[14]

Indeed, how did perceptions of the campaign – and the prospect of an invasion – differ for these oft-forgotten Luftwaffe branches: the air signals units, the anti-aircraft units, the paratroopers that the flying arms left behind collecting dust? Then, when flying multiple daily sorties for months on end, what spurred on the Luftwaffe's airmen of all ranks and designations to keep venturing over the 'pirate island': *Führer* or Fatherland, revenge or desperation? As the final victory over Britain became ever more elusive, how did they view the inconclusive campaign? Above all, what did the Battle of Britain mean for the entire Luftwaffe and Nazi Germany at large – and how seriously did they think they were going to win?

For as long as these questions are unanswered, our understanding of how the Battle of Britain unfolded will remain lopsided. The true objectives, operational culture and battle cohesion of the Luftwaffe in the campaign cannot be fully comprehended on a military level if its combat mentality, experiences and morale on a more personal level are not also understood. Eighty-five years after the Battle of Britain, a comprehensive history on everyday life and death for the entire Luftwaffe during its ten-month *Luftschlacht um England* is yet to be penned. Thus, *Eagle Days: Life and Death for the Luftwaffe in the Battle of Britain* rectifies this by providing a typical operational history of the German air force during the iconic campaign, but adopting a holistic approach that encompasses the rank and file of the infamous organization.

Eagle Days traces the historical, societal and political conditioning that the men of the Luftwaffe were exposed to regarding the British – as well as determining how its recruitment process, military training and previous campaigns affected their mindset heading into the battle. This panoptic view of the Luftwaffe should also be extended to the aerial campaign as a whole. The Battle of Britain was far more than sun-soaked tussles over Kent: it was a bitter transnational struggle in which both sides were fighting for *their* view of European security and freedom, though with vastly different intentions. The offensive action taken by its fighter pilots and *Kampfflieger* ('bomber crews') in the skies above Britain had a direct bearing on the morale of not just the Luftwaffe personnel who concurrently protected the Reich from British air raids, but also of the German public.

Furthermore, the campaign also inspired the German and Austrian airmen of the future, who watched the Luftwaffe's sorties in the *Luftschlacht um England* with amazement – whether they had already signed up for the German air force, or were young schoolboys desperate to emulate their flying heroes once they became of age. Drawing upon everything from wartime Luftwaffe field letters, diaries, combat reports, operational memoranda and post-war memoirs, to Nazi speeches, newspapers, magazines and newsreels, this book goes deeper than the dogfights to explore the fears, ambitions and convictions that gripped the men who rained death on Britain amid its 'Finest Hour'.

Eagle Days achieves this because it never loses focus on the complex human condition. We fly alongside the queasy young Luftwaffe recruits on their first perilous raids over Britain; we witness the agitation of German paratroopers anxious to be dropped over the country. Equally, we hear from the family men who ache to be in the arms of their wives once more; we examine the abhorrent Nazi ideology spurring on some German flying personnel to strike at Britain's 'Jewish plutocrats'. Thus, the Luftwaffe is observed as a sum of its parts – from the courageous aviators who wanted to fly, to the vengeful Nazi combatants who saw the subjugation of others as their birthright, and to those who fell somewhere in between.

This, in turn, transforms the Luftwaffe's historical role during the RAF's 'Finest Hour' from a cartoonish antagonist to a multidimensional, flawed yet formidable opponent – making it even more remarkable and crucial that Fighter Command managed to prevail against it. Taking off from Anglo–German relations in the inter-war period, roaring full throttle towards the *Luftschlacht um England*, before landing in German post-war memory of the momentous campaign, *Eagle Days* tells the Luftwaffe's story in the Battle of Britain through a far more unique lens than its bombsights and gunsights alone.

PART I

AGITATION

1

Corps of Vengeance

As his icy blue eyes flashed maliciously within his porcine face, it was evident that Hermann Göring's tripwire had been triggered; Marshal of the RAF Sir John Salmond knew he could do nothing but exasperatedly watch the explosion. A representative of Imperial Airways, he had merely hoped to discuss the working relationship between the German airline Lufthansa and British airline pilots, whom the former had been training in blind flying.[1] Accordingly, Salmond had been personally received in Berlin during the early spring of 1935 – now, though, he looked towards the accompanying Reich Air Ministry* interpreter, *Oberstleutnant* Ulrich Kessler, to make sense of the portly windbag in front of him. Kessler's staid face wavered slightly as he confirmed the redirection of the conversation to a far more uncomfortable subject: the Treaty of Versailles' enduring ban on German military aviation.

Salmond quickly interjected that Germany had already been theoretically granted equity back in December 1932. The affronted Bavarian replied that the Germans did not require other nations to bestow them with aeronautical parity. With a

---

* *Reichsluftfahrtministerium* (RLM) in German.

supervillain's aside, however, he added that the 'little air force' of his own that he had been constructing was the real equity that he had hankered after for the past fifteen years. Kessler froze upon hearing Göring's confession, initially hesitant to convert it from the German swirling in his brain to the incriminating English – but the future commander-in-chief of the Luftwaffe was adamant that Kessler should pass on his exact words. 'A little one?' Salmond enquired, raising one suspicious eyebrow. Göring shrugged. 'Well, I would call it little.'

Defeated by his bone-deep proclivity for gloating, he sheepishly admitted to Hitler straight afterwards that the game was up: the British knew about the Luftwaffe. Yet, a month before Salmond's visit, Göring had promised his young airmen that 'The whole Luftwaffe would soon be brought out into the open.'[2] Already, his 'little air force' boasted 900 flying officers and 300 anti-aircraft officers, who commanded around 17,000 men.[3] As the Luftwaffe fighter ace Adolf Galland testified, 'In February 1935, Göring put in an appearance at Schleissheim and to everyone's delight showed off the first Luftwaffe uniform.'[4] Johannes Steinhoff reminisced of the period that 'We were proud. We were pampered. We were the "guarantors of the future". Göring's air bases were conceived in typical Third Reich style – very modern and comfortable, almost luxurious.'[5]

Such was the success of the commander-in-chief's blandishments that, when Göring privately claimed to a group of Luftwaffe junior officers in 1936 that 'You will one day be my corps of vengeance,' Herbert Rieckhoff – later the *Geschwaderkommodore** – of the *Kampfgeschwader* ('bomber wing') KG 30 confirmed there 'probably wasn't a single person in attendance who wasn't totally motivated to give his utmost in support of this regime'.[6]

That same year, their brimming enthusiasm would be truly put to the test when Göring deployed his new air force in the Spanish

---

* Like a *Staffelkapitän* ('Squadron Captain'), *Geschwaderkommodore* ('Wing Commodore') was a command position rather than a rank.

Civil War (1936–9). Before we advance towards detailing the Luftwaffe's story in the Battle of Britain, we must neither undersell the immense operational advantage and experience it accumulated in the immediate years leading up to the campaign, nor lose sight of how German perceptions of their British rivals were shaped on the long, arduous road to 1940.

After Hitler formally recognized General Francisco Franco's far-right Spanish Nationalist government in late September 1936, a dedicated expeditionary force consisting of Luftwaffe and German *Heer* ('army') personnel – the *Legion Condor* (Condor Legion) – was mobilized on 29 October 1936 to assist Franco's forces against the Second Spanish Republic. The Germans tried to keep the existence of their Condor Legion a fiercely guarded secret, although Britain was already aware of their collusion with Franco by September 1936.[7] As Rieckhoff noted, only the crème de la crème of the Luftwaffe were selected for the intervention in Spain, known as *Übung Rügen* ('Rügen Exercise')*:

> They had to report to General [Helmuth] Wilberg's liaison office in Berlin. There, they were given an ID card stating that they had been released from the German Wehrmacht, and were given behavioural measures – they were not allowed to carry out any correspondence, not to inform their wives or parents about their command, nor to talk about Spain after their return – and were then shipped to the Iberian Peninsula without being told their destination.[8]

The depth of this deception could be seen in how the German airmen making their way to Spain donned civilian clothes and were given elaborate false backstories to explain their journey. Galland later testified that his own group of Luftwaffe pilots

---

* The intervention had initially started with *Unternehmen Feuerzauber* ('Operation Magic Fire'), a *Deutsche Lufthansa* airlift to assist the Nationalists on 29 July 1936. *Winterübung Rügen* ('Winter Rügen Exercise') referred to a large maritime transfer of German volunteers to Spain – but the entire intervention became known as *Übung Rügen*.

were to say that 'they were all on a "Strength Through Joy"* leisure trip to Genoa'⁹ if their ship was stopped and searched. Upon arrival in Spain, the Legionnaires were kitted out with Spanish uniforms and Luftwaffe aircraft were repainted with the Spanish saltire (*crux decussata*) to maintain the façade.

The Legion started clocking up combat flying hours on the aircraft that would later form the backbone of the Luftwaffe in the Battle of Britain. Three hundred Heinkel He 111 B-1 bombers had been ordered by the Reich Air Ministry, with the first batch entering Luftwaffe service in January 1937; the aircraft's first operational deployment took place two months later during the Battle of Guadalajara.¹⁰ Walter Lehweß-Litzmann – who later became the *Kommodore* of a Luftwaffe bomber wing – enviously devoured letters from his brother Peter, who had been transferred from the bomber wing KG 55 in Ansbach to the Condor Legion. Peter reported that:

> A lot of [He] 111s have already fallen on their faces, not due to enemy action, but due to some kind of engine malfunction coupled with stupidity. I have already developed an unusual level of bad luck, as something always happens to me when I have a new machine, i.e. one that has just been partially overhauled. Once, when I got one from Seville, I landed quite peacefully in Burgos, rolled out, then the landing gear broke on the right. Then, as a result of the ground loop, the left one, the spur, the pot, also broke. Everything is a wonderful mess.¹¹

Nevertheless, flying such aircraft enabled Luftwaffe mechanics to fine-tune some of the shortcomings that could only be revealed under the strain of operating within a warzone. The Dornier Do 17 bomber had also entered frontline Luftwaffe service in early 1937, though it was not as numerous as the He 111.

---

* '*Kraft durch Freude*' ('Strength Through Joy', KdF) was a Nazi leisure organization which coordinated recreational trips and tourist activities for German workers and their families.

Largely carrying out bombing and reconnaissance missions in Spain, it was favourably received by its aircrews. Do 17 pilots were especially taken with its excellent handling qualities at low altitudes, where its radial engines thrived best.[12] The raid in which German bombers would be most infamously deployed in Spain, however, was the bombing of the Basque city of Guernica on 26 April 1937.

The Luftwaffe's failure to properly assess the cultural significance of the city beforehand – and, thus, their inability to anticipate the international outrage that the raid would provoke – demonstrated its early propensity towards a reckless and cavalier bombing policy.[13] The raid killed between 200 and 250 people, although the highly inflated figure of 1,700 dead was quickly circulated.[14] Then, when rumours propagated that the Germans had been involved in the city's mutilation, it afforded the British the perfect opportunity to lambast their rival's morality. On 29 April 1937, an article in the *Dundee Courier* newspaper seethed about 'The Guernica Massacre':

> Throughout history [the Germans] have as a people gloried in aggressive war carried on by highly trained armies beyond their own frontiers. To-day they know that war will come home to the cities and towns of Germany, and that it will be precisely of the kind that Guernica and Eibar have experienced... Germany does not fear either the Red Army or the army of France. But its fear of the air forces of every surrounding country is real. Experimenting in Spain will not be entirely one-sided in its lessons.[15]

The article concluded that the bombing of Guernica 'marks a step further in the steep descent of modern scientific civilization to daemonic savagery' and 'will rank as the most completely a-moral action yet recorded in connexion with modern war'.[16] Indeed, even some Luftwaffe personnel had grown uncomfortable with their actions. *Major* Diethelm von Eichel-Streiber recalled that 'I had hated to bomb villages in Spain and asked to be transferred to a fighter unit. I was refused. So, I returned to Germany and reported to General [Erhard] Milch, who 'told me to let him know when I had my pilot's B2.'[17]

In light of the heightened Anglo–German tensions as the Condor Legion spread its deadly wings over Spain, it became of even greater importance for the Luftwaffe to maintain an air of international cooperation with the RAF. In June 1937, a German delegation travelled to the British city of York in order to participate in the International Air Meeting that was held there. Wolfgang Falck, then the adjutant of the *Jagdgeschwader* ('fighter wing') JG 132 based at Döberitz in Brandenburg, reported on the event to the Reich Air Ministry. He made particular note of the genuine warmth that the Luftwaffe had felt from their British counterparts:

> We Germans were treated with extraordinarily warm hospitality and courtesy and were lavishly entertained, so that a very friendly relationship quickly developed between the English and the Germans; at first, it was almost more cordial than between the individual groups of Germans who had previously been strangers to each other. The hospitality of the English is admirable. We were invited to everything imaginable – even the newspaper with reports on the meeting was delivered to us free of charge.[18]

Falck spoke especially fondly of the fact that 'With the officers of the RAF, with whom we met over the course of the days, we were immediately united by a bond of the warmest camaraderie and, in addition to the recognition that was always paid to German aviation, there was a warm and loud wish from all sides for closer and common cooperation and friendship between English and German aircraft pilots.'[19] Capitalizing on this diplomatic momentum, Göring received Marshal of the RAF Hugh Trenchard, 1st Viscount Trenchard, a month later in Berlin. At Döberitz, Falck recalled the continued geniality between the two sides:

> As part of a large reception, Sir Hugh first presented us with a photograph of the Richthofen *Staffel*, which had been taken by

an agent during the First World War. Then followed the photo collage of an air battle between the old Richthofen *Staffel* and an Allied unit, which Trenchard had provided with very comradely, handwritten comments that showed his respect for the German fighter pilots. Both gifts hung in the officers' mess in Döberitz for years until everything fell to pieces towards the end of the Second World War.[20]

Thus, by the summer of 1937, it appeared as if the flyers on both sides had forged a firm mutual understanding that defied the turbulence of the wider geopolitical situation. But, as always, the clamorous Göring had to go and put his foot in it. Initially, he had gone on the charm offensive towards Trenchard: 'You are well known in Germany,' he informed him warmly, 'and I have a high regard for the air force you have created.'[21] Then, during the banquet held in Trenchard's honour at Schloss Charlottenburg, Göring allegedly remarked to him that 'It will be a pity if our two nations ever have to fight.'

What he said next, however, made Trenchard's lip curl with distaste. 'Your airmen are very good,' Göring sighed to the Marshal. 'It's a pity they haven't the machines we have', before he concluded that Germany could 'make the whole world tremble'. A furious Trenchard allegedly replied: 'You must be off your head. You said you hoped we wouldn't have to fight each other. I hope so too, for your sake. I warn you, Göring – don't underestimate the RAF.'[22] Trenchard dropped his serviette, walked straight out of the banquet and never conversed with Göring again. Despite the frosty exchange, the planned visit of the Reich Air Ministry's State Secretary Erhard Milch and his delegation to Britain in the autumn of 1937 still went ahead.

Yet, the fraught diplomatic relations were hardly eased by the British press making constant references to 'General Milch and his technical henchmen.'[23] Ahead of the autumn Luftwaffe visit to various RAF airbases and aircraft factories, the air correspondent of a British newspaper even had to reassure its readers in October 1937 that 'The [Royal] Air Force can be trusted to hide what it doesn't want people to see.'[24] It was just as well, too, given that Milch went on to make a number of RAF personnel uncomfortable with a single goading question: 'Now, gentlemen, let us all be frank. How are you getting on with your experiments in the detection by radio of aircraft approaching your shores?'[25] Thankfully for humanity, the affronted RAF officers kept *schtum* on the precious radar technology that would later assist Fighter Command against the Luftwaffe in the Battle of Britain.

Following this string of faux pas, Göring attempted to smooth over the damage. While attending an opera for the Berlin Hunting Exhibition in November 1937, the Luftwaffe's commander-in-chief leant over to catch the ear of Sir Nevile Henderson, then the British ambassador to Germany. Göring exclaimed to Henderson that it was 'Inconceivable even to imagine that there should ever be war between men who got on so well and respected each other so much as the British and German airmen.'[26] For as long as politics was kept out of the picture, this was perhaps true. But, with National Socialism exponentially rotting both the core and conscience of the Luftwaffe, the chances of a peaceable resolution were slimming dramatically.

Moreover, if the RAF had found out just how aggressively the Condor Legion was testing out its military capabilities in the concurrent Spanish Civil War, that possibility would have dwindled to zero. In 1938, the Luftwaffe debuted a new aircraft in Spain that would later epitomize *Blitzkrieg* ('lightning war') in the Second World War: the Junkers Ju 87, a *Sturzkampfflugzeug* ('dive-bomber') or 'Stuka' for short. Tipping over into a white-knuckle dive of over 300 mph, this harbinger of death enjoyed great success during the Spanish Civil War. By swooping down at an angle of 60–90°, the Ju 87's near-vertical approach meant that the trajectory of the released bomb could be placed far more accurately than in traditional horizontal bombing. Its suitability for conducting tactical precision strikes against enemy infrastructure and communication hubs quickly became evident.

A number of urban Spanish targets, from ports such as Valencia to major cities like Madrid, became the victims of Ju 87s testing their accuracy. One of the most infamous examples came in May 1938, when four Spanish villages and towns – Albocàsser, Ares del Maestre, Benassal and Vilar de Canes – were indiscriminately bombed by Ju 87 pilots during the springtime advance in Aragon. Within Benassal, a large church was entirely gutted by 500 kg bombs; the Luftwaffe then snapped photographs of the smouldering ruins on the ground and from the air, hoping to establish the bomb patterns and extent of the damage.[27] The final major German aircraft that received its baptism of fire in Spain – albeit to a lesser degree – was the inimitable Messerschmitt Bf 109. Equipped with all the modern fighter entrails of a retractable undercarriage, trailing edge flaps, and an enclosed cockpit, it was trialled using cannon fire to pack a harder punch.

The Bf 109, Ju 87, He 111 and Do 17 would later be deployed to tremendous effect by the Luftwaffe during the opening years of the Second World War; the former, of course, would rise to become Fighter Command's foremost rival in the Battle of Britain. In addition to testing out its aircraft and armament, it gained important air–land integration experience in Spain through the collaboration of the Luftwaffe fighter and bomber crews, anti-aircraft units, *Heer* and Nationalist infantrymen, signals units and intelligence corps. Attacking both tactical and strategic targets, too, was added to the Luftwaffe's deadly résumé, while another crucial takeaway for the fighter pilots that later proved exceptionally important during the Battle of Britain was the adoption of a new flying formation.

In the early years of the Second World War, the British flew in a stiff 'Vic' formation of three: one fighter, the section leader, flew out in front like an arrowhead, with two wingmen flanking him from behind on the left and right. Yet the '*Schwarm*' – literally 'swarm', but later known as the 'Finger Four' to the RAF – was already being developed by the Luftwaffe during the Spanish Civil War. Four staggered fighters were split into pairs ('*Rotten*'),*[28] a system which was found to be more flexible and made it easier for all four pilots to keep an eye on the horizon for enemy fighters.[29] This was quickly adopted, with Bob recalling how 'I was transferred to Königsberg in East Prussia in 1939 in order to form the I. *Gruppe* ['Group'] of the fighter wing JG 21. The *Gruppe* was equipped with Me 109 Cs and Ds. At that time, we were already flying in squads with two planes each and in swarms with four planes each, whereby three swarms always formed a *Staffel* of twelve planes.'[30]

By the time that the Spanish Civil War ended on 1 April 1939, the operational experience garnered by the Condor Legion was substantial. The Luftwaffe's fighter pilots had claimed 386 aerial victories; its flak units chalked up a further fifty-nine aircraft shot down.[31] All-in-all, roughly 20,000 Luftwaffe personnel had seen combat in Spain, with tours lasting between six and twelve months on average.[32] Most favourable of all, they had sustained relatively light fatalities: in total, between 300–400 Germans had been killed, although 50 per cent of these deaths were attributed to 'careless driving on the torturous winding roads in the mountains of Spain'![33] Having successfully installed Franco as Spain's new leader, the returning Condor Legion received a hero's welcome in Germany.

On 6 June 1939, Göring declared amid his speech in their honour that 'The young and newly resurrected Luftwaffe tried to show and prove that it could fight and win again, and that is the great tradition that it has inherited from the World War.'[34] The international significance of what they had achieved could be seen in Göring's conclusion that 'Our opponents have seen

---

* As part of the swarm – known officially to the Luftwaffe as the *Schwarmgefechtslinie* ('swarm battle line') or *Schwarmwinkel* ('swarm angle') – the *Schwarmführer* ('swarm leader') was positioned slightly out in front, supported by a *Rottenflieger* ('wingman') on one side. On his other side was the *Rottenführer* ('pair leader'), who had his own wingman flying on his other flank, located slightly behind the main three.

the impressive strength of our proud Luftwaffe.' Hitler chimed in with his own sincere praise, expressing his delight that 'I can finally greet you boys myself. I am so happy to see you here before me. And, above all, I'm very proud of you.'[35] The intervention demonstrated the lengths that the Luftwaffe – and, more broadly, the Nazi leadership – was willing to go in order to gain the operational edge over their international rivals.

Despite the controversy and criticism that the Condor Legion garnered, the German air force considered the lessons of the Spanish Civil War to be well worth learning. It appeared to confirm that, whatever might happen in the future, its Nazi masters now appeared to truly realize the full potential of the young Luftwaffe. This had been further reinforced by the fact that in addition to the Rügen Exercise in Spain, the Luftwaffe had also been accruing operational experience through serving as the aerial muscle for Hitler's rapacious policy of *Lebensraum** ('living space'). This parasitic expansionism intended to secure the lifeblood of the German *Volk* by supplying the Third Reich with vital resources, manpower and materiel from external territories.

The expansion of Nazi Germany's borders had initially started with legally recovering the Saarland from British and French occupation in a League of Nations-observed plebiscite from January 1935, with over 90 per cent of Saarlanders electing to be united with Germany.[36] The emboldened *Führer* then publicly announced that the Wehrmacht would swell beyond 500,000 men through reintroducing military service; it was followed by the Third Reich's remilitarization of the Rhineland on 7 March 1936.[37] Only a handful of flying *Staffeln* ('squadrons') and anti-aircraft batteries supported the operation in the Rhineland, largely ferrying aircraft to repaint their frames in the Reich

---

* Though the concept was rooted in late nineteenth-century ideals of imperial expansionism, it became most infamously associated with the Nazi intention to seize *Lebensraum im Osten* ('Living Space in the East' – namely Poland and the Soviet Union).

colours and replace their registration numbers.[38] Nevertheless, the Luftwaffe was poised to be one of Hitler's more persuasive deterrents in providing the necessary hovering menace to enable his geopolitical ambitions.

By 1938, the Luftwaffe declared that its purpose was to 'Protect the German Reich and Fatherland, the people united under National Socialism, and their *Lebensraum*.'[39] This was very much reflected in its heightened presence over Austria during the *Anschluss* ('annexation') on 12 March 1938. Group Captain John Lyne Vachell, the British air attaché to Berlin from 1937 to 1939, despatched a report on 29 March 1938 in which he commented that 'Some 300 to 400 aircraft must have taken part'[40] in the *Anschluss*, even if it did not yet seem to him that the Luftwaffe was fully ready for war:

> Aircraft played a very prominent part in the entry of German armed forces into Austria. The first aircraft to land was that of *Herr* [Heinrich] Himmler, head of the SS, who, with fourteen of his men armed with submachine guns, was prepared to take the Vienna airport at Aspern by force if necessary – but it had, in fact, already been taken over by the Austrian Nazis. The first aircraft of the German air force arrived shortly afterwards, at 5 a.m. on 12 March, and carried a staff which took control of the airport. This aircraft was followed at short intervals by formations of fighters, bombers, reconnaissance aircraft, which dropped pamphlets over Vienna, and aircraft conveying troops.[41]

After the successful *Anschluss* in his homeland, Hitler now used the excuse of three million ethnic Germans in Czechoslovakia to move towards the Sudetenland region of that country.[42] This decision, however, catapulted the Germans right into diplomatic strife with the British and the French, leading to crunch peace talks. The Munich crisis of September 1938 – in which the British prime minister Neville Chamberlain, the French prime minister Édouard Daladier, and fascist Italy's leader Benito Mussolini later conceded the Sudetenland to Hitler – particularly stoked intense fear of war in Luftwaffe circles.

One Luftwaffe airman based at the Aviation Technical School in south-east Berlin captured the tension on 27 September 1938: 'It's do or die. We are facing a difficult time. But we have our *Führer*. We have a godlike trust in our *Führer*.'[43] After meeting with Luftwaffe and Reich Air Ministry personnel, such as general engineer Wolfram Eisenlohr, *Hauptmann* Huhneber and *Oberst* Wendland in October 1938, the British aeronautical engineer Sir Roy Fedden commented that 'The rank and file [were] profoundly grateful to Mr Chamberlain for his courage and pertinacity on behalf of peace.'[44] Göring, meanwhile, continued to bury his head in the sand regarding the possibility of provoking war with Britain.

Following the Munich crisis, he berated Kessler for having destroyed some documents belonging to the Luftwaffe air attaché Ralph Wenninger which pertained to the Godesberg Memorandum, in which Hitler had demanded that Czechoslovakia should concede the Sudetenland to Nazi Germany.[45] A strained-looking Kessler confirmed to him that 'Yes, I burnt them; I had to, considering the way the political situation looked to us.' Göring replied with a dismissive wave of the hand:

> Never mind; you were wrong. [Nevile] Henderson left me just a few minutes ago. He tried to work my tear glands and play the sentimental tune. He tried to persuade me to use my friendship with the *Führer* to make me give in. He threatened war with England. I told him that England would be smashed in that event. Now Henderson is off to my 'friend' [Walther von] Brauchitsch* to threaten him; he knows threats don't go with me. Whatever made you think that England would go to war with us?[46]

Kessler retorted that 'there is not a man in England who believes

---

* Brauchitsch was commander-in-chief of the German *Heer* ('army') until Hitler took over this position himself on 19 December 1941.

that they have anything to gain by war. Perhaps that idea existed before 1914. Today they know that they would not be richer by impoverishing us. But they do have their prestige to lose.' He denied being an Anglophile when Göring described him as such, admitting that 'I have come to see that England, and not France, stands in the way of Germany's becoming a great power.' Yet he warned Göring that 'England cannot afford to be pushed around by Hitler':

> The man in the street is not interested in Czechoslovakia, nor does he hate Germany. But he is interested in fair play and in protecting Britain's prestige. I have watched the Hyde Park orators talking about the imminent war – I also experienced the relief when Hitler did not invade Czechoslovakia the day after the Heinlein ultimatum, and I saw people on the streets and in the restaurants kissing each other when Chamberlain announced that he would fly to Germany to see Hitler.

For now, though, the Anglo–French policy of appeasement papered over the gaping fissures in European diplomacy that had appeared by late 1938. Kurt Scheffel, who later served in the Battle of Britain as a Ju 87 'Stuka' pilot, recalled the first Luftwaffe propaganda flights which took place over the Sudetenland on 5 October 1938. He remembered passing through Graslitz (Kraslice) and Heinrichsgrün (Jindřichovice), noting that 'people everywhere were enthusiastic about the successful entry of the German troops.'[47] On 8 October 1938, a junior doctor with a flak regiment – who would later transfer to the bomber wing KG 2 during the Battle of Britain – wrote a letter from Eger, now Cheb in the Czech Republic:

> Then came the invasion of Eger. It was beautiful – it is indescribable how people welcomed us with jubilant hearts and literally showered us with flowers. As we stood, a woman held out her hand to me and asked if we would have to retreat from the Czechs again, and when I said firmly that the troops were marching

in enormous numbers and that we were the Wehrmacht and not the *Freikorps* ['Free Corps'], she broke down in tears and said she would be able to bring her husband and children back from hiding in the forest. Many women cried for joy and decorated us with the most beautiful flowers – they distributed food and tobacco.[48]

For Siegfried Fischer, who would soon become a Luftwaffe dive-bomber pilot, 'This was, then as now, the greatest day of my life. It made me a German citizen.'[49] He claimed that 'Three-and-a-half million Sudeten Germans, almost as many inhabitants as Norway has today, were oppressed by a Czech majority in the so-called democratic state of Czechoslovakia.' He added that 'The German occupation in 1938 was not felt to be an occupation at all. Rather, we waited in happy anticipation for the German troops finally to liberate us.' Amid this seemingly 'peaceful' takeover, however, the scars of German-inflicted violence were evident in the Sudetenland. The junior doctor who was then stationed in Czechoslovakia observed how:

> On the border that we crossed with the first troops, you could clearly see the shot-out windows and walls of the border sentry post. A bridge had been blown up by the Czechs, but our sappers have already built a new one and we continued on. The Czech soldiers had to abandon everything because of the rapid advance, and they themselves often fell into captivity. There was a shooting in a village and members of the Wehrmacht were killed. Artillery was immediately deployed and where resistance arose it was immediately broken. The Jewish and Czech businesses are closed and the owners have taken refuge. The people breathe a sigh of relief.[50]

This demonstrates that even before the Second World War broke out, some of the future Luftwaffe personnel who would serve in the Battle of Britain had already borne witness to – and, in some cases, even approved of – German violence and antisemitic persecution in the territories they occupied. Hans-Ekkehard Bob, on the other hand, tried to maintain that it was an entirely peaceful transfer of power: 'The occupation of the Sudetenland, which at that time belonged to Czechoslovakia, was planned, and we were told to fly fighter escort for Ju 52 transport planes. Our planes were equipped with machine-guns, but we had neither ammunition nor ammunition belts. Fortunately, there was no shooting then.'[51]

What is little known, however, is that the Luftwaffe immediately set to using the newly acquired *Lebensraum* in the Sudetenland as a testing site for its Stukas. Scheffel recalled that 'The inspection of the bunker line near Kaaden on 17 October had given rise to the desire to attack such a bunker from the air with bombs in order to determine the effect. For this purpose, a bunker was chosen that was located in an open field near Mokotil [now Mukoděly], 9 km south of Podorsam [Podbořany].'[52] Assigned as a security officer, Scheffel was tasked with ensuring 'that the area around the bunker was clear of people when the *Staffeln* attacked'. On 21 October 1938, a large crowd of *Heer* and Luftwaffe personnel gathered to watch the bombing trial:

> At the appointed time, the bunker was vacated and the first aircraft attacked. It was quite clear that only hits directly next to or on the bunker could have any effect. The bombs that had been thrown as part of the first payload were very close to the bunker, but it was still standing. *Unteroffizier* Grobe of the 1. *Staffel* scored the first effective hit. This was noticeable because there was no large mushroom of smoke shooting upwards; rather, the explosion was widespread. *Leutnant* Unbehaun's next bomb must have been laid just as closely. *Staffelkapitän** *Hauptmann* Ott was the last to throw himself into his slanted dive. There was a direct hit. When the smoke cleared, you could see that the bunker was no longer there.[53]

---

* *Staffelkapitän* ('Squadron Captain') was a command position rather than a military rank.

Scheffel added that 'The bunker was completely torn apart, with its remains lying about eight metres away from the original location. There was a five-metre-deep hole where the bunker had been. The *Heer* officers were very impressed by the accuracy of the Stukas.' These early *'Blumenkriege'* ('flower wars'), where Germany annexed regions with large populations of ethnic Germans and German speakers, elevated Hitler to a godlike level in Germany by the late 1930s. But, most crucially, the seeds of *Blitzkrieg* were being sown in the occupied Sudetenland nearly a full year before their terrible bloom in Poland. Thus, from its victory in the Spanish Civil War to enabling Hitler's *Lebensraum*, Göring's 'little air force' of 1935 was now all grown up – and, like any contumacious teenager, it was ready to unleash its full might and fury upon anyone who opposed it.

2

The Line Was Dead

During a conference held in Hitler's New Reich Chancellery study on 23 May 1939, *Generaloberst* Erhard Milch was disconcerted by the *Führer* he saw in front of him. The State Secretary face – already no stranger to a pained grimace – dropped even further as he realized that 'Hitler had changed a little in comparison to the earlier years. He was much more convinced of his own virtues than he was earlier. I recalled his modesty was not so apparent anymore. Then he patted himself on the back about his successes. He talked for a long time about his successes: Austria, Sudetenland, Czechoslovakia.'[1] The latter, referring to Hitler's most recent acquisition of Bohemia and Slovakia in March 1939, had further boosted the *Führer*'s confidence that he could throw his weight around without provoking war with Britain and France.

By the end of the spring, however, the mounting discord with the Allies over the territorial dispute of the Polish Corridor in Danzig (Gdańsk) had become a thorn in Hitler's side that he could not ignore. He allegedly ranted at the conference about how 'Britain did not wish that he would clear up the question of the Corridor with Poland.' Milch claimed that Hitler 'Then abused the British. He said the British wished to have the world to themselves and they wouldn't even let poor little Germany have the Corridor.'² Yet he alleged that the *Führer* 'Always emphasized that he wished to direct his policy in such a manner as to work with Britain':

> He always said, 'Even if Britain now condemns my action and threatens me, I shall know how to deal with it. There will never be war with Britain, even if I have to give in politically. But the British are very good at bluffing. So am I. We shall see who is better at this game.'³

The minutes of the May 1939 conference, on the other hand, gave a far more realistic assessment of what would happen if the Germans pushed things too far. 'The war with England and France will be a life-and-death struggle,' they outlined tensely. 'The idea that we can get off cheaply is dangerous; there is no such possibility. We must burn our boats, and it is no longer a question of justice or injustice, but of life or death for 80 million human beings.'⁴ That Britain continued to be of particular concern could be seen in a follow-up note on the conference:

> The effort must be made to deal the enemy a significant or the final decisive blow right at the start. Considerations of right and wrong or treaties do not enter into the matter. This will only be possible if we are not involved in a war with England on account of Poland.⁵

By the late summer, though, Hitler's bluffs would soon give way to belligerence. On 22 August 1939, his private retreat 'Berghof' in Berchtesgaden, Bavaria, was stuffed to the brim with all the commanders of the *Oberkommando der Wehrmacht* (OKW or 'Wehrmacht High Command'), along with every other relevant officer. Numbering around forty in total, the men were told about the situation with Poland: 'In the next few days orders will be given as to whether you have to report or not [for the invasion]; everybody has to be ready; you have to mobilize and you have to take up from this starting point. However, everybody has to wait, as diplomatic negotiations are still going on.'[6]

Göring, as Milch recounted, said that 'Through a Swedish gentleman, Count Dahlehus, he was still negotiating with Britain, hoping to avoid war. I am still convinced today that Göring did not wish the war to break out. Then, when it was clear that war with Poland would break out, I advised Göring to go to England personally, of course with Hitler's permission, in order to try and save what could be saved.'[7] Nothing could be salvaged, however, once the Germans irretrievably broke Britain's trust on 1 September 1939, when 1.5 million Wehrmacht soldiers poured into Poland.[8] Peter W. Stahl, who later became a Ju 88 bomber pilot, recalled how that fateful day had begun for him like any other:

> As an experienced weather pilot, I should have known what would happen when the clouds on the horizon, in this case politically, began to gather more and more. Because there was a lot of talk about the danger of war in those July and August days of 1939. I have always been an incorrigible optimist. I had every reason to be as such, too. Our little house in Deep, close to the waters of the Baltic Sea, seemed to be on another planet. I had married young and had the dream job of the time as a civilian test pilot. Life was indescribably beautiful...[9]

On the morning of 1 September, Stahl was due to fly to Berlin from the little house – but, before he had even left for work, he was already plotting how to get back to his beloved wife. 'I was already imagining how I could buy time to come home earlier than expected (honking cheekily in front of the house) and folding my arms behind my back in anticipation of the welcoming kiss, hiding a package with a little something from Berlin in my pocket.' He envisioned teasing his sweetheart with a silly game – 'Right or left hand?' he would playfully ask – though she would naturally be spoilt irrespective of the answer given.

But, that morning, Stahl happened to turn on the radio to check the time. The foreboding music that unexpectedly passed over the airwaves made both of their stomachs drop.

'And then it comes,' he later wrote mournfully. '*Achtung, Achtung*! German Radio is now broadcasting... the statement that the *Führer* made to the world at 6 a.m. this morning!' Stahl and his wife looked at each other anxiously; he carefully put down the spoon he was using to boil an egg, a single vein twitching in his temple, and they listened intently to the radio bulletin. 'After a wordy introduction, the fateful sentence finally came: there has been firing back since 5 a.m. this morning! My wife suddenly turned pale right down to her hair. Her question hangs in the room: 'Then we have war now?' Stahl described how 'My limbs have become heavy. Breakfast no longer tastes good. And I still can't believe it. I still flew to Berlin, came back on time and was happy about the little thing I was able to bring with me. And as usual we went to the beach to swim. That was the day the war began.'

He was swiftly made a reserve *Unteroffizier* of the Luftwaffe despite being a civilian, and as an instructor of 'blind flying' – using instruments only and not relying on visual clues – Stahl was immediately thrown into 'teaching young Luftwaffe soldiers and officers the secrets of adverse weather conditions and night flying'. Stahl was put on constant 'navigation flights with students on the Junkers Ju 52 around Greater Germany, around and around – I hardly ever get home. Gradually you become very sleepy'. He reflected when writing his memoirs that 'I recently came across my old diary again. There stands my younger self in front of me, sometimes a bit strange, but it is the image of the young man who was dropped on top of this stubborn bullshit war without his involvement and who (of course) had to make sure that he didn't fall off.'

In post-war memoirs, it was naturally in the best interests of Luftwaffe veterans to emphasize their lack of agency regarding the outbreak of war, so as to distance themselves from the aggressive conduct of the Nazi regime. Nevertheless, traumatized by the relentless bloodshed of the Great War and post-war hyperinflation, the broad German reception of these new hostilities was genuinely lukewarm at best. As Dennis Showalter, Peter Pechel and Steinhoff wrote, 'In silence, thousands of men and women watched the tanks and trucks roll by.'[10] Fresh from serving as a fighter pilot in the Spanish Civil War, Eduard Neumann simply noted that 'We always hoped that it would never come to that.'[11] Even the malicious attempt of the Nazis to present the Poles as the aggressors via a Silesian radio station did little to inspire the vengeful outrage they hoped to stoke among the Reich's citizens.[12]

However, not everyone in the Luftwaffe was crestfallen by news of *Fall Weiss* ('Case White'), the plan for the invasion of Poland. Karl Bruns, an airfield construction manager, painted a far more optimistic portrait by enthusing in his diary: 'The impudent attacks by the Poles are said to have been avenged. German troops crossed the Polish border at 5 a.m.; German *Staffeln* started taking off from their bases! Serious and adjusted to the historical moment, everyone enthusiastically sings the national anthem. The fight for freedom for Germany's future has begun!'[13] Yet, it is perhaps more likely that because Bruns and his colleagues were serving in a less combative Luftwaffe role, they had been more easily seduced by the Nazi presentation of Case White as a necessary crusade.

He documents of his own Luftwaffe construction unit, for instance, that 'with hot heads and feverish hearts, we sit in the officers' mess in Hagenow and hear the *Führer's* call!' However, Bruns' account fails to hide the dumbstruck response of the flying arm to the war: 'The youngest officers of the airborne units and old medal-decorated First World War officers sit around me with serious faces.'[14] Even if Case White was a shock to the German airmen, however, it was nothing compared to the impact on the Poles, for whom the paralysing thunderbolt of *Blitzkrieg* now forked across their homeland. Their heartbreaking realization of what was transpiring came when, as the Stuka pilot Paul-Werner Hozzell recalled:

It had been customary with the German and Polish railroad personnel on duty to wish each other 'good night' after the last scheduled train had passed via the railroad telephones installed at both sides of the river. They had known one another for years. In the night from 31 August to 1 September there was no such call from the German side. When the Polish personnel tried to ring up their German colleagues, the line was 'dead'. The Germans had switched it off. In view of the extremely tense situation this was a danger signal, and the Poles acted accordingly.[15]

Indeed, some Luftwaffe personnel were less shocked at the outbreak of war after having witnessed the German build-up at the border. A *Soldat* ('soldier') serving at Finsterwalde airfield in north-eastern Germany wrote on 30 August 1939 that 'You must have expected a letter earlier, but there was a postal blockage here, there were no longer any scheduled trains – all military transport. There will probably be no more vacations before the war… in Berlin things must now be very busy with ration cards, etc. Hopefully you have stocked up enough.'[16] He added that the 'trucks rolled in day and night by us. Old men that are over sixty years old were drafted…'

The Germans had launched Case White to seize what they saw as German territory that had been ceded by the Allied Powers to Poland following the First World War – particularly the Danzig corridor, which isolated East Prussia from the rest of Germany, but also West Prussia, Posen and East Upper Silesia. In addition, Poland was the logical stepping stone after Czechoslovakia towards securing additional *Lebensraum im Osten* ('living space in the East'), thus setting up the possibility of a future invasion of the Soviet Union. This was not immediately pressing, however, due to the Molotov–Ribbentrop pact from 23 August 1939, in which Nazi Germany and the USSR had agreed to carve up Polish territory between them.

Germany's eastern border with Poland also offered the invaders a bounty of oil, grain and minerals – and, viewing the Polish as racially inferior, the Nazis equally anticipated that forcing the local population into slave labour would sustain Germany's eastern expansion. Keen to avoid getting bogged down into a military quagmire such as in the First World War, the Germans knew they needed to seize Poland quickly. *Blitzkrieg* ('lightning war'), it was hoped, would avoid the bloody attrition that had characterized trench warfare between 1914 and 1918. It was an adaptive, tactical means of waging warfare that had its roots in a much older tradition of Prussian military thinking and *Bewegungskrieg* ('mobile warfare'), which had seen a more practical application in the First World War.

The incorporation of these older tactics into a new, slick means of mechanized warfare that was advanced by General Heinz Guderian came to mesh favourably with the *Führer*'s desire for a rapid new approach. The Luftwaffe had initially been intended as a 'Risk' air force, one whose theoretical destructive power aimed to serve as a deterrent to other nations, making it too risky to consider fighting against Germany. However, the Spanish intervention had bloodied the Luftwaffe, and it had proven to itself that it could successfully engage in all forms of aerial warfare.[17] Combining highly mobile *Panzer* ('tank') divisions on the ground with fearsome air support from the Luftwaffe, the shock-and-awe tactics utilized in *Blitzkrieg* were designed to quickly punch through the weak spots within enemy strongholds, reeling armies and fleeing citizens.

The Luftwaffe called upon the 1,600 aircraft of *Luftflotten**[18] ('air fleets') 1 and 4 to wipe out the Polish Air Force (PAF) on the ground before supporting the *Heer*'s advance: a central thrust straight through Poland, flanked by a pincer movement in the north and south.[19] To counter them, the Polish wielded

---

* By 1939, the Luftwaffe was split into four *Luftflotten* ('air fleets'), each of which was commanded by a general. The *Fliegerkorps* ('Aviation Corps') were then assigned across the *Luftflotten*, leading down to the *Fliegerdivisionen* ('flying divisions'), *Geschwader* ('wings'), *Gruppen* ('groups' – not to be confused with the RAF's larger Groups) and finally the *Staffeln* ('squadrons'). *Luftflotte* 5 was added in April 1940 amid Operation 'Weser Exercise' in Norway and Denmark, provisionally tasked with attacking Scotland and the northern half of England during the *Luftschlacht um England*.

935 aircraft in total – just under half of which were fighters or fighter-bombers.[20] The main PAF fighter was the PZL P.11, a high-wing monoplane that dated back to 1931; flying 100 miles per hour slower than the Bf 109, it would become outclassed in Case White.[21] Nevertheless, Bob recalled how the Germans still felt the sting of the Polish fighter pilots early on in the campaign:

> I gathered my first experience in fighting operations in the Poland Campaign through escort missions for bomber planes. In the beginning we had rather strong air fights with Polish biplanes that were quite slow but on the other hand very manoeuvrable. For that reason, only a few air victories could be obtained. Afterwards we were removed to the west of Germany and received the Me 109 E.[22]

Watching their homeland being mercilessly lacerated by the Luftwaffe stoked a raging fire within the PAF fighter pilots; from Łódź and Warsaw, they made their courageous final stand.[23] Against the odds, they succeeded in bringing down dozens of German bombers with their antiquated fighters. Their lionhearted resistance was reflected in the fact that they ended up losing just forty-eight more aircraft over Poland than the Germans; this same frenzied tenacity would later serve the Polish fighter pilots well in the Battle of Britain.[24] Despite this spirited defence in Case White, however, the PAF's weak communications, vastly inferior numbers and largely obsolete aircraft ultimately meant that the Germans secured overwhelming air superiority after just five days.

By the middle of September 1939, the PAF had been suppressed for good. The Luftwaffe now hammered away at enemy artillery, army formations and Polish shipping in the Baltic Sea. German confidence was further strengthened by the Soviet invasion of Poland on 17 September 1939, with a member of the Luftwaffe Construction Company 3.XII writing a week later that 'Russia has fully aligned itself with Germany.

That was another little glimmer of hope. As long as they aren't at loggerheads, you can still hope for a happy ending.'[25] A *Flieger* based with a supplementary Luftwaffe personnel unit at Stendal airfield in northern Germany expressed his hope that the final Soviet–German demarcation of Poland 'should significantly fortify our strength and position against England'.[26] Yet, the *Flieger* could not entirely dismiss the threat that Britain still posed:

> It's just not entirely conceivable that the British blockade will end. But I firmly believe that we will probably use the strongest possible means to deploy ourselves whenever the opportunity arises. Of course, there are great difficulties for us – but, either through the air war or the submarine war, England will be hit hard. In the meantime, the German pilots in Poland had the best opportunity to gain the 'experience' that is necessary to use any weapon. We will certainly not lack aggressiveness, commitment and courage, as proven by the many exploits and attacks of the German bombers and '*Stukas*', as well as the newest exploits of our young submarine arm.[27]

Behind the scenes, however, tension was festering between the blustering Göring and his headstrong fighter pilots. Steinhoff had been commanding an experimental night-fighter *Staffel* in Bonn – whose job was to shoot down Bristol Blenheims that were airdropping British propaganda leaflets over the Reich[28] – when he was summoned to Berlin towards the end of September 1939 for a conference on night-fighting. Arriving at the Reich Air Ministry on *Wilhelmstraße*, the young Steinhoff peered up at the imposing structure. Its crisp lines, sandy bricks and neatly studded windows revealed little of the intimidatingly sumptuous interior. 'I will never forget that meeting,' he later recalled. 'We all sat around an enormous round table. Everything in the room seemed to have oversized medieval dimensions. The huge chandelier hanging from the ceiling was made out of a wagon wheel. The chairs were covered with thick yellow leather.'

Steinhoff added that 'The masterpiece was in the middle. The back of that chair had to have been at least six feet high. That was where Göring sat.'[29] When Göring finally appeared at the meeting, he 'lit up a Virginia cigar and lectured for at least half an hour. What he had to say sounded like the script of a movie about the First World War: biplanes looping, attacking from below, flying so close that the pilots could see the whites of their enemy's eyes.'[30] Steinhoff noted how 'I couldn't stand it any longer. After all, I was the only one present who had actually flown any night missions.' Thus, he 'dared raise one finger' to halt the commander-in-chief's barnstorming monologue:

> Göring looked at me in astonishment. He pointed his cigar at me, giving me permission to speak. I simply said that everything had changed. We flew at much higher altitudes. We wore oxygen masks. Furthermore, we had no navigational aids and the cities had been blacked out. The only way we could spot the enemy was if he just happened to be picked up by a searchlight. That's as far as I got. Göring gestured with a wave of his cigar that I had said enough. 'Young man,' he said, 'young man, you still have a lot of experience to gain and a lot to learn before you think you can have your say here. Now why don't you just sit back down on your little rear end.'

Flustered and furious, the humiliated Steinhoff took his seat. 'At that moment,' he later recalled, 'I realized that the man was obviously an amateur, and I began to hate him.' These first stirrings of discontent, however, were soon drowned out by how quickly Poland was being conquered. A Luftwaffe serviceman in an air signals unit wrote on 11 October 1939 how 'We were deployed in the East for three weeks. But we weren't allowed to take a direct part in the fight either because, as the troops' victorious advance became more and more rapid and the front became shorter, we were withdrawn.'[31] The future Stuka pilot

Hans Deibl, who was finishing school in the autumn of 1939, recalled:

> Maps hung in all the classrooms on which colourful pinheads impressively marked the advance of our troops. Everyone gathered around the radio when fanfares to the song 'Swastikas Will Soon Flutter Over All the Streets' sounded and another special announcement was delivered, such as: 'Our troops have reached the city limits of Warsaw.' The rapid advance of the German Wehrmacht, later referred to as the '*Blitzkrieg*', fascinated us on the one hand, but on the other hand it also confirmed our concerns that we would definitely miss the war, as it looked as if all the military objectives of this campaign would be achieved in just a few months.[32]

On 6 October 1939, as the last embers of Poland's defence burnt out, Adolf Hitler made a so-called '*große Friedensrede*' ('great peace speech') to the Reichstag* that outlined the next steps towards securing peace in Europe.[33] The *Führer* described the German seizure of Poland in Case White as signalling that 'One of the most senseless acts of Versailles is thus eliminated.'[34] Altogether, it had taken the Wehrmacht just thirty-six days to successfully execute Case White. This rapid advance had partly been assisted by the fact that General Wolfram von Richthofen, who had once led the Condor Legion, deployed over Poland the destructive lessons he had learnt during the Rügen Exercise in Spain. On 8 October 1939, a Luftwaffe analysis report noted that:

> General von Richthofen announced that he had observed an outstanding morale effect of the high-explosive bombs in Poland, as had previously been the case in Spain. The

---

* After the Reichstag fire of 1933, the Nazis transferred the Reichstag from its former location at the *Platz der Republik* to the nearby Kroll Opera House in Berlin. Any references to the Reichstag after this year refer to the latter location.

acoustic effect of exploding bombs on people is greater than the impression caused by later inspecting destroyed houses, etc. The troops' morale tolerates sustained artillery fire more easily than repeated explosive bomb attacks.[35]

Of the Luftwaffe air raids on Kraków, it was pointed out that 'Only after the first air raid were a few protective trenches dug on the outskirts of the city and makeshift air raid shelters were set up for some villas, although these were not significant in terms of numbers. The overall impression was that the population was unprotected.'[36] It was further observed that 'Air raid shelters were not found in the airbases or the towns visited, with the exception of individual cases in the residential area of Kraków. When flying over Polish territory, cover trenches were occasionally observed. No useful information could be obtained about the behaviour of the population during air raids. Apparently, no significant preparations have been made for air protection for the civilian population.'[37]

In addition to the vulnerability of the local population, the report highlighted that 'Polish-dropped munitions were found which corresponded to the state of our bomb development in 1918 and which also externally closely resembled the old German P.u.W.* explosive bombs' from the First World War. Despite the technological imbalance between the two sides, however, the Luftwaffe's ransacking of the PAF's resources began almost immediately after Case White. The Luftwaffe's clear hand in exploiting the local Polish and Jewish population was evident in a Luftwaffe report on the administration of Poland during 1939:

> In order to capture and exploit the Polish LW [Luftwaffe] equipment, the Aircraft Staff East was set up in October 1939 and deployed in Łódź […] soon, around 1,500–2,000 Polish

---

* P.u.W = *Prüfanstalt und Werft der Fliegertruppen* ('Aviation Testing Facility and Shipyard') bombs.

and Jewish workers were made to repair the war materiel taken by the German Luftwaffe. Their work was initially carried out via a *Hauptmann* Fuchs, financed with the sum made available in advance; later, sufficient funds were available through the sale of the acquired equipment.[38]

To directly fund itself, then, the Luftwaffe not only pillaged the PAF's resources, but it also sold off the parts it did not want – all while forcing Polish and Jewish slave labourers to repair their own country's equipment and facilities for German use. Yet, in spite of its clear domination, the Luftwaffe had felt its first deaths keenly. Peace remained the desired option, and the airmen hoped it would now be their reward for the successful culmination of Case White. A *Soldat* who was serving with a Luftwaffe munitions unit during the Battle of the Bzura near Warsaw wrote on his return to Germany on 9 October 1939 of how:

> We also suffered heavy losses. I want to inform you how many losses we had in the company and in the whole regiment. Eight dead and twenty-six wounded in the company and 400 casualties in the whole regiment. That is a large number in relation to the days of struggle.[39]

The same day, a *Gefreiter* attached to the Luftwaffe Heavy Flak Unit 492 in Mannheim penned wistfully that 'Perhaps the world's leading men will come to an understanding. It's all pointless to destroy everything, and afterwards everything comes to nothing.'[40] Nevertheless, while the first major battle may have been won, the war was only just beginning. Hitler had previously claimed in August 1939 that 'There is no actual rearmament in England, just propaganda';[41] he had banked on British reticence to fight Nazi Germany over Czechoslovakia as being indicative of its weakened military capacity.

Yet, Case White triggered France and Great Britain's respective pledges to militarily assist Poland if it were invaded by a European power, and both countries had declared war against Nazi Germany on 3 September 1939. To make things even worse for the Luftwaffe, signs that Göring would prove to be a meddlesome force in the air war against Britain were already beginning to surface at that early stage. Following the Anglo–French declaration of war, Rieckhoff claimed that the commander-in-chief's first hot-headed telex declared that:

> 'The unrestricted air war against England is hereby opened.' The second telegram referred to the first and ordered: 'Air Fleet 2 immediately opens the air war against England. To do this, it will attack the British aircraft carrier Hermes in Sheerness [northern Kent] with a chain (three aircraft) of Ju 88s.' Previously it was said that the sky would darken if England was attacked. Now, on Göring's orders, three aircraft were to open the air war! In any case, Air Fleet 2 ensured that this order was withdrawn before it was carried out.[42]

Many Luftwaffe personnel simply kept the faith that the war with France and Britain would soon dissipate in the face of Poland's crushing defeat. Some took comfort from the old adage that it would all be over by Christmas, with the Stendal-based *Flieger* writing hopefully towards the end of Case White: 'We assume that the German soldier will sit under the Christmas tree in Germany yet!'[43] As had happened with the older generation in the First World War, however, Germany's soldiers were heading closer to a rip tide than the yule-tide.

3

How Insane Is This War!

Soaring high in the skies of Scotland, Willibald Klein found himself cursing the recent outbreak of hostilities with the island nation. Serving as a radio operator and aerial gunner on an armed reconnaissance He 111, his crew were scouring for secrets over Lerwick in the Shetland Islands. As he looked down, 'The people – many people in the streets – waved up at us as if it were deepest peacetime.'[1] The local residents then identified the foreboding airframe as foe, not friend. Suddenly, their heartiness gave way to horror. Desperately trying to defend the exposed archipelago, a British flying boat now engaged with Klein's He 111; the bullets scattered, the civilians fled. 'We knew that under those roofs in the old buildings, human life was pulsating and hearts were beating with fear and anxiety,' Klein recalled sombrely.

He added that his entire aircrew felt: 'How insane is this war! Here we are firing away above the houses and those poor, peaceful people must suffer for it.' Meanwhile, in post-*Anschluss* Austria, a young airman named Johan Rausmayer was admonishing the British in a letter to his American pen pal Earl Hansen. 'England, who reigns a quarter of the world, fears Germany, as we only want to get back our colonies, our territories,' he ranted in broken English on 1 December 1939. Expressing that 'We hope that we can fight to France or fly to England,' Rausmayer vehemently wished 'England the death' in his letter.[2] These fascinating accounts display how differently just two airmen in the Third Reich could feel about the early air campaign against Britain: one desperate for it to end, the other hungry for it to begin.

Sidestepping the crestfallen British prime minister, Neville Chamberlain, the *Führer* had set his sights on the intractable Winston Churchill in his supposed 'Peace Speech' of 6 October 1939. 'Mr Churchill may be convinced that Great Britain will win,' he sneered in his address. 'I do not doubt for a single moment that Germany will be victorious. Destiny will decide who is right.'[3] For the moment, however, destiny was dragging its feet. The *Sitzkrieg* ('sitting war') – known to the British as the 'Phoney War' and to the French as the *drôle de guerre* ('strange war') – now began, stretching from autumn 1939 until the early spring of 1940. Its name stemmed from the relative inaction of the French army and the British Expeditionary Force (BEF) against the German *Heer* following the Polish campaign.

During the Phoney War, the Luftwaffe started agonizing over whether a concerted attack against Britain would be required to break the deadlock. A year beforehand, as Rieckhoff recalled, *General der Flieger* Helmuth Felmy – commander of Air Fleet 2 – 'had spoken out in serious words in a detailed justification against the air war with England and pointed out the weakness of the Luftwaffe, the strong forces of the Empire, and also that Germany could do nothing more than pinpricks'.[4] Then, Felmy had determined in his *Studieplan* ('Study Plan') 39 of 13 May 1939 that successful attacks against British industry would still be doubtful even by 1942.[5] *Oberst** Hans Jeschonnek concurred with Felmy's conclusion, warning that 'Terror attacks on London as the stronghold of the enemy defence would

---

* Promoted to *Generalmajor* on 14 August 1939.

hardly have a catastrophic effect or contribute significantly to a war decision.'[6]

This was in direct contrast to *Oberstleutnant* Josef 'Beppo' Schmid's findings in his *Studie Blau* ('Study Blue') of July 1939: he conversely stated that 'Operations could be carried out with even greater effect [over London] due to the greater density of population.'[7] Schmid headed the Luftwaffe's Air Intelligence Department 5; as a participant in the Munich Beer Hall Putsch of 1923, he wielded great political favour with the Nazis.[8] Some of the information that Schmid had utilized in Study Blue was outdated by around four to five years, especially regarding Britain's aircraft industry.[9] He produced a number of chimerical reports in 1939, and Study Blue became the benchmark for Luftwaffe wartime intelligence even though Schmid spoke no foreign languages, did not fly himself, lacked the air war experience of Felmy, and consistently misinterpreted his intelligence data.[10]

Despite the flawed methodology in Schmid's hyper-optimistic reports, however, they found great favour with Göring – particularly as they often pandered to him and indulged his belief that a strategic air campaign against Britain was achievable.[11] Hitler, too, knew that contingency plans were required should Case White provoke war with Britain. As such, on 31 August 1939, he issued 'Direction No. 1 for the Conduct of the War': this determined that 'In conducting the war against England, preparations are to be made for the use of the Luftwaffe in disrupting British supplies by sea, the armaments industry, and the transport of troops to France.'[12] He further identified that 'A favourable opportunity is to be taken for an effective attack on massed British naval units, especially against battleships and aircraft carriers.'

Hitler crucially stipulated, however, that 'attacks against London are reserved for my decision', although 'Preparations are to be made for attacks against the British mainland, bearing in mind that partial success with insufficient forces is in all circumstances to be avoided.'[13] Thus, the idea of conducting a dedicated aerial offensive over Britain had already started materializing by the late summer of 1939, even if there were caveats in the need for adequate manpower and resource allocation. As Rieckhoff noted of this fateful year:

> The year 1939 was all about preparing for a battle with England. This was the last war game of the naval command in Oberhof, in which the *Kriegsmarine* [German navy] wanted to justify its strategic demands, which culminated in the conquest of northern France and Brittany and the acquisition of bases in Norway. This was the major planned exercise by Air Fleet 2, in which the organizational and supply needs for the war against the islands and a procedure for throttling British supply traffic were to be determined. Finally, there was the Air Fleet 2 manoeuvre in August 1939, during which the ground organization in East Frisia was checked for its suitability as a drop-off area against England.[14]

The possibility of attacking Britain from above, then, consumed the Luftwaffe's thinking towards the end of 1939. Indeed, Hitler's 'Principles for the Conduct of the War against the Enemy's Economy' identified in November 1939 that 'Great Britain is the driving spirit and the leading power of our enemies.'[15] It should be noted that the next sentence of this document is often translated into English as 'The *conquest* of Britain is, therefore, the prerequisite for final victory.'

Yet, the actual German word used in Hitler's order is '*niederringen*', which translates more closely as to 'overpower', 'defeat' or 'floor' an opponent.[16] Semantically, then, it is important to recognize that by November 1939 the *Führer* was not pressing for an active 'conquest' of Britain via an invasion. Indeed, the *Kriegsmarine* had been most disparaging of the Luftwaffe's capacity to wage a dedicated campaign against the British mainland. On 10 October 1939, the general index of a War Diary from the Maritime Warfare Command – one of the *Kriegsmarine*'s most senior command authorities – documented that:

> *Ob.d.M.* [*Oberbefehlshaber der Marine,* 'Supreme Commander of the German Navy'] calls for capacity to increase the submarine construction programme to 20–30 *U-Boote* per month (previously 10), possibly at the expense of the other parts of the Wehrmacht. Since the Luftwaffe will not be ready for major attacks on England before autumn 1940, the *Kriegsmarine* initially has to bear the burden of the war against England alone.[17]

Although the *Kriegsmarine* naturally had a vested interest in advocating for its own expansion ahead of the Luftwaffe's, it was true that the German air force was only really equipped to keep itself occupied with reconnaissance flights over Britain and a scattering of dogfights in the Phoney War. Poor weather curtailed much of its activities during this period, though there were a few key successes, with *Hauptmann* Werner Mölders of the fighter wing JG 53 scoring ten aerial victories during the Phoney War.[18] As the year came to a close, though, the most significant interaction between the RAF and the Luftwaffe was a clear German victory in the Battle of the Heligoland Bight.

On 18 December 1939, twenty-two Vickers Wellington bombers from the RAF's Bomber Command were en route to attack *Kriegsmarine* ships in the major German naval base at Wilhelmshaven on the North Sea coast. At the same time, *Hauptmann* Wolfgang Falck – who was now based in Jever as a *Staffelkapitän* attached to *Zerstörergeschwader* ('Destroyer Wing') ZG 76 – had been leading his *Staffel* of Bf 110s (also known as Me 110s*) in a training flight near the island of Borkum. *Leutnant* Hermann Diehl was using their formation to demonstrate to a *Kriegsmarine* officer how the new *Freya* radar

---

* Although the Germans did sometimes use 'Bf' (*Bayerische Flugzeugwerke*) and 'Me' (*Messerschmitt*) interchangeably to refer to the Bf 109 and Bf 110, these two aircraft were technically developed for Bayerische Flugzeugwerke AG. After the company was reconstituted as 'Messerschmitt AG' on 11 July 1938, all aircraft thereafter received the 'Me' designation. Contemporary German documents typically favour the term 'Bf' for the 109 and 110, hence the use of these designations in this book.

system could detect incoming aircraft. Suddenly, an unauthorized blip made Diehl's heart flutter; he had picked up the incoming wave of Wellingtons.

At first, there was a nail-biting delay in communications, even though the naval radar also picked up the British aircraft twenty minutes later.[19] But, with eight minutes' advance warning, Luftwaffe fighters successfully intercepted the Wellingtons and shot down twelve of them, constituting a loss rate of around 50 per cent for the RAF.[20] As the future night-fighter Paul Zorner recalled, 'This defensive success was due to a completely new, top secret technology, which experts called radio measurement technology, and which made it possible to locate an enemy far beyond visual range using electromagnetic waves.'[21] This engagement marked the first truly successful use of German radar to intercept British aircraft in the Second World War.

Falck himself scored two aerial victories in the Battle that had been assisted by *Freya*, which served to strengthen his interest in radar detection. According to Zorner, Falck had 'immediately recognized the importance of the new technology and began to systematically deal with it from the perspective of integrated air defence'; he would later be credited as the father of the German night-fighters.[22] Zorner further recalled that it was 'thoroughly exploited by German propaganda as the "Air Battle over the German Bight"', and it subsequently dominated Luftwaffe propaganda during the early portion of the Phoney War. That the engagement remained in the German consciousness could be seen when a female resident of Berlin later wrote to a loved one on 11 January 1940 that 'The radio is on – it's the news. Our planes have again shot down four English bombers over the German Bight.'[23]

For most of the Phoney War, however, the Luftwaffe circled around Scotland's Firth of Forth, the Orkney Islands and the Shetland Islands. 'When Winston Churchill told the English people that fog and rain would ensure that England was protected from the German Luftwaffe, German aviators smiled and, with dozens of flights over the English* island kingdom, taught Mr Churchill better,'[24] an early Luftwaffe propaganda report on a patrol over the Shetland Islands smugly noted. In lieu of intense dogfights, the Nazi press often focused on the reconnaissance flights as a means of demonstrating that the Luftwaffe was not just sat twiddling its thumbs:

> A flight to England is no child's play. Our reconnaissance scouts know how to speak about it – not only of victorious battles with English fighters, of successful attacks on the ships and picket boats heavily armed with anti-aircraft guns, but also of fighting against storms, fog and cold, and even of the wrestling against the waves of the North Sea. Day in and day out, the grey eagles roar along, aggressive, ready for battle, and determined to use any machine or man. The will to win is their power and strength, their ability to outperform the enemy, and unsurpassable German workmanship. You hear that again and again in all the stories our pilots tell us.[25]

Nevertheless, the Phoney War dragged on drearily for many Luftwaffe personnel. One *Soldat* in a reserve Luftwaffe flak section in Linkenheim – wrote on 15 October 1939 that he had 'been spending my days deep underground in the *Westwall*' (Siegfried Line) and that 'There is nothing to be done. You just kill time as best you can with more or less meaningful service. It's actually impossible to read anything decent in the cramped bunker. You can imagine how the mind here is rapidly approaching absolute zero.'[26] The day before, a *Leutnant* in the 3. *Staffel* of the fighter wing JG 54 had itched for the British to make a move:

> The second Englishman in our *Gruppe* was shot down yesterday. I was there for the first time, but I didn't have any dogfights,

---

* German propaganda overwhelmingly referred to 'England' and the 'English', even if the reports were on operations conducted over Scotland, Wales or Northern Ireland.

and yesterday we were just doing sports because our *Staffel* has been taking a break for a few days at the moment. A few minutes later we were at our berth and ready to go. A whole squadron had been reported, but that turned out to be a mistake. Only one machine was seen and shot down. We hope that the English will eventually accept the consequences of their many speeches and finally attack.[27]

An *Unteroffizier* serving in the II. *Gruppe* of the bomber wing KG 55 wrote similarly on 19 November 1939: 'It's a bit boring for us anyway. I've been waiting for a few days now for the opportunity to finally fly to the west. Hopefully it will work out today. Of course, I would prefer to fly in the north, where I could bring the miserable assassins the reward they deserve...'[28] An *Obergefreiter* air signalman serving with Air Signals Regiment 3 wrote five days later: 'How I would have liked to have been in Poland too! But as it is we lie in the swamp and dirt of the comms zone. It's cruel when you have nothing to do.'[29]

To break the monotony and uncertainty of the Phoney War, then, the appetite for a fight was growing among some Luftwaffe personnel. An *Obergefreiter* stationed with Luftwaffe Flak Regiment 2 confided to his Berliner parents on 14 December 1939 that his training unit hoped 'to get away soon – preferably to the front. We are all of this opinion'.[30] In addition, Luftwaffe recruitment propaganda exerted a powerful effect over the impressionable young men despite the lull in military action. Deibl, the future Stuka pilot, recalled that by January 1940, 'I was determined that I wanted to become a pilot. After a call in the press that was specifically aimed at high school graduates to fulfil their duty and volunteer for the Luftwaffe, my last concerns were dispelled.' He added that:

> In accordance with the text of the advertisement, I already saw myself as an officer of the 'proud Luftwaffe' in a stylish uniform.

The advertisements were richly illustrated. They showed fighter planes in an impressive dive or a knife-edge pass; enemy aircraft were not depicted in them. The colourful images implied happy, unmolested, fair-weather flight in the midst of war.[31]

Now finding himself queuing in a freezing corridor of Vienna's *Stiftskaserne*, he nervously asked himself, 'Did everyone who waited here so patiently have the same dream in mind as I did? At the same time, I noticed that everyone waiting was significantly taller than me, not to say that I was the shortest in line. I had never noticed the fact that I was only 166 cm tall [5 foot 5 inches] as clearly and at the same time as frighteningly as I did now.'[32] Deibl was also intimidated by the hypermasculine depiction of the perfect Luftwaffe serviceman in Nazi propaganda, fretting that he would not match up to the *Führer*'s idealized eagles. As he shuffled along, Deibl's eyes drifted to another Wehrmacht recruitment poster looming above those who were answering the call:

> I was nowhere near the blond giant with the bright blue eyes and the angular face on the Wehrmacht advertising poster I was just standing next to in the queue. Right next to it, a second poster depicting a group of advancing Wehrmacht soldiers covered the shabby grey-green paint of the corridor wall. Strangely fixated, these men stared into the distance with steely boldness. Where exactly? A lot of confusing stuff was going through my head, and I found the situation increasingly strange and depressing. Soldiers rushing back and forth with briefcases clutched tightly to them didn't even glance at us.[33]

Having volunteering amid the wintry Phoney War, young, keen men like Deibl would soon be required by the Luftwaffe when the weather cleared in the spring. Nevertheless, he was hardly going to swan into the proud organization: for both Luftwaffe officer candidates and NCOs, only five per cent of applicants passed the entrance examination that got them through to the interview stage.[34]

If they made it that far, the real work began when they came face to face with the Luftwaffe's formidable aviation psychologists. The demanding nature of these interviews are revealed in a report submitted to the Americans after the war by Dr Siegfried Gerathewohl, who was a senior psychologist to the Luftwaffe and in charge of an aircrew selection centre during the war. Firstly, as Gerathewohl outlined, there was the 'Exploration by Questioning', which was:

> A penetrating, detailed question scheme whose aim is determined by what has been discovered previously, which embraces a true anamnesis of the personality and which clarifies the relationship to the outside world and especially to the field of aviation. In the investigations are included interest in flying, relationship to things and experiences, attitude towards risks, danger and death.[35]

That the Luftwaffe was especially advanced in this area could be seen in the way that even its recruits were unnerved by being subjected to this emerging science. During his Luftwaffe recruitment assessments in the spring of 1940, Deibl recalled how 'In the afternoon we were handed over to the psychologists, as we felt. Because this science was still young and frightening back then. People automatically associated this discipline with mental illness and no one wanted anything to do with it.'[36] Deibl then recounted:

> I was led to a long table, at the end of which six stern-looking gentlemen were sat. They took turns asking questions that hit me like rapid fire: 'Can you play an instrument? What colour is the Siemens company poster? Are you in favour of the death penalty? Do you masturbate? Have you had sexual intercourse? Someone in your family is murdered – what would you do to

the murderer? When was the current Pope elected?' In this confusing mess, the men took turns at a swift pace. It seemed endless and annoying, with the gentlemen busy taking notes. It was later explained to us that this storm of questions was not aimed at the correctness of the answers, but rather to test the ability to react and resist stress.[37]

The next phase included a more general conversation, where 'the form is unrestrained, free and unbound', while another phase contained 'the analytical exploration (psycho-analysis)', in which 'the form and structure aim at the discovery of the contents of the deep psychological aims. It serves to uncover unconscious relationships.' This phase, however, was designed solely to screen out any candidates where 'emotional malfunction or disturbance is suspected' and who had 'abnormal behaviour reactions because of inner conflict tensions', as 'a tendency towards accidents can develop in this manner'.[38] Only if the subject was deemed stable enough was the 'Aggressive Exploration' carried out:

> The patient should be shaken and the core of his personality broken. The shock effects are intended to stimulate the patient emotionally and to release symptomatically important effects. The release of abnormal symptoms occurs in cases of affective disturbances, like a catharsis. This method is best carried out by analytical treatment. It is important that the psychologist maintains an appropriate control and that an explanation of the facts as well as a reconciliation be made.[39]

Finally, the 'Pedagogic Exploration' was carried out, which focused on 'charitable sympathy and the readiness to help. The examinee must recognize the readiness of the psychologist to help and be aware of his unquestionable authority. The discussion and explanation are made with suggestions but not with restraint.' Thus, it is apparent that Luftwaffe personnel were subjected to rigorous psychological testing in the early years of the Second

World War, which may partially explain why their general breakdown rates were relatively low in the following years.[40]

If a Luftwaffe candidate was able to successfully navigate these daunting odds and obstacles, an even more intensive training scheme followed. The two year basic course for officer candidates began with four months of training that included elements such as infantry tactics, formal training, military discipline, small unit leadership and the legal and administrative fundamentals of the Luftwaffe.[41] Basic training was followed up with a nine- or ten-month training course in which officer candidates specialized within a certain branch of the Luftwaffe, whereupon additional courses would be undertaken that were specially tailored to their role. Training, then, was of a remarkably high and fastidious standard during the Phoney War; but, once the winter thawed, it would be time to put theory into practice.

For the Luftwaffe flak crews tasked with defending the Reich at the turn of spring 1940, the war already seemed to have resumed, although the feeble RAF Bomber Command attacks did not typically give the Luftwaffe much cause for concern in this period. At the time, Nazi Germany counted 197 heavy and 48 light flak batteries across the country, leading to an increased strength of 450 heavy batteries by the summer of 1940.[42] Some of the flak crews felt compelled to explain to their families that they were doing everything they could to stop the British air raids. Erich Dohl – a trained mechanical engineer who worked as a rangefinder in a Luftwaffe anti-aircraft unit – tried to assure his wife and daughters in this vein at the end of February 1940:

> Last night, a plane flew over our position at low altitude. The searchlights were all on, but they couldn't see him through the heavy and low clouds. We didn't have permission to fire, which means it was a test illumination. However, I can tell you that there are enough searchlights and of course guns here in our area. The planes also have anti-aircraft guns just like us.

So, if the evil enemy really comes, he will definitely be brought down by us or our planes. So many can't miss – and many hounds soon catch the hare.[43]

Some Luftwaffe personnel relished the prospect of being unleashed against Britain in a way that the Phoney War had not yet permitted. The *Unteroffizier* serving in the II. *Gruppe* of the bomber wing KG 55 wrote hungrily that 'Hopefully there will now be a nice spring and summer so that we can at least do a lot. Have mercy on the Englishman: he will have to pay for his misdeeds a hundred times over. It's a shame that such a beast exists among the peoples. But we will ensure that it is completely destroyed...'[44] Thus, by the dawn of spring 1940, some of the Luftwaffe's personnel were ready to dish out their displeasure over the 'warmongering' British if required.

As the Phoney War entered its final months, some members of the German public wholeheartedly supported the idea of Britain being subjected to an aerial offensive. A resident of Hann. Münden in north-west Germany wrote on 21 February 1940 that 'The English have done a lot again. Hopefully we will soon bomb their own country properly. It's a terrible bind, our poor boys who fall into their hands...'[45] The paralysing fear of being subjected to an air raid could be seen in the way that a resident of Innsbruck in Austria recalled how 'We had an air raid alarm at 3.30 a.m. today, and I was so frightened that I got a terrible headache and couldn't sleep at all, even though the all-clear was immediately given again...'[46]

In such a paralysed war situation, apparent breakthroughs in the air war were keenly documented by citizens of the Reich. After the Luftwaffe raided Scapa Flow on 16 March 1940, a woman also living in Innsbruck wrote the next day that 'We are now hearing on the radio about the new Luftwaffe attacks in England. There is a lot of hope here for a possible understanding ever since the Finnish–Russian peace negotiations...'[47] A doctor living in Premnitz in north-east Germany, meanwhile, wrote on

20 March 1940 that 'We are very excited about the latest political events: the Finland–Russia peace, the *Führer*–Mussolini meeting, and the government crisis in France. We were very pleased by the great success of our air raid on Scapa Flow.'[48]

A resident of Hemmerde near the Ruhr region of Germany penned just over a week later that 'Now, I'm just curious to see how the war develops in the spring. Our Luftwaffe will probably have to do the main work. But what use are all the assumptions? If the weather just gets better, we'll see whether we can hope to end the war this year. A month doesn't matter if we can only see the Englishman being beaten up on his island. He has to pay for this anyway...'[49] By the spring of 1940, then, it was becoming increasingly evident to both the German public and the Nazi war leadership that the Phoney War had to change.

Hitler had determined at the beginning of the *Sitzkrieg* that any future invasion of France and the Low Countries was intended 'to defeat as much as possible of the French Army and of the forces of the Allies fighting on their side, and at the same time to win as much territory as possible in Holland, Belgium and Northern France, to serve as a base for the successful prosecution of the air and sea war against England.'[50] During a conference on 23 November 1939, he expanded upon this requirement by noting that:

> We have an Achilles heel: the Ruhr. The progress of the war depends on the possession of the Ruhr. If England and France push through Belgium and Holland into the Ruhr, we shall be in the greatest danger... certainly England and France will assume the offensive against Germany when they are armed. England and France have means of pressure to bring Belgium and Holland to request English and French help.[51]

Hitler further claimed that 'In Belgium and Holland, the sympathies are all for France and England... if the French Army marches into Belgium in order to attack us, it will be too late for us.' He implored, then, that 'We must anticipate them… we shall sow the English coast with mines which cannot be cleared,' identifying that 'England cannot live without its imports. We can feed ourselves.' The *Führer* concluded that 'The continuous sowing of mines on the English coasts will bring England to her knees. However, this can only occur if we have occupied Belgium and Holland…'[52]

Yet, on 10 January 1940, initial German plans for an invasion of France and the Low Countries were spectacularly foiled in what became known as the Mechelen incident. After *Major* Erich Hoenmanns gave *Major* Helmut Reinberger an unofficial lift in his Messerschmitt Me 108 '*Taifun*' to a staff meeting in Cologne, the duo crash-landed in Belgium near Mechelen-sur-Meuse – but Reinberger had top secret documents on his person that outlined the proposed invasion of the Netherlands and Belgium. He tried twice to burn his incriminating papers (including asking for a match from a local farmer), but the charred documents were discovered by the Belgian authorities. Upon finding out about the debacle, Hitler snapped that 'It is things like that which can lose us the war!'[53] Nevertheless, this allowed the Wehrmacht to fashion a new version of the plan, which had been code-named *Fall Gelb* ('Case Yellow'), that now departed significantly from the original plans captured in Belgium.

As the wintry Phoney War finally cleared for good, the stage appeared set for a decisive manoeuvre to be carried out against France and the Low Countries during the mid-spring of 1940. A new Luftwaffe recruit based at Kaltenkirchen, near Hamburg, wrote in March that 'We're supposed to get away from here at the beginning of April. Hopefully we'll get to pilot school. There, one has a worthy occupation at least. We notice next to nothing about the war. But something must come in the near future. Hopefully it will turn out well…'[54] Sure enough, the Luftwaffe soon closed in on its next victims; but, to the surprise of everyone involved, they were not British, French, Dutch or Belgian.

4

A Lobster Dinner

'All of us in the Luftwaffe thought we'd be off to England one way or another,' the German air signalman Reinhold Runde pondered thoughtfully. 'No one dreamed that it would be Denmark and Norway.'[1] In early April 1940, German intelligence had reported that Britain and France were going to invade Norway to deprive Nazi Germany of the rich mineral deposits located there, such as copper, iron, cobalt, molybdenum and nickel.[2] But, if the Germans could beat the Allies to this forecasted punch, Scandinavia could serve both as a mineral treasure chest *and* as an additional base of operations for the Luftwaffe, as well as allowing the Kriegsmarine to further consolidate its presence in the Baltic Sea.[3] Thus, on 9 April 1940, the Wehrmacht launched *Unternehmen 'Weserübung'** – 'Operation Weser Exercise' – against Denmark (*Süd*) and Norway (*Nord*).

The importance of briefly recounting the Scandinavian campaign is that, prior to the Wehrmacht's invasion of France

---

* In addition to these two countries, Sweden had been considered for occupation to forge a Nazi stronghold in northern Europe. Sweden, however, was allegedly spared due to Göring's impassioned plea for the Wehrmacht to not invade his much-loved adopted country: it was the homeland of his first wife, Carin, and he had spent many happy years in Stockholm during the inter-war period.

and the Low Countries on 10 May 1940, it saw the first major aerial clashes of that year between Luftwaffe and RAF aviators – further building the former's confidence when the latter ultimately failed to defeat them. In addition, Luftwaffe airmen of all branches gained vital experience in navigating and flying over their first major water-ringed country, as well as proving that they could successfully provide air support to amphibious Wehrmacht landings. On the opening day of Operation Weser Exercise, the German assault force numbered 12,250 men; 3,400 of them were destined for Denmark, but since the other 8,850 men were earmarked for Norway, it was clear which of the two countries was considered to be the harder target.[4]

Given that Denmark already shared a land border with Germany, it was not too difficult for the *Heer* and Luftwaffe to apply the same dizzying *Blitzkrieg* success that they had carried out in Poland. Crossing the border with little resistance on 9 April 1940, the *Heer* quickly seized Copenhagen and completely overran Denmark in just twelve hours. The attack had taken the city's inhabitants completely by surprise, with the naval guns that defended the port all blocked up with grease and unable to fire. Moreover, the ground fire was so weak that it barely grazed the incoming German aircraft.[5] The ease with which Operation Weser Exercise South was carried out could be seen in the way that the heart of the Danish capital was secured by German soldiers on bicycles![6]

While the *Kriegsmarine* landed more soldiers on the coast and neighbouring Danish islands, the Luftwaffe's key breakthrough had come two hours into the invasion, when it landed its *Fallschirmjäger* ('paratroopers') onto the airfield at Aalborg. Capturing this harbour city meant that it could serve as a vital halfway house between the Reich and Norway, as well as another vantage point from which British shipping and coastlines could now be harassed. The airfield was only lightly defended, with Luftwaffe intelligence estimating that it wielded just fifteen obsolescent Hawker Furies with which to defend itself.[7] Lacking flak defences and further rocked by the Luftwaffe after it landed additional shock troops – who immediately seized key bridges, communications hubs and garrisons – Aalborg and Vordingborg quickly fell into German hands.

The Luftwaffe's role of softening up Danish airfields, along with the bomber wing KG 4 'General Wever' dropping leaflets over Copenhagen telling the residents 'Do Not Resist', proved sufficient in light of the Danish army being so easily overrun on the ground.[8] This was a powerful example of where the Luftwaffe's painstakingly cultivated image as an invincible air force was highly successful. Eager to prevent Copenhagen from going up in flames as Warsaw had done the previous year, King Christian X swiftly conceded defeat.[9] The fact that Denmark crumpled so easily appeared to confirm that Case White had not just been a fluke: Hitler's Wehrmacht was truly a force to be reckoned with by the spring of 1940.

However, the Luftwaffe did not take the threat of British air raids lightly during their military consolidation in Denmark. Wilhelm Johnen, later one of the most decorated German night-fighter pilots, recalled that 'The British not only flew by day – they also attacked important military objectives in Denmark by night. The RAF pilots flying over the North Sea at night already displayed a skill which forced the Luftwaffe leaders to take their activities seriously.'[10] Nevertheless, the utter speed of Operation Weser Exercise South's execution could be seen in how Karl Bruns was summoned almost overnight from a merry, wine-filled evening in Schwerin in north-east Germany to expand the Luftwaffe's new airfields in Denmark:

> It was 10.30 in the evening when the telephone suddenly rattled. Krüger from the construction office calls me. That all-knowing fellow has actually managed to track me down, he knows us well! After a short while, the connection with the *Luftgaukommando* ['air district command'] was established.
> 'Bruns here!'

'[*Oberbaurat**] Wittneben here! Bruns, you have to stop your work in Schwerin immediately. You will get new tasks in Denmark.'

'Yes, but –'

'No buts! You will report to me in Hamburg at eight o'clock tomorrow morning. You will receive an order to expand two airfields; you will find out everything else tomorrow! That's all!'[11]

Bruns now found himself frantically scrabbling to get his team together in time to get to Aalborg; one of his colleagues received just an hour's notice.[12] Norway, on the other hand, was going to be a harder nut to crack than Denmark: in addition to lacking a land frontier with Germany, its jagged topography presented multiple challenges to the Luftwaffe. With the *Kriegsmarine* forced to navigate the perilous nooks and crannies of the 1,200-mile coastline, the increased harassment from Allied air forces over Norway necessitated a swift campaign that required all three Wehrmacht branches to work closely in tandem. As Walter Lehweß-Litzmann noted:

> Since transport problems were expected when conquering Norway's extensive territory, transport machines were primarily needed. The majority was provided by *Lufthansa*, which supplied its Junkers Ju 52s, Junkers Ju 90s and Focke-Wulf 200s, including crews attached to transport and supply units. Their actions were coordinated on our staff by the Director of *Lufthansa* and internationally known aviator, Baron von Gablenz, with the rank of *Oberst*.[13]

Five hundred Ju 52 transport aircraft were assigned to the campaign over Norway, some of which were crammed with paratroopers from *Generalleutnant* Kurt Student's 7. Air Division.[14] In addition, seventy reconnaissance and coastal

---

* *Oberbaurat* = Chief Building Surveyor.

aircraft were committed to scope the situation over Norway, with a hundred Bf 109s and Bf 110s deployed to provide air cover. In total, the X. Air Corps mustered 1,000 aircraft for the invasion.[15] The Luftwaffe had recognized the educational potential of Operation Weser Exercise North, with 340 of the transport aircraft pilots having been selected from the pilot training schools.[16]

The Luftwaffe's paratroopers took the airfield at Stavanger on 9 April 1940 as part of the commencement of Operation Weser Exercise North. A document later filed at the Nuremberg trials by the Norwegian government, entitled 'Germany's Crimes Against Norway', recalled how the invasion 'brought war to Norway for the first time in 126 years. For two months the war was carried on throughout the country, causing destruction amounting to 250,000,000 *kroner*.'[17] Compared to the Danish walkover, however, the struggle put up in Norway proved to be challenging for the Luftwaffe, with a planned paratrooper drop over Oslo later being aborted due to unfavourable weather conditions. This allowed the Norwegians to hang on for a bit longer, as they retained control of the rail and road communications which connected Oslo to Trondheim further north – the lifeline that kept the strategic port of Narvik alive for the natives.[18]

Nevertheless, Oslo soon fell after Ju 52s were able to get through and drop off a German infantry division over the largely uncontested Fornebu airfield.[19] In addition to the main cities, a whole host of Norwegian towns and villages were targeted by the Luftwaffe – from Elverum, Nybergsund and Kristiansund, to Molde, Namsos, Steinkjer and Bodø. The unremitting brutality of some German airmen during Operation Weser Exercise was seen when Bodø hospital was bombed on 27 May 1940: the Norwegian government reported to the Red Cross that the patients who were being urgently evacuated from the burning building were strafed by the Luftwaffe aircraft that had attacked the hospital.[20] In addition, some German airmen who had been captured by the Norwegians had orders on them which noted that 'all movements on the roads were to be attacked.'[21]

With the increasing cauterization of *Kriegsmarine* sea transports by the Allies, though, mounting pressure was being placed on the X. Air Corps to airdrop the necessary men and materiel to reinforce German positions in Oslo, Kristiansund, Stavanger, Bergen, Trondheim and Narvik.[22] The Battle of Narvik, in particular, saw the RAF desperately attempt to stop the Luftwaffe from supplying these vital reinforcements, initially affording crucial breathing space to Allied forces on the ground.[23] On 18 May 1940, *Oberleutnant* Karl Tyezka* was briefing his paratroopers ahead of being dropped over the town: he recalled how 'We take turns trying to take in some of the beauty of Norway's mountains through our small windows. We only see rough, jagged mountain peaks, some of which are covered with snow – and this enjoyment doesn't last long, either.'[24]

As the Luftwaffe transport aircraft climbed up to 2,500 metres, it seemed as if it were gliding across a fluffy carpet of clouds; there was rarely a gap, although stunning views of Norway's dramatic landscape occasionally peeked through the wisps. Tyezka added that 'We've been flying for over two hours now, filled and thrilled by the austere, sparse beauties of this flight. But the thoughts are already lingering for a time in Narvik. Our actual goal: "Are we going to get through, can we still manage to be of any real help? What awaits us there?" These questions should soon be solved for our aircraft.' Drinking in the languid views, Tyezka's ponderings were rudely disturbed:

> Suddenly, one of my people shouts: 'An He 111!' I try as best I can to look out of our small window and actually see an He 111 roaring along in the opposite direction. We then give back in to our thoughts. The man at the window shouts again: 'An

---

* Tyezka later had his surname Germanized to 'Tannert' in February 1942, likely because of its Polish origins.

enemy aircraft!' A brown biplane is heading towards us from the sun. We are blinded, and the distance is too great to determine exactly what kind of warbird it is. We have hardly sat back down on our seats when a long burst of fire tore through the air. Our rear gunner fired. We were still listening to the shots when they were already hailing into our machine. A stream of bullets riddled the machine from behind, starting at the front. Some men jump up: 'I've been hit, I'm wounded!' come the shouts.[25]

Tyezka attempted to reassure his agitated men, though the transport aircraft now jerked around erratically as it took evasive manoeuvres against the enemy fighter, which had now been identified as British. As he passed around the aircraft, checking who had been wounded, his blood suddenly froze as the jarring horn to jump blared out. Ochre flames were now licking at the fuel tank; if they did not jump now, they would be blown sky high. The men's anxiety further spiked when they were unable to open the hatch – 'Damn, the door is closed!' – but it finally gave way with a desperate tug.

Acrid smoke was spewing into the aircraft, stinging the eyes of the paratroopers as they stumbled blindly towards the exit. Tyezka was grateful to observe, however, that his men's training instincts kicked in during the jump: 'Thank God, every man kept his calm during these critical encounters and survived the situation with iron nerves.' His eyes darted around as he drifted further down towards the Earth. 'Wherever may we be – where should we land?' he mulled in his head, before a 300-metre-thick cloud formation quickly forced him to pull his legs up for fear of clipping the edge of a mountain.

His relief was palpable when he came through the other side and 'Gradually, the earth shines through again.' Tyezka's account over Narvik ended up being published in a Luftwaffe propaganda book in 1941, so one must be discerning when reading such heroic accounts. Nevertheless, they were hardly the only Luftwaffe personnel with a perilous story of encountering

British fighters above Norway. In late April 1940, a band of He 111s from the bomber wing KG 54 took off from Celle in north-west Germany for Trondheim, located over 600 miles away.

Obergefreiter Kurt Hiekel, a flight mechanic on one of the He 111s, recalled the first half of the leg-numbing journey: 'my pilot William Czech* from Berlin was sweating when he landed on the narrow airfield in Aalborg, which was overcrowded with bombers. I also felt a little uneasy in my stomach, because the brakes were barely able to cope with the overload.'[26] Taking off from their halfway base at Aalborg on 25 April 1940, Hiekel's crew settled in for the last leg of the long flight to Trondheim...

> We had now been heading towards the combat area in a formation for over two hours. I was sitting on the folding seat next to *Unteroffizier* Czech, behind our observer, *Unteroffizier* Theodor Kutsche, who was a teacher in his civilian career. Even further back, with his back to the direction of flight, our radio operator, *Gefreiter* Siegfried Fernes, was dozing.

Unbeknown to the Heinkel aircrew, however, eighteen British Gloster Sea Gladiators had been summoned to Trondheim at the same time. Despite having no ground crews, fuel or spare ammunition, and buffeted about by snow drifts, the Gladiators set out that same day in the desperate hope of catching any passing Luftwaffe aircraft. As the German bomber approached Trondheim, Hiekel recalled that his only major concern was if their temperamental He 111 would survive the journey:

> We flew north past Oslo into the Trondheim area. The visibility was good, so that the city could be clearly seen from a height of 6,000 metres. This time the two engines of our plane, heavily

---

* It was no doubt unfortunate for a Luftwaffe airman to have both the English form of 'Wilhelm' as a first name and the surname of an occupied nationality during the war!

loaded with fuel and bombs, hummed without disturbance. This was by no means normal; we considered our Heinkel, the 'Gustav', to be a defective bomber, because we had already been through all sorts of troubles with her.[27]

But, as their heavily laden Heinkel groaned on, the men's stupor was shaken by the dreaded words over the radio: 'English fighters!' Hiekel immediately threw himself into the nose of the bomber to look through the air gunner's front window, but did not see anything at first. The radio operator then went to fire at the fighter that suddenly popped up, but 'Our right engine was already producing a trail of smoke.' Hiekel scrambled up to pull the lever to lock the landing gear, in case a drop in oil pressure affected it.

The flight mechanic ordered Czech to turn off the smoking engine and the Heinkel crew attempted to turn back, but the bomber quickly began to slump from the drag on the 'dead' engine side. The pilot then instructed Kutsche, the observer, to make an emergency bomb drop without setting them to detonate. Unfortunately, Kutsche had panickily jettisoned them as live bombs; Hiekel admitted that 'I preferred not to look down to see where they fell.'

Try as they might, the flagging Heinkel simply could not withstand the rest of the journey on just one engine. Czech briefly thought about where they could make an emergency landing – but the damaged Heinkel made the decision for them. They were forced to ditch the bomber in the icy waters of the Skagerrak strait between the Jutland peninsula in Denmark and the east coast of Norway:

> We got into our dinghy and started vigorously paddling southwards. Gradually, our aircraft sank away behind us and we were completely alone on the wide expanse of water. There was no land to be seen. We were quite afraid that we might be captured by British submarines that appeared and were operating in the sea area against German maritime supplies. We fired

several flares. After a few hours, when it was already getting dark, we were finally able to draw the attention of a German *Kriegsmarine* coastal protection boat to us.[28]

Thus, against all odds, the British Sea Gladiator pilots had truly lived up to their aircraft's name by shooting down five of KG 54's He 111s.[29] Although the Luftwaffe were gaining vital combat experience in flying over greater stretches of water, the dangers of becoming submerged in their freezing depths were thus becoming evident to the Luftwaffe aircrews deployed in Operation Weser Exercise North. Lehweß-Litzmann claimed that 'The "success" we achieved was far from assured, even though we occupied Oslo and five other port cities.'[30]

Towards the end of the month, the Luftwaffe required a further concentration of their attacks against Allied naval and land forces at Namsos and Åndalsnes. Consequently, as Lehweß-Litzmann further noted, the Luftwaffe 'flew attack wave after attack wave with all available aircraft from our air bases located in southern Norway, Denmark and northern Germany'. To counter them, 'The Norwegian Air Force, including training and reserve aircraft, only had around 200 aircraft – divided into an army aviation unit and a naval aviation unit – as well as a barely developed anti-aircraft defence system. These forces inflicted only minor casualties on our massive effort.'[31]

By early May 1940, the Luftwaffe's relentless attacks on both Allied troops and the Royal Navy forced their retreat from Namsos and Åndalsnes. On 12 May 1940, however, the Allies rallied and switched their attention to Narvik, with British, French and Polish forces eventually ousting the Germans from the town on 28 May.[32] Yet, with the Germans concurrently waging Case Yellow and then *Fall Rot* ('Case Red') against France and the Low Countries, the Allies found that they could not salvage the situation in Scandinavia.

In frustration, they withdrew from Narvik on 8 June 1940; two days later, Norway sank under the swastika. The Luftwaffe had not come out of the campaign unscathed, with 260 aircraft destroyed over the sixty-two days; the British, in comparison, lost 169 aircraft.[33] In a reflection of the treacherous weather conditions and challenging topography in Norway, nearly eighty of those Luftwaffe aircraft had been lost through accidents.[34] In addition, as Lehweß-Litzmann testified, 'The few Norwegian pilots heroically threw themselves at our bombers and destroyers and engaged them in short dogfights.'[35]

The RAF, too, had kept hold of air parity long enough to vastly complicate the Luftwaffe's aerial operations over Norway, allowing the Royal Navy to inflict considerable blows upon the *Kriegsmarine*.[36] Klein conceded that 'Soon after Narvik, most of the old [Luftwaffe] crews had been killed,' and were 'replaced by inexperienced trainees'.[37] Yet, Operation Weser Exercise constituted an important moment in which the RAF had been unable to fully suppress the Luftwaffe.

Even more worryingly for Britain, the sea-faring nation, the Luftwaffe's aerial support had enabled the other Wehrmacht branches to launch successful amphibious landings, while the German air force had also managed to airdrop elite *Heer* units to capture Norway's islands. During the Battle of Britain, a captured *Leutnant* was later overheard by British intelligence claiming that his pilot brother 'was especially proud of having transported the celebrated Austrian mountain troops to Narvik'.[38] Sometimes, the Luftwaffe's pertinacious Ju 52s even went on one-way trips, bellyflopping onto nearby icy lakes to keep the Austrian mountain troops in the fight.[39]

The Luftwaffe's accumulated experience continued to grow, with Klein claiming that as a rear-gunner on an armed reconnaissance He 111 bomber, 'I was so experienced [over Narvik], no fighters could get the better of me.'[40] This is not to say that operational naivety did not impact other aircrew members, however, with Klein describing a hairy moment over Narvik in which his pilot – a former Lufthansa employee – misunderstood Klein's order to 'Pull!' (fly up) and 'Press!' (go down) as a Spitfire was breathing down their necks! Nevertheless, the German occupation of Norway and Denmark, in Brun's words, 'surged through all

German hearts like an electric shock'.[41] Scandinavian riches swiftly filled German pockets, with Lehweß-Litzmann recounting how:

> In Oslo, we got a few rooms in the largest hotel on *Karl-Johanns-Gate*, the main street that leads to the harbour on one side and to the royal castle on the other. The hotel employees were hired to take care of the food and they immediately started preparing a lobster dinner, which for us was a real feast that we didn't know back home in Germany.[42]

The boost to morale within the Luftwaffe could be seen in the Stuka pilot Paul-Werner Hozzell's recollection that 'We flew at low level above the waves in the direction of the Norwegian coast. Somehow, a happy feeling came up between the aircrews. Pressing the button of the radiophone, I heard my men singing. They sang the song of the *Kriegsmarine* because all of us felt so close to the sea now: "*Wir sind Kameraden auf See*" ("We Are Comrades at Sea").'[43] Most concernedly for the British, then, the Luftwaffe were now acclimatizing to conducting aerial operations over water-framed nations. Yet, as Lehweß-Litzmann notes, Operation Weser Exercise also had disturbing implications for the lengths to which the German armed forces were willing to go for the *Führer*:

> The sudden occupation of Denmark and Norway marked the first serious violation of neutrality in the Second World War. Actually, it should have caused movement, concerns, queries, reservations or even protests in the officer corps, whose older representatives had already taken part in the illegal occupation of neutral Belgium a quarter of a century ago. However, I don't remember noticing such reactions, perhaps because I didn't think about it myself.[44]

Hitler's blatant disregard for international neutrality was further demonstrated by the updated plans for Case Yellow, which were being refined during the opening weeks of Operation Weser Exercise. Were a future invasion of the Soviet Union to take place, he was concerned that Germany's western flank could be harassed by Anglo–French forces if they were supported by neutral Belgium and the Netherlands. The *Führer* was not fussed about violating Belgian neutrality, tarring the Belgians with the same brush as the French due to their mutual defence negotiations.[45] However, he was initially less convinced about the strategic benefits of breaching Dutch neutrality: in October 1939, he had objected to an early proposal for the *Heer* to advance through the Maastricht strip during Case Yellow.[46]

With Case Yellow being postponed twenty times from 12 November 1939 up to 10 May 1940, however, Jeschonnek was slowly able to convince Hitler of the need to suppress the potential threat of the RAF operating from Dutch airfields.[47] In addition, the updated Case Yellow plans following the disastrous Mechelen incident highlighted that marching through the Netherlands could be strategically advantageous. Instead of a predictable German thrust straight through Belgium into France, the *Heer* would simply conduct diversionary attacks at France's heavily protected Maginot Line while advancing through the sparsely protected Ardennes forest. By seizing the Netherlands and Belgium at the same time, Allied soldiers stationed in the latter would have a key northern avenue of retreat cut off *and* would also be separated from their counterparts in northern France.

Hitler eventually concluded that the 'Breach of the neutrality of Belgium and Holland is meaningless. No one will question that when we have won.'[48]

The Luftwaffe knew Case Yellow would take every ounce of strength it had, with a report by the bomber wing KG 1 'Hindenburg' noting on 8 April 1940 that the aim would be to 'destroy as many aircraft as possible – primarily fighters – along with the fixed installations of the airborne organization and the supply of the enemy air force'.[49] But, for as long as France and

Britain stayed in the fight, Hitler's mania for more *Lebensraum im Osten* would remain threatened. If the Weser Exercise had been the equivalent of a castling move in chess, Belgium and the Netherlands would now be the final pieces that would unexpectedly put the French, and therefore Great Britain, into checkmate.

5

Poor *Poilu*!

10 May 1940. During the witching hour, hundreds of German bomber crews across the Reich began to run their engines and go through their pre-flight checks. Near the Netherlands, a band of He 111s and Junkers Ju 88s from the bomber wing KG 4 were already soaring close to the Maas estuary – but, with the order to penetrate Dutch airspace not before 03.50 a.m., the early birds circled around the Channel to kill time.[1] Eventually, at the crack of dawn, six wings of He 111s and Do 17s – around 300 bombers – attacked more than twenty airfields scattered across northern France, Belgium and the Netherlands.[2] A total of 4,000 aircraft had been designated for the campaign by the Luftwaffe[3], 2,750 of which were earmarked for the French aspect of Case Yellow.[4]

The typical *Blitzkrieg* opening of knocking out as many enemy aircraft on the ground as possible ran like clockwork: by the next day, the Dutch and Belgian air forces had been almost obliterated, while the French *Armée de l'Air* and the RAF were left reeling. Having seized tiny Luxembourg on 10 May, the Germans were met with a valiant but futile attempt by the Dutch to delay the Wehrmacht's advance. The Dutch Army Aviation Brigade had been bombarded so intensely in the preliminary *Blitzkrieg* that it was only able to operate ten of its original 220 aircraft on 13 May 1940.⁵

The situation looked even graver when paratroopers from Student's 7. Air Division were dropped over the Netherlands. They immediately seized important defensive positions and vital objectives like the Moerdijk bridge that lay on the way to Rotterdam.⁶ One of the Luftwaffe paratroopers who had landed, *Hauptmann* Karl Lothar Schulz, acknowledged that 'The Dutchman fought bravely, harder than we expected from a people who had not waged war for over a hundred years.'⁷

Nevertheless, the Dutch troops were horror-stricken as hundreds of parachuting troops floated through the air like dandelions. As more German troops scrambled out of gliders, lorries, transport aircraft and even seaplanes, the Dutch soon realized that this was an overwhelming assault.⁸ Their main forces surrendered on 14 May 1940, though the request from a Luftwaffe air signals unit at Waalhaven in Rotterdam to postpone the bombardment of the city tragically did not reach Air Fleet 2 in time.

So, the Luftwaffe sent in a hundred aircraft to attack Rotterdam.⁹ Fifty-seven He 111s dumped high-explosive bombs on the city centre that ignited the innards of a margarine warehouse and broke Rotterdam's water mains. A total of 980 people were killed in the raid and a further 78,000 people were made homeless.¹⁰ A Swiss magazine, the *Zürcher Illustrierte*, featured an exclusive article entitled 'Rotterdam – Queen of the Meuse – Victim of the Air War', which published the pictures that a Swiss man had taken shortly after Rotterdam's destruction:

> Various circumstances made him hesitate to show his photos to the public until today. We publish it with the conviction that truthful reporting is our task. Our Swiss informant lived in Rotterdam. He lived there on 10 May, the day the war broke out. On 11 May he was no longer safe in his office from the stray machine-gun bullets of the German paratroopers, who

had penetrated the city. The heavy bomber *Staffeln* were already flying over the city all the time… the bombs had done their job. As far as Gouda, 24 kilometres away, charred remains of paper had fallen incessantly from the sky without wind, witnesses to the fire in Rotterdam. On that day, Holland capitulated – the soldiers had stood at the border for fourteen months – and many collapsed, shaken.[11]

Schulz, on the other hand, recalled the gleeful wonder of the Luftwaffe paratroopers as they gazed at the German fighters and bombers roaring above the beleaguered city: 'Brightly illuminated by the sun, they pass us as if pulled forward by invisible rubber cords. They are much faster than us. Shortly afterwards, when the Messerschmitt fighters and destroyers literally whizz past us and below us, I have to, in view of this wonderful picture, think of the funny, popular joke: "When Hermann lets his planes fly, the birds walk!"'[12] Peter W. Stahl watched as his bomber colleagues returned from attacking Rotterdam on the second day of Case Yellow:

> Crews come and go at the command post. Oil-stained combis, yellow life jackets, fur boots and yellow covers over the hoods. I hear snatches of conversation: 'Lots of fighters over the target' – 'Falling on ships in the harbour' – 'Convoy in the canal' – 'Rotterdam is burning…'[13]

Despite the Luftwaffe's evident domination, however, any combat zone naturally came with its own risks. When one bomber landed, Stahl recalled that 'The entry hatch opens. I see a bloody hand feeling its way down. Helpers pull the seriously wounded gunner out of the floorpan. He immediately collapses and has to be transported away with a *Sanka* ['ambulance'*]. Neck shot.'[14] Back in Rotterdam, meanwhile, the paratroopers kept a keen eye

---

* '*Sanka*' is short for '*Sanitätskraftwagen*', or an ambulance.

out for British aircraft. Schulz noted that 'The enemy pressure on our northern bridgehead became ever stronger. But our men held out in the houses, some of which were already burning, and did not give up the bridgehead.'[15] He added of the first night in the area:

> As expected, the Englishman came. Every five minutes, bombers flew to the *Plas*, threw flares down and, illuminated by their glow, high-explosive bombs were dropped elsewhere; then they fired machine-guns and cannons into the area from a height of almost 50 metres. But light flak was soon on the scene, repelling the British and forcing them to high altitudes. During the British bombing raid, six of my people were wounded and a building filled with ammunition was hit. There was a wonderful fireworks display. All the ammunition and a lot of flares went off. Days later, our feet were still treading on exploded cartridges and bullets that were scattered everywhere.[16]

Once again, the RAF had failed to counter the Luftwaffe effectively; yet another European country easily folded under the Wehrmacht's insurmountable will. Belgium, too, was faring poorly under the German barrage. The Belgian Army Air Force had possessed a mere 250 aircraft by the spring of 1940 – just 20 per cent of which could be classed as modern.[17] In the opening days of Case Yellow, the seemingly impregnable Fort Ében-Émael on the Belgian–Dutch border was breached after Luftwaffe bombers created a disorientating smokescreen, before waves of paratroopers were released over the fortress. Adolf Galland, whose fighter wing provided cover for this imaginative raid, recalled that:

> The sector where they were to be dropped lay deep inside Belgium, and the action could not be reconciled with the general plan of German operations, as far as this was known to us. What dropped from the transport planes and sailed down

to earth were – dummies. On landing, these invasion dummies set off a complicated mechanism which produced a good imitation of battle noises. The Belgians were deceived and flung considerable forces into the supposed danger area. Their absence from important defence positions was of great advantage to the attacking Germans.[18]

Capitalizing on the Belgian confusion, hordes of Ju 52s then towed over the fort gliders containing more than seventy men, who promptly landed and stormed the fortress.[19] Slinging explosives down the barrels of the artillery guns to disarm them, the German soldiers then used flamethrowers to pin the entire Belgian garrison of 750 men in the fort's corridors until more *Heer* reinforcements arrived.[20]

The airfields at Liège, Maastricht and Brussels were captured in turn, and although Belgium limped on until 28 May 1940, it fell into Nazi hands after eighteen days of fighting. By then, only France continued to hang on during Case Yellow, albeit by the most rapidly fraying of threads. By 12 May 1940, the Luftwaffe had wiped out around half of the *Armée de l'Air* and the RAF's aircraft in France.[21]

While General Joseph Vuillemin desperately tried to conserve what he had left of the French air force through dispersing it across the country, he only succeeded in further diluting its defences. Although the *Armée de l'Air* had managed to increase its number of modern aircraft from 500 to 1,100 during the Phoney War,[22] its main fighters at this time – the Bloch 152 and Morane-Saulnier MS 406 – lagged behind the Germans' Bf 109 by 50 or 75 miles per hour respectively.[23] Provided with close air support from the Luftwaffe, the *Heer*'s infantry and panzer divisions punched through the Ardennes forest.

Here, the lack of French airpower was felt keenly. Slicing through the Meuse line at Sedan on 13 May – where the 1. *Panzerdivision* had been heavily assisted by Stukas breaking down the morale of the French troops – the Luftwaffe's flak batteries and fighter pilots successfully fended off Allied attempts to dismantle the Meuse bridges. The Luftwaffe's fighter pilots were at their most deadly on 14 May, declaring it to be the *Tag der Jagdflieger* ('Fighter Pilots' Day') after they brought down more than ninety British and French aircraft and destroyed 56 per cent of the RAF's Fairey Battles and Bristol Blenheim light bombers.[24]

The Luftwaffe's successes in the skies over France bolstered the confidence and morale of the German troops down below. On 19 May 1940, a soldier with a *Heer* reconnaissance unit wrote:

> We are only now getting to know our loyal companions, the proud Luftwaffe, and what they are good for. We like to see the Messerschmitt fighters and the Stukas the most. The iron conveyor belt of the panzer units rolls in front of us. We have also crossed the Maginot Line and nothing... you can believe it, as they say on the radio, the German troops are advancing according to plan. Now I'm experiencing it myself. I want to close today with the words: together with God for Germany's final victory...[25]

On 20 May 1940, the French coastline at Noyelles-sur-Mer in northern France was finally spotted by the advancing *Heer* at Abbeville; the two halves of the Allied armies were now 50 miles (80.5 km) from one another. Capturing Amiens on 21 May, the Germans closed in on the 1.7 million Allied soldiers who were trapped in a coastline pocket in northern France.[26] Such was the coercive effect of the Luftwaffe during this period that it sometimes held sway over the local population in a way that German boots on the ground did not. One *Heer* signalman with the German infantry wrote on 22 May 1940 that:

> Most of the civilian population had fled, but everywhere was overtaken by the German troops and they were now returning

to their villages in endless treks. One encounters terrible misery there. You ask yourself again and again why did the people just flee, and when you ask them: 'Oh, the German planes, the bombs!' It was only the fear of the German Luftwaffe and their unrivalled Stukas that drove them away.[27]

On 26 May 1940, the Luftwaffe set its sights on harassing the retreating Allied soldiers who were scrambling to be evacuated from the small port of Dunkirk. Knowing that the BEF had reached the end of the line in France, an audacious seaborne evacuation – Operation Dynamo – was underway. Royal Navy ships, together with about 850 British civilian vessels, attempted to rescue as many Allied soldiers as possible from two concrete moles at Dunkirk's outer harbour. On both 29 May and 1 June, the German fighter pilots outflew Fighter Command in terms of sorties conducted and the number of aircraft they had amassed.[28]

There is an entrenched narrative surrounding the 'miracle of Dunkirk', however, that Fighter Command had faced a much hardier and more tireless opponent than itself. In reality, the Luftwaffe was already feeling the strain as it headed into the air battle above the beaches. The Bf 110 *Zerstörer* ('destroyer') – a twin-engined fighter that had previously seen great success in Poland, Denmark and Norway – numbered around 350 at the beginning of Case Yellow; by the start of Dynamo, its units had lost more than 40 per cent of the operationally ready Bf 110s.[29] The two weeks of combat losses leading up to Dunkirk had also drained the Luftwaffe's bomber strength considerably, while serviceability rates were severely affected by the use of forward airfields.[30]

This is not to denigrate the remarkable accomplishment of the RAF in holding back the Luftwaffe long enough for 338,226 Allied soldiers to be evacuated by 4 June 1940. Instead, it is to accurately depict the accumulative losses that affected the might and mindset of the German air force a month prior to the Battle of Britain. In addition, the psychological impact of more frequent and intense dogfights with modern British fighters over Dunkirk – particularly the revered Supermarine Spitfire Mk Ia, which had a slightly faster top speed than the Bf 109 E-3* – should not be underestimated. During the nine-day air battle, it is estimated that the RAF lost around a hundred aircraft in exchange for ninety-seven Luftwaffe aircraft shot down, albeit with twenty-eight Bf 109s and thirteen Bf 110s that were damaged but could be repaired and made operational again.[31]

The redoubtable Spitfire and its stalwart comrade, the Hawker Hurricane, posed new dangers to the Luftwaffe bombers and their fighter escorts that would soon carry on into the Battle of Britain. Certainly, the Bf 109 performed better than both the Hurricane and Spitfire at higher altitudes, had a more impressive rate of climb and was more comfortable in a steep dive. Another key advantage was its fuel-injected Daimler-Benz 601-A engine, which gave it a certain technological edge over the Rolls-Royce Merlin engine fitted within the British fighters. Unlike the latter, in which the engine's float chamber was prone to flooding with fuel under negative g-force, thus forcing the carburettor to cut out, the Bf 109's fuel-injected engine reduced its chances of stalling and kept it in the fight for longer.[32]

Nevertheless, both the Hurricane and Spitfire were normally able to outturn a Bf 109 and especially the Bf 110, with both aircraft wielding their own unique advantages. Spitfires were swifter, lighter and more agile; Hurricanes contained a more stable gun platform, were produced faster, and were easier to throw back up

---

* The Spitfire Mk Ia's top speed was 354 mph at 18,500 feet, though it is often cited as being 364 mph. Mitch Peeke points out, however, that the latter figure was more likely to be the maximum speed without combat loadings. The Bf 109 E-3, on the other hand, reached a maximum speed of 348 mph at 17,500 feet. Such figures naturally varied at different altitudes and among variants, so some mild fluctuation in statistics is to be expected. No sources, however, put the Bf 109 as being faster than the Spitfire in this period. See M. Peeke, *1940: The Battles to Stop Hitler* (Barnsley: Pen & Sword, 2015), 179.

after sustaining combat damage, as they were less complicated to repair. The RAF presented an unprecedented challenge for the Luftwaffe: although the fighter pilots the Germans had faced in previous campaigns had not lacked courage and tenacity, their aircraft had often been outclassed in speed, armament or manoeuvrability – possibly a combination of the three. What Fighter Command lacked in combat experience and the offensive initiative was made up for by its modern fighters and greater parity with the Luftwaffe.

For the Luftwaffe, Dunkirk hung somewhere between a powerful assertion of its military prowess and bitter disappointment at Hitler's infamous *Halt* order, which prevented them from trying to knock the Allies entirely out of the war. Bob recalled that 'It was planned to exterminate the British Army at Dunkirk through the Luftwaffe. It was our task to protect the bomber attacks against the British fighting planes.'[33] He willingly conceded, though, 'It is a fact that we did not succeed in preventing the British army from escaping to Great Britain. The British soldiers escaped in large and small boats to the British island and their great number of boats (thousands) made an effective attack nearly impossible.'[34]

Galland similarly noted that 'Although Dunkirk was a heavy blow for England and had a political rather than a military effect on her French ally, for Germany it was nothing like a total victory.'[35] Indeed, it is important to document the exhaustion that Luftwaffe personnel were feeling after flying up to ten sorties a day during Case Yellow. One Luftwaffe airman later recounted the scene that greeted him on 26 May, and although it was for an air war propaganda book, the recollection nevertheless demonstrated the bloody struggle heading into Dunkirk:

> Above us, beside us, below us, ever closer, ever better aimed, the heavy shells of the enemy flak explode. You can already hear it over the engine noise of your own machine. In a moment, we have to go in. Every bomb is not only dropped, but also carefully

and with great precision. You see everyone. The other machines too. That gives us new courage in this hellfire of tightly closed defence batteries. Poperinghe, Bailleul, Armentieres, Ypres. This fiery square is flown over again. We still have bombs. And Dunkirk burns to starboard ahead. Oil tanks have been in bright flames for days.[36]

The airman also spoke of the grisly injuries he sustained in the raid, noting that 'Something is burning on the left side of my face. It is not worth mentioning. The cockpit was hit and what appeared to be shards of glass are smeared across my face. The firing guns spray around us again and again, and the banging never ends.'[37] For the German airmen forced to bale out or who lost their comrades the harsh realities of Case Yellow were at odds with the lionizing propaganda of the Reich. Yet, for other Luftwaffe airmen, the disarray among the Allied forces simply confirmed their belief that their enemies could not hold out for much longer. The Stuka pilot Helmut Mahlke recalled that:

> As we flew back to base, I couldn't get the horrific spectacle that was Dunkirk out of my mind. An absolute inferno! Those poor beggars down there desperately fighting for survival. Just as well that at our height we were spared the gory details. True, the enemy was also doing his utmost to shoot us down. Just how much lead had been aimed my way again today I dreaded to think. But those men down there, trying to escape but packed together so tightly on board their ships that they could hardly move, didn't stand a chance if set upon by Stukas. Why didn't they lay down their arms? They'd lost this battle. There could be no argument about that.[38]

Indeed, the impressive performance of the Luftwaffe over Dunkirk could be seen in the gushing praise that German soldiers on the ground heaped upon their airborne comrades. An *Unteroffizier* serving as part of the *SS-Verfügungstruppe* (Combat Support

Force)[39] near Hazebrouck in France wrote on 30 May 1940: 'We can understand why the enemies are outrageously afraid and therefore angry at our planes. They ruthlessly drop their bombs on the ordered targets, and the Stukas rush at the enemy at full speed. Our planes are so good that our own troops, who are in the front lines, are afraid of *our* planes. You could be mistaken!'[40]

A *Gefreiter* of the *Heer*'s Infantry Regiment 85 penned on 5 June 1940 that 'The day before yesterday, German bombers and fighters flew over our positions in the direction of Paris. When we heard the Wehrmacht report yesterday, we knew that they had once again done a great job. I am convinced that they will soon be doing their work on our front as well. In fourteen days, we will be in Paris and the war in France is coming to an end…'[41] Two days later, a *Gefreiter* from Artillery Regiment 5 wrote how:

> Everything is still fine here, and I think we're going to continue, let's go to Paris! About two hours ago, I witnessed a wonderful spectacle here. About fifty German Stukas attacked the enemy over the Channel. Wonderfully precise and accurate, they attacked their targets and delivered their deadly and ruinous bombs on target. Machine after machine sped down with an eerie roar and sirens, and soon all we could see were flashes of smoke and fire. It was about time, too, because the French artillery was firing like mad, and the shells were still hissing over our safe shelter. So far, we have had few dead and wounded. One always has to be surprised that nothing more happens, we almost believe in divine providence ourselves…[42]

Even though its ordeals were keenly felt across the Wehrmacht, Case Yellow had turned out to be a monumental success for the Germans. The Luftwaffe did not just bolster the morale of the *Heer* in a more abstract form during the Fall of France: in this case, it was the 'flying artillery' which literally tipped the balance from death to life for the German soldiers on the ground. These impressive displays of interservice coordination reaffirmed to the Wehrmacht that its tactical synergy was unparalleled in efficiency and effectiveness, thus appearing to confirm that by midsummer of 1940 it was an invincible force.

This feeling was potent enough that the prospect of successfully invading Britain was already starting to percolate within some German soldiers by the end of 'Dynamo'. In a bloodcurdling glimpse of how certain SS members were already psyched to storm Britain's shores, for instance, the SS-*Unteroffizier* wrote on 7 June 1940 that 'We all still firmly believe in taking part in the deployment against the "island".' Then the war would be over too...'[43] Indeed, the idea that the Germans ought to finish what they started permeated the minds of some Luftwaffe personnel as well. 'It has come to the point where we can settle accounts with this wretched bunch,' as a Luftwaffe *Feldwebel* of the bomber wing KG 55 'Greif' wrote triumphantly on 30 May 1940. 'I saw England during the attack on Dunkirk. Well, this promised land will also know how to appreciate our acquaintance.'[44]

However, the Luftwaffe's commander-in-chief had a different priority in mind: revelling in the Allied retreat. On 5 June 1940, Göring spoke proudly in his Daily Order of how 'I thank you from the bottom of my heart, my soldiers and comrades, for your achievements. I know that they were superhuman, because this glorious victory could only be achieved if you – each and every one of you – gave everything you had.'[45] That same day, Case Red (the final half of the Battle of France) commenced, with the Germans advancing over the Somme to subdue the remaining Allied units in the resultant Battle of France. Some Luftwaffe personnel had a relatively uneventful experience during Case Red, with an air signals *Obergefreiter* writing on 10 June 1940 that:

> Although so much has happened in the last month, and I am still 'in the country', I actually have little to do and exhaust my patriotic activity by listening to the radio. During the day we lie in the sun or in the water, we play football – at night, we let the French throw bombs at us. They don't hurt us either, because

our airfield is beautifully located in the valley; next to it flows an extremely valuable river that fills the whole valley with fog at night so that the evil enemy never knows where to look for us. But it has to be that time soon.[46]

He then complained, 'Isn't it ridiculous: here you can count the bombs and "play" war, and on the front they die every hour. We feel like rags, we don't like the word 'air news' anymore. We can only eavesdrop on what's going on. Well, let's hope that the war will end sooner this time...'[47] The French put up a courageous fight during Case Red, inflicting 5,000 casualties a day on the Germans from 4 to 18 June 1940 – more than double those of Case Yellow[48] – but, by the end of the Western campaign, they had lost 90,000 of their own men.[49] One *Wachtmeister* ('sergeant') of the Luftwaffe's Flak Branch 776 wrote on 15 June 1940 that 'It's also wonderful to be in the ranks of such an army, even though sometimes heavy chunks of metal come flying in.'[50] He added that:

> Everyone expects the war to end in August and the French to surrender this month. At the moment, the enemy's fortifications are once again being bombarded with heavy barrages, and what the Stukas didn't destroy during the day, the artillery makes it completely ready for attack. It's now 3 a.m. and there's no chance of sleeping. The 'bon comrade' ducks his neck and will not fully agree on the leadership of his army. They have terrible losses. Poor *poilu!** This is not how they imagined the clout of the '*Boches*'. Yes, Adolf only has a lot of work to give. We are approaching the greatest victory in world history. God willed that![51]

With the French government successfully exiled, the Germans occupying Paris on 14 June, and French defences launching

---

* '*Poilu*' was a nickname for the French infantrymen from the First World War onward, akin to 'Tommy' being synonymous with the British soldier.

their final efforts on 18 June, it appeared to certain Luftwaffe personnel by the summer of 1940 that the Wehrmacht was divinely protected – and that Hitler was its infallible messiah. Yet, it is easy to forget that the Germans were still taking losses from Bomber Command on the Reich home front during the French campaign. On the last day of the battle at Dunkirk, for instance, a resident of Hamburg wrote how:

> The English have already visited six times in fourteen days. Friday to Saturday before Mother's Day,* they were there for the first time. There were thirty-three dead and seventy injured, and an oil factory and a hotel on the *Reeperbahn* were destroyed; not much else happened. It was nice fireworks. I watched in the garden until the first pieces fell down, but then I went back to bed. I haven't been in the basement yet, it's of no value. Those on the Reeperbahn also got it in the basement.[52]

Although he was not too concerned by the damage that the British wreaked, the Hamburg resident expressed his concern at the lack of response from the German defences, noting that 'On Thursday night they dropped quite a bit again, even though it was pitch black. Almost nothing was heard from the anti-aircraft guns, only the bombs sometimes – boom, boom, boom. Not much happened. It was just outside of Hamburg…'[53] In addition, even after Dunkirk had fallen, bombs from the French *Armée de l'Air* still occasionally fell deep in Reich territory.

On 7 June 1940, a housewife in Constance – located just 40 miles away from Zürich in Switzerland – wrote how 'This week we *Konstanzer* ('Constancers') were awakened from our sleep for the first time by an air raid alarm. French planes

---

* In Germany, *Muttertag* ('Mother's Day') always falls on the second Sunday of May, unless it interferes with Pentecost. It had been observed in Germany since 1922, but was declared an official German holiday in 1933 by the Nazis, who sought to consolidate the 'Aryan motherhood cult'.

dropped bombs and only hit the Swiss border at Tägerwielen. The cowards probably had nowhere else to put them, and couldn't bring them home, but thank God they filed them into open ground and couldn't do much damage.'[54] Other members of the German public saw beyond the success of Dunkirk and recognized that France might not be the final destination for the Wehrmacht.

A German housewife who was living in Tübingen pondered in a letter of 4 June 1940: 'The hardest fight is certainly still to come, because the English are fighting back tenaciously. I made a bet with *Herr* Plumm that the war would be over by October. He, who was always so hopeful and optimistic, doesn't want to believe in it. Yet, if it continues like this, I'll definitely believe it. But England must initially experience it first-hand...'[55] There was a growing sentiment among some German civilians that Britain should repent for the smattering of bombs that had been dropped on German territory so far. One resident in Göttingen wrote fervently on 5 June 1940:

> I wish from the bottom of my heart that England now gets a full turn. I'm also interested in whether the *Führer* offered peace or whether, based on the situation, it was simply said that the Western powers still don't want peace. And now the fight goes on to complete annihilation, according to the report.[56]

Even as the French campaign wound down, then, there were some Reich citizens who felt that the BEF's smack on the nose at Dunkirk had not been adequate punishment for the Tommies. The civilian from Hann. Münden wrote on 18 June 1940: 'Now France has already laid down its arms. It is in fact overwhelming, hard to believe. I hope that thousands of bombs will be thrown at England that night, they must also be made aware of the war!'[57] The Nazi leadership and the Luftwaffe, on the other hand, mainly hoped that after the Dunkirk evacuation, the interfering 'pirate island' had finally walked the plank for good.

PART II

CONFRONTATION

6

A Future Without Albion

Walter Knappe had been to hell and back. As a Luftwaffe navigator, he had participated in the ruthless bombing raids that razed the helpless Rotterdam to the ground in mid-May 1940. Everything had gone to plan in that particular sortie – but his karma would soon come over Dunkirk. Flying high above the burning port on a routine sortie, Knappe's blood suddenly chilled to ice; his crew's aircraft had separated from the main German bomber formation. French bullets rattled the vulnerable aircraft, igniting both engines into almighty fireballs. Somehow, the stricken aircraft limped on and finally bellyflopped in Belgium.

'When we landed,' Knappe recalled, 'we were enveloped in a heavenly calm.'[1] It seemed that the three German airmen had found their own miracle of Dunkirk. However, as heart-bursting gratitude flooded the airmen at surviving against all odds, the euphoria soon vanished: bullets ricocheted violently off the downed aircraft, shredding its burnt-out shell. Knappe instinctively began to sprint away with his radioman and pilot as the incandescent locals descended upon them. It soon proved fruitless. Hands aloft and heads bowed, they submitted to their fate as prisoners of war.

Suddenly, a wounded howl rang out during the tense standoff. Knappe clutched at his shoulder, which now hung down like meat on a butcher's hook; one of the Belgian locals had shot a bullet clean through it.

'Hot blood ran down into my clothes, and I staggered,' he described vividly. 'The civilians rushed us, robbed us of everything we had, and cursed us.' Swiftly taken into French custody, Knappe watched from the sidelines as his countrymen fought to finish the carnage they had started. He held no grudge against his captors; on the contrary, he later spoke warmly of the communal bond that he felt he had forged with other wounded French, British, German and North African prisoners in the hospital.

Yet, as is the case with any incarcerated man, Knappe yearned to be free. And, with the German capture of Dunkirk, his wish was swiftly granted. Knappe soon found himself on a hospital train heading for home: 'I cannot describe the feeling I had as I neared the German border as a wounded soldier coming home from captivity.' In an uncanny echo of the adoring treatment that his enemies received from the British public after the Dunkirk evacuation, the incapacitated Knappe was equally stunned at the German population's reaction to the returning casualties. 'We were speechless,' he said, 'when the first flowers were put into our hands.'

Following the successful culmination of Case Red by late June 1940, the German Wehrmacht was riding high on the wave of success: the new conflict appeared to have departed significantly from the grisly attrition that characterized the Great War. 'This war is so completely different from the last trench warfare,' as the housewife from Constance had written to her serving son during Case Red. 'It's always moving forward, and you have to be amazed at what our troops are able to achieve.'[2] Yet, it was not just the remarkable application of *Blitzkrieg* that the victors celebrated, which had been executed to tactical perfection: it was also the vengeful catharsis of conquering some of the nations deemed responsible for Germany's economic, military and political dilapidation after the Treaty of Versailles.

Thus, the end of the campaign in the Low Countries, but especially in France, signalled to the Germans that historic justice had finally been served for their martyred forefathers in the Great War. It appeared to confirm to the young Luftwaffe men that they were marching on the right path in this war – and, indeed, on the right side of history. This was reinforced by the grateful letters they received from their fathers back home, as the Stuka pilot Helmut Mahlke recalled:

> On 22 June 1940, in the same railway carriage in which France's Marshal Foch had received the German surrender of 1918, the German–French Armistice terms were presented and signed. It was not only the same railway carriage; it had also been taken to the very same spot in the Forest of Compiègne where the 1918 surrender – leading to the infamous Treaty of Versailles – had been signed. This symbolic gesture was not lost on Germany's older generation, those who had experienced the end of the First World War and particularly those who had fought in that conflict. It brought them great satisfaction; or at least that was the impression I got from one of my father's very infrequent letters.[3]

Carmine wine and amber liquors flowed deep into the night; the finest charcuterie adorned the tables as the Luftwaffe's men eagerly acclimatized their palates to the local cuisine. As an *Unteroffizier* with a Luftwaffe airfield construction battalion wrote from Alsace on 30 June 1940: 'All the inns and cafes are extremely busy. They are all German soldiers who can now suddenly live well with their few pennies, interspersed with a few curious civilians.'[4]

Sometimes, the understandably visceral hostility from the local populations shook certain Luftwaffe personnel to the core. *Oberst* Johannes Fink, who was *Kommandant* of the Do 17 bomber wing KG 2 in Arras, recalled how 'In 1940, I was billeted in Charleville. I found children's books showing pictures of German soldiers cutting off children's heads. I was deeply shocked. The house belonged to a doctor's family. We had no equivalent to this in Germany.'[5] Others, however, were more than happy to laud their perceived German superiority over the recently vanquished French.

A *Feldwebel* in the II. *Gruppe* of the bomber wing KG 55, for instance, wrote rudely on 29 June 1940: 'We've been stuck here in France for some time now. I've already seen Paris. I'm hardly pleased with the "*Grande Nation*", because in general, they've probably stopped developing since the World War. In any case, they cannot remotely compete with us, and the cleanliness leaves a lot to be desired...'[6] A Nazi propaganda book entitled *Hölle über Frankreich* ('Hell over France') fervently reinforced this view: 'The German Luftwaffe was up to tackling all sorts of new problems. What's more, in close cooperation with the *Heer*, it had developed a fighting style that could be called perfect.'[7]

As with the Polish campaign, some young Luftwaffe personnel back in the Reich were anxious that the chance to fight over France had passed them by. Heinz Rökker, later one of the Luftwaffe's most successful night-fighter pilots, was undertaking further training in late June 1940. He recalled how 'Successful fighter pilots like Mölders and Galland were our role models. Therefore, in our youthful thirst for action, we feared that the war might be over before we could prove our commitment.'[8] A *Stabsfeldwebel* with the Luftwaffe transport column 9/VII in France echoed this sentiment in a letter home from 19 June 1940, as Case Red entered its final week:

> We are beginning to have very strong doubts as to whether we will be relocated at all. I won't believe it until we're on the road. Judging by the general situation, it seems almost unbelievable. Of course, we don't know anything. Well, let's wait and hope for the best. If I'm later released to go home from here, I'll have to

feel ashamed, and I can't ease my conscience either. See, we often think about that [... we] blame ourselves so much inside that we are doomed to be with the waiting troops and not with the fighting ones. It's difficult for a person like me to be able to do so little.[9]

Despite his abject disappointment at missing out on the fighting in France, the *Stabsfeldwebel* could hardly contain his unbridled joy at the outcome of the campaign: 'Isn't it fabulous what our soldiers are doing under our God-given *Führer*? It just borders on the impossible. It's the same for all of us – you can hardly believe it. The boldest predictions are exceeded.'[10] The *Stabsfeldwebel* ultimately tried to make sense of the ongoing war by concluding, 'May this great experience, which is given to us, purify and strengthen us inwardly, so that we never become weak in fulfilling our duty. That is my oath...'[11]

Part of the Luftwaffe's military prowess in Case Yellow and Case Red stemmed from the rigorous training and mounting combat experience. By the summer of 1940, the average German airman was twenty-six years old and had served for nearly five years.[12] As a 1936 information guide for joining the Luftwaffe stated, 'The aviator is also supposed to be a soldier first, and the way to this goal is paved by the proven, albeit rough methods of a Prussian–German soldier's upbringing.'[13] Yet Rökker, who had been transferred to the Luftwaffe military academy in Gatow, Berlin on 4 July 1940, criticized this need for 'zombie-like obedience'[14]:

> Our supervisory officer, *Leutnant* Nolte, must have gone through a tough training course himself, because he was a sadistic 'grinder'. This means that when we were drilling with a gun for alleged offences, he gave us completely nonsensical orders, e.g. Repeatedly 'Lie down – stand up... crawl,' running around the shooting range five times, etc. This aroused our anger. When we were lying on the ground again, we had to stop a comrade from attacking our tormentor with a gun.[15]

Hans Deibl, later a Stuka pilot, recalled similar treatment on being transferred to the same academy in Berlin, which he reached a week before Rökker. Deibl spoke of how the theoretical lessons at the military academy were 'cramming and cramming again. Sometimes we were so overtired that we had trouble following the lessons, which progressed at breakneck speed and demanded the utmost of us with regular tests'.[16] Nevertheless, their harsh treatment during their training meant that Deibl 'later felt a strange lack of fear during my operations. Today, I believe, it resulted from this instilled harshness in us, especially against ourselves'.[17] He described how:

> In 1940, the training of the Luftwaffe was still very precise and hard and, as admitted by the enemy nations, had reached the highest level of all air forces known at the time. The basis for this was the establishment of a special training inspectorate in the Reich Air Ministry, which was headed by the chief himself, Hermann Göring. According to his wishes, the flight instructors had to be particularly dynamic and enthusiastic soldiers who were only allowed to recruit the best and most suitable young men.[18]

Between late June and mid-August 1940, Johannes Kaufmann – later a Bf 110 *Zerstörer* ('Destroyer') pilot – took a group of young airmen for a six-week course at Salzwedel in northeastern Germany. He described their training as 'intensive and uninterrupted', noting that 'At this relatively early stage of the war flying training differed little, if at all, from the days of peace. There were no signs as yet of the increasingly drastic cuts in the training schedules brought about by the fuel shortages of the later war years. It was simply that everything had to be done in a much shorter time.'[19]

These trainees would soon be needed, given the losses that the Luftwaffe had sustained in the French campaign. By the end of Case Red, it had sacrificed 30 per cent of its bomber force,

30 per cent of its dive-bombing units, 40 per cent of its transport aircraft, and close to 20 per cent of its precious Bf 109 fighters.[20] Around 50 per cent of the aircraft it still wielded had suffered damage on operations, while more than 15 per cent of its Bf 109 pilots had been lost.[21] On 3 July 1940, the newly qualified flight instructor Paul Zorner was bemused by the over-enthusiastic greeting he received at Zeltweg airfield in Styria, Austria:

> Immediately after landing, I reported to the head of training, *Oberleutnant* Braun, who had his office in Flight Control. At that time, it was the only brick building in the eyrie. Braun greeted me with a smile. '[This is] *Leutnant* Zloch,* transferred from the Lönnewitz C-School to the Zeltweg C-School!'
> 'Welcome, we are eagerly awaiting you!'
> 'I didn't expect such a reception, *Herr Oberleutnant*. Is that because there is such a big shortage of teachers?'
> 'Well, you'll soon notice that!'[22]

Zorner may have only been twenty years old, but with 231 hours under his belt as an assistant flight instructor, he was already fully qualified as a '*C-Schule*' teacher.[23] In German pilot training, 'C' aircraft were the most technically difficult to fly of three flight classes (A, B and C). They weighed over 5,000 kg (11,000 lb) and normally contained multiple engines and seats, with students required to fly at ever greater distances, speeds and altitudes.[24] *Oberst* Heinrich Benz, the *Kommandant* of Zeltweg and the *C-Schule*, expressed his cynicism regarding his fresh-faced new flight instructor. 'How are you supposed to work as a "C" teacher?' Benz asked Zorner, raising one sceptical eyebrow. 'And that's supposed to go well?'[25]

While Zorner was undeniably precocious, the fact that the Luftwaffe was so keen to use him in this role – rather than flinging

---

* Zloch's father later changed his name to Zorner, as Zloch is a Czech–Polish surname.

him onto the frontline – reflected the deeper need to replenish the experienced flight instructors who had been killed over France. Indeed, high-quality training remained of paramount importance given that some Luftwaffe airmen had lost their lives by simply neglecting to use their oxygen masks at high altitude. Dr Georg August Weltz – a Luftwaffe medic and doctor who later headed up the Institute for Aviation Medicine at Munich – recorded that death from anoxia (severe lack of oxygen) had been alarmingly frequent among German flyers during the Battle of France.[26] These findings were augmented by research undertaken by Professor Franz Büchner, who headed the Institute for Aviation Pathology in Freiburg from April 1940.[27]

Büchner, who specialized in hypoxia and its effects on organ systems, received for dissection all of the brains of German aviators who had died of anoxia.[28] Such was his renown in the field that he was later nominated for a Nobel Prize in 1963 for his pioneering research into the influence of oxygen deprivation on the human body.[29] All other brains from Luftwaffe personnel killed in action, meanwhile, were transferred to Professor Hugo Spatz at the Kaiser Wilhelm Institute for Brain Research. As a result of Büchner's anatomical research, it was determined that the heightened rate of anoxia occurred when flying at an altitude of 7,000 metres, or nearly 23,000 feet.[30] To prevent this serious issue, Luftwaffe personnel were subsequently ordered to put on their oxygen masks as soon as they hit 4,000 metres (13,000) feet.[31]

Possessing air superiority, then, did not render one immune to the everyday risks and challenges of operating in an aerial combat zone – something which Helmut Mahlke soon became painfully aware of. On 30 June 1940, the VIII. Air Corps was reassigned to the south and south-east of Cherbourg in anticipation of targeting Allied shipping and their convoys in the English Channel. Mahlke's *Gruppe* was assigned a new landing ground near Falaise. 'To disguise the move,' he said, 'the aircraft flew to the new base separately and at low level. They took their chief mechanics with them, crammed into the rear cockpit alongside the gunner.'[32] After their earlier departure, Mahlke was coming to land when the twisted remains of an aircraft at one end of the airfield suddenly caught his eye; he urgently touched down and scrambled out of his Stuka to speak with the advance ground party.

One of its dejected personnel 'confirmed my worst fears: *Oberleutnant* Wolf-Dietrich Herbst from Emmagrube had hit the ground while on his final approach'. Along with Herbst, 'The senior *Oberleutnant* of my *Staffel*, together with his gunner, *Unteroffizier* Franz Kollmitt from Tollack in East Prussia, and chief mechanic, *Unteroffizier* Erich Schug from Neuwied on the Rhine, had all been killed in the crash. The pilot must have tried to take the final curve too steeply, or with insufficient speed. The machine had stalled and gone straight in.' Their deaths especially stuck in Mahlke's throat in light of the onerous campaign they had just fought together:

> After all that they had been through with us during the recent operations and air battles in France, the loss of the Herbst crew – and their chief mechanic – hit us particularly hard because it had been so unnecessary and could have been so easily avoided. It just went to show how quickly the slightest lapse of concentration, even by the most experienced of pilots, could lead to disaster. It was a sombre gathering that paid its last respects to three good comrades as they were laid to rest in Falaise cemetery.

With such careless lapses in concentration often provoked by the strain of war, the men desperately needed a break from operations. Yet, on 18 June 1940, Winston Churchill – Britain's newly installed prime minister – immortally declared: 'What General Weygand called the Battle of France is over. I expect that the Battle of Britain is about to begin.'[33] Deichmann claimed after the war that the Luftwaffe was 'surprised to be ordered to prepare for war against England, because we all thought the war was over'.[34]

In late June, a *Soldat* in a Luftwaffe supplies unit in Berlin was cheered by the fact that 'Various comrades who were with me on leave brought good news: for example, factories that used to manufacture armaments are beginning to convert to peacetime articles. So, we are actually all pretty confident that it won't take too long...'[35] For some wartime factories to be planning a post-war future, then, further illustrates the initial confidence that Britain would soon back down. Nevertheless, the German press created a shrewd contingency for any further military action. On 28 June 1940, the *Fehrbelliner Zeitung* newspaper proclaimed:

> The final battle with England thus enters its decisive stage. The English have already received a small foretaste of the forthcoming battle from the attacks of the German Luftwaffe. These undertakings are aimed exclusively at military installations and war industries, while the British planes, on their nocturnal flights into German territory, make the defenceless civilian population the target of their bombings. The day will come when the British people will recognize with horror how criminally their air force has brought disaster upon England.[36]

This priggish sentiment rang hollow, given that the Luftwaffe had already reduced multiple European cities such as Warsaw and Rotterdam to smouldering ruins. Nevertheless, it was a clever propagandic twist that prophesized two favourable outcomes to the German public: whether the British found themselves snivelling at the negotiating table, or crushed within their own island by the Luftwaffe, the so-called rule-breaking *Luftpiraten* ('air pirates') would end up getting their comeuppance one way or the other. A similar approach can be traced in the propaganda book *Hölle über Frankreich*:

> Victory! Like prayers of thanksgiving, the *Deutschlandlied* ['Song of the Germans'] and the 'Song of the New Reich' rose up to the blue sky of Normandy. The pilots stood there, tense

figures. Fire glowed in their eyes. They were silent and looked straight ahead. Stared at the hell they'd been through over France. Looking into a future that will be more beautiful than the present. A future in which it is not possible for a France seduced by England to campaign against Germany – a future without Albion!³⁷

By placing emphasis on Britain's culpability for having seemingly led France astray, the Nazi propagandists sought to justify any future military action against the 'real' enemy. In addition, the highly deliberate choice of the ancient Greco–Celtic name 'Albion' ('white land') evoked connotations of 'perfidious Albion'. From the Romans to Napoleon, and indeed to the Second Reich, 'Albion' had long been the thorn in the sides of the world's greatest empires; now, it was the Third Reich who needed to deadhead the British.* Teade Sysling, a Dutch boy who grew up during the Nazi occupation of the Netherlands, later recalled how:

> From the very first day of the German occupation, the Nazi propaganda machine tried to establish the notion that the Germans were in Holland to protect their brothers from the perfidious Albion. To the Germans, Albion sounded more sinister than England or Great Britain. They often used Albion (Latin for white – Cliffs of Dover!†) in their propaganda as an emphasis of the evil that the English enemy represented. In their

---

* The Romans initially referred to Britain as 'Albion' which became superseded by 'Britannia' – from the Latin term for the British Isles, *Insulae Britannicae* – by the 1st century AD. See E. H. Warmington, 'Albion', in S. Hornblower et al. (eds.), *The Oxford Classical Dictionary*, 4th ed. (Oxford: Oxford University Press, 2012), 49; A. Geiser, *Das perfide Albion* (Bielefeld and Leipzig: Velhagen & Kasing), 1915.

† There is debate over whether the name 'Albion' truly derived from the White Cliffs of Dover, or if this is simply a mythological origin. Nevertheless, it has long been used to describe Britain as a whole, often being used poetically from the mid-thirteenth century onwards. See R. Barber (ed.), *Myths and Legends of the British Isles*. 1999. (New York: Barnes & Noble, 2000), 3.

scenario, Germany was Holland's natural friend with a common enemy, the English.[38]

At the same time, the Nazi propagandists were careful to not state the specifics of how 'a future without Albion' would be achieved.[39] Certainly, the war did not yet feel over for some Luftwaffe personnel who were protecting the Reich from Bomber Command. A *Soldat* serving in a Luftwaffe supply unit in Schwerin, northern Germany wrote of a British air raid on 24 June 1940: 'It was a horrifying sight when it started around 1.45 in the morning... now it is becoming increasingly clear to us what it meant when our superiors initially said that we were soldiers at the front. We never really wanted to believe that, but now we see the light.'[40]

Although British and German cities would not come fully under fire until late August 1940, both sides sporadically harassed the other's industrial targets and port cities in June and early July. The port city of Kiel on the Baltic Sea experienced one of its earliest air raids on 2 July 1940, where military targets, submarine construction shipyards and oil bunkers located within the suburb of Mönkeberg were all attacked.[41] In this particular raid, however, eleven people lost their lives – mainly in the neighbouring post office building.[42] An SS report stated that immediately after the 2 July attack, 'A special case of anxiety psychosis was reported from Kiel, where on 4 July, a rumour that spread like an avalanche about a major attack imminent in an hour caused an uproar throughout the city.'[43]

It was recorded in Kiel after the July 1940 air raid that there was 'increasing nervousness, particularly anxiety and depression, among women and children; a general decrease in the desire to work as a result of growing exhaustion; and, in isolated cases, even the occurrence of symptoms (headaches, colds)'.[44] Likely adding to the residents' nervousness was the fact that German provisions for civilian air raid defence were often inadequate in the early years of the war, as the Wehrmacht and armaments industries failed to allocate enough construction machinery and building materials to the scheme, such as steel and concrete.⁴⁵

In addition, some German civilians proved to be a liability in a manner that threatened the safety of their cities during these early British air raids. One resident of Kiel, Elsbeth Benz, wrote to her son of how 'People behave like bastards – they stand on the balcony and turn on the light despite the alarm.'⁴⁶ Indeed, so many people were ignoring the German blackout that in 1940 Hitler brought in the disincentive of cutting off the offender's electricity for at least eight days; it could even be used as a weeklong punishment for entire districts that contained high numbers of offenders.⁴⁷

The Bomber Command air raids over Nazi Germany gave the wartime leadership yet another reason as to why Britain needed to be quickly subdued. But in some spheres of the Luftwaffe, there was a growing confidence that the British would not prevail. A *Stabsfeldwebel* in the Luftwaffe transport column 9/VII had dismissively written on 28 June 1940 that 'I wouldn't want to be an Englishman and live on the threatened island. The warmongers are still making big speeches, but for how much longer. Now they are all alone with their money sacks, and no matter how many beautiful words and promises they utter, no helpers are drawn to them.'⁴⁸

Some, like the Luftwaffe technical officer Werner Bartels, even anticipated that German boots would one day thunder onto British soil: '[Galland] and I were the first pilots to walk down the *Champs Elysées* after France capitulated. We made plans to be the first ones to walk down Bond Street in London as well.'⁴⁹ Even more concretely, one *Leutnant* paratrooper – based at the parachute school in Wittstock/Dosse in north-east Germany – wrote on 24 June 1940: 'Having taken part in the Polish campaign and the great offensive in the West from home, I am now with the paratroopers and am waiting to be deployed against England. Hopefully something will come of it!'⁵⁰

Thus, for some of the Luftwaffe's paratroopers to expect to be dropped against the defiant island rebuts Deichmann's generalization that 'We all thought the war was over.' It must be recognized, too, that a small number of Luftwaffe personnel had already experienced what it was like to conquer British territory. On 28 June 1940, Luftwaffe bombers attacked the harbours of St Peter Port (Guernsey) and St Helier (Jersey) in the Channel Islands, killing forty-three people and injuring forty-four more.[51] Two days later, *Hauptmann* Liebe-Pieteritz was spearheading a formation of four Do 17s from *Aufklärungsgruppe* ('Reconnaissance Group') 123 (F), when he observed that the airport at Guernsey had become a ghost town.

Hesitantly touching down to gauge the response from the islanders, and holding his sidearm in a death grip, Liebe-Pieteritz was immensely relieved to discover that the airport had largely been abandoned.[52] Dropping official conditions of surrender, the Germans had quickly occupied Guernsey on 30 June and then Jersey the day after. A *Leutnant* attached to Air Fleet Command 6 later recounted the occupation in a letter from 20 August 1940:

> We took the islands of Guernsey and Jersey in a surprise attack and were the first German soldiers to set foot on English soil. You should have seen the stupid storytellers of the English! With another naval officer, I was the first to hoist the imperial war flag on the island of Jersey, which English women had to sew according to our specifications at night. The governor of the island gave the occupying German troops a sumptuous meal in the best hotel in Guernsey and we enjoyed it, as the English government paid for everything. In general, we were very surprised at the attitude of the English population, who could almost be called fawning and submissive.[53]

It is hardly surprising that the vulnerable local populations of the Channel Islands – with just over 90,000 civilians across

Jersey, Guernsey, Alderney and Sark – quickly yielded to the Wehrmacht's advance.[54] Yet, it is evident that their subjugation bolstered the morale of the German invaders, who now knew how it felt to have British currency jingling in their pockets and stolen English cigarettes hanging from their lips.

In the week before the Channel Islands were unexpectedly occupied, the Luftwaffe had already started equipping its men for a potential aerial campaign over Britain. On the final day of fighting in France, Günther Rall recalled how his newly relocated home unit in Jever had to 'exchange our aircraft – the Messerschmitt Me 109 E-1 and E-3 – for new versions of the E-4 series with more powerful armament and the possibility to stow additional equipment for the pilot'.[55] There was, however, something that stood out about the new accoutrements they received:

> We also swap our somewhat impractical brown flight suit for a two-piece flight suit consisting of a light jacket and comfortable trousers, with large patch pockets in which all sorts of life-support supplies can be stowed away – from pocket-knives and doses of chocolate, to a fishing line with a hook. New life jackets are issued, rubber dinghies, luminescent water dye packs, flare pistols and ammunition to be carried in leather loops around the lower leg. Aha. It is therefore assumed that in the near future, we could fall into the water more often than before.

Rall had correctly deduced that 'The next enemy we are supposed to bring to their knees with our new sea rescue equipment will be England,' as reflected by the specialist training that they quickly undertook over German waters: 'the four days in Jever pass with surveillance and navigation flights over the German Bight'. Yet, although the Luftwaffe had already had experience in flying over water, most of its fighter and bomber pilots had become accustomed to the narrow, splintered fjords of Norway – not the watery fortress of the English Channel.

As Rieckhoff noted, 'In the Baltic Sea and in the narrow triangle between Heligoland and the mouth of the Elbe, however, one could not learn to navigate because one saw new landmarks every ten minutes at the latest. There was a strong aversion to flying at sea. The airmen weren't used to it. The unreliability of the engines frightened them. The sea rescue service was inadequate and only reasonably useful for flying boats.'[56] Already, fear of losing one's bearing over the rolling waves sunk into the Luftwaffe's psyche, as Rall documents in more detail:

> Anyone who has ever flown across the sea in a single-engine plane knows that one never really feels comfortable doing so, especially when the horizon and the water surface merge all around into a featureless blue-grey, and that is often the case here. One loses a feeling for the altitude and visibility; you have no landmarks for your flight path. If you don't master dead reckoning and don't have at least a few blind-flying hours to your credit, you're in acute danger. But even the more enlightened unconsciously listen a little more closely to the running of the engine; they glance more often at the oil pressure, the oil temperature and the fuel gauge.[57]

Yet, even during the preparatory sea flight training for waging a campaign against Britain, a dangerous schism was appearing between the blasé Göring and his sacrificial eagles. 'It's a pity that the *Herr Reichsmarschall* doesn't fly in my flight,' Rall noted bitterly. 'Perhaps he would have reconsidered his sonorous assertion that England was no longer an island…'[58] With no signs of 'Perfidious Albion' yielding after the fall of France, navigating the choppy English Channel was no longer just an unnerving prospect for the Luftwaffe aircrews: it had become a prerequisite for victory.

7

Flags Up! Hearts Up!

Now then – England. We know that we haven't really smelt gun smoke yet; we also know that the fighter pilots of the Royal Air Force are at least equal opponents, but that doesn't scare us. On the contrary: we are eager to fight. The British are not invincible; [we] must have caused terrible trouble over France. Burning shipwrecks are bobbing in the Channel. Every day, German Stukas attack the enemy's merchant fleet, which cannot be effectively protected by the naval flak in Dover, nor by destroyers or enemy hunters. A few more days, and the strait will be impassable. Then an essential prerequisite for the invasion will be created.[1]

According to Günther Rall, that was how the Reich's burgeoning air war over Britain was *supposed* to go by early July 1940.

Given how forcibly the BEF and RAF had been ejected from northern France, the supreme confidence that was now wielded by many of the young fighter pilots was understandable. In a front-page article embossed with the headline 'The Greatest Military Campaign of All Time', the *Baruther Anzeiger* newspaper declared in its bumper issue on the French campaign that 'Only one enemy remains: England.'[2] In the same issue, the newspaper published a poem by Georg Weßler entitled 'Now We Want to Storm Albion!'[3] The sense of rhyme and rhythm is lost when translated into English, but the sentiment was unequivocal:

> The German armies swept forth, breaking down all resistance of their naked defences. The German eagle flew ahead – the *Panzerhussar*s, engineers and infantry stormed. They have overrun Holland, carried death to Flanders, and British pride lies shattered in Dunkirk's dunes. The steel flood roared on, over the Somme, Marne and Seine; the rivers drank French blood and bitter tears of misery.[4]

The last stanzas of the poem called for 'Flags up! Hearts up! It rings from all the towers: "we will force fate and break it – now we want to storm Albion!"' Thus, by the beginning of July 1940, the Nazi press was already beginning to brandish combative language regarding an invasion of Britain. Indeed, the *Briesetal-Bote* newspaper spoke about the possibility of German troops invading Britain as early as 2 July 1940. In an article entitled 'Fear and Confusion in London', it claimed that:

> While, after the London radio broadcast, the only interest of the British public is to be assured that the German attack will be 'completely thwarted', any attempt at reassurance just raised new concerns. Every speaker so far has felt compelled to point out that it is entirely possible for German troops to land in England.[5]

Yet, although the invasion expectations of some younger fighter pilots were in line with the Nazi press, their more experienced superiors were not quite so convinced that fighting the island would be a fruitful endeavour. The radio operator and rear gunner Willibald Klein had little appetite for any further heroics: 'We weren't even the type of soldier who flew and fought in order to be shot dead, hoping that posterity would erect a monument for our sake or write a beautiful poem about us.'[6] Admittedly, the post-war need to distance himself from any glory in serving a criminal regime may have slightly coloured Klein's comment.

However, after encountering fiercer resistance than expected over Dunkirk, the seasoned fighter ace Adolf Galland recalled how 'The German pilots thought that the [Nazi] propaganda about England being unprepared was wrong. It annoyed us. We found the English well prepared. Not an easy job. We were not hopeful of victory.'[7] It is important to recognize, of course, that these despondent comments all came *after* the war, when Sealion had long been rendered as nothing but a tantalizing hypothetical scenario.

Yet, even with the megalomaniacal catalyst of holding much of north-western Europe in his failed artist's hands, by late June 1940 the *Führer* himself was not all that enamoured with the prospect of invading Britain. Hitler has often been portrayed as having a sneaking, almost sentimental admiration for the British that made him waver over invading them. The Luftwaffe fighter pilot Hans-Ekkehard Bob averred that the *Führer* had 'a soft spot for England, that he was an admirer of England and that he didn't want to humiliate the English and so on'.[8] Galland further noted after the war:

> I had the occasion to talk to him without any other company and I told him, we will have the opportunity to attack London – and he said, 'stop it, stop it! I don't want to hear this – the whole attack on England is against my opinion, against my will. I would like if I could stop it. The English population is of such high class, and they are so similar to the Germans, that I hate to fight England.'[9]

However, it was arguably not so much the countryfolk that Hitler truly admired, but rather the historical blueprints of imperialist expansion and colonial exploitation that the British

Empire offered. In addition, unlike his racist abhorrence for his Eastern European rivals, he appreciated what he called the 'high class' Anglo-Saxon roots of the British. Accordingly, his first priority after the capitulation of France was to determine 'whether England is prepared to end hostilities'.[10] But, with the Wehrmacht's prolific continental victories failing to intimidate Britain into suing for peace, the Nazi leadership recognized that alternative provisions had to be made.

On 30 June 1940, Göring issued a directive to prepare Air Fleets 2, 3 and 5 for conducting a fresh series of aerial operations above the impudent Brits:[11] these early opportunistic raids would help the Luftwaffe acclimatize to their potential new target.[12] Consisting of sweeping armed reconnaissance sorties, scattered bombing raids and flurries of fighter activity, the Luftwaffe prodded around the island's airspace to uncover any discernible strengths and weaknesses within its operational response.[13]

Two days after Göring's directive, Hitler requested a full Wehrmacht survey into launching an amphibious invasion of Britain, propagated by the OKW Chief of Operations Staff, *Generalleutnant* Alfred Jodl and signed by *Generaloberst* Wilhelm Keitel, the head of the OKW.[14] The Sealion plans would require the German *Kriegsmarine* and Luftwaffe to deploy soldiers of the *Heer* onto the British coastline to occupy the country by force. In a directive regarding 'The War against England' on 2 July 1940, Hitler determined that:

> A landing in England is possible, providing that air superiority can be attained, and certain other necessary conditions fulfilled. The date of commencement is still undecided. All preparations to be started immediately.[15]

The *Heer* was required to estimate the numbers of the surviving British Army, while the *Kriegsmarine* was ordered to identify 'possible landing points for strong *Heer* forces (25–40 divisions) and estimate the strength of English coastal defences'.[16] The document warned, though, that 'All preparations must be undertaken on the basis that the invasion is still only a plan, and has not yet been decided upon.' It would only stay a plan, too, unless the Luftwaffe could first seize air superiority – and, eventually, air supremacy – over the stricken island.

By July 1940, even the infamous SS grew concerned about the uncertainty hanging over an invasion of the British Isles. The *Unteroffizier* of the SS Combat Support Force pondered on 2 July 1940: '... will we still be deployed to England? We don't know much about the general situation at the moment, firstly because we're not interested in it and secondly because you can only hear one German station down here (or upstairs) very weakly.'[17]

The concurrent Battle of the Atlantic, however, had reinforced to the Germans just how reliant the island nation was on imported goods. In May 1940, the Germans had shrewdly agreed an oil pact with their future Axis partner, Romania.[18] Securing a frequent delivery of 200,000–300,000 tonnes of Romanian oil per month, Germany's regular oil supply was thus seemingly assured.[19] On the other hand, in July 1940 Britain began to feel the loss of oil deliveries from Romania.

So, if additional pressure could be put on British maritime targets in and around the English Channel, perhaps this could force a concession from the defiant island without the need for a risky amphibious landing. The Luftwaffe had buzzed around the English Channel since late June 1940; now, in early July, it made a stinging beeline for Allied shipping, convoys, ports, docks and other maritime targets.

Thus, the *Kanalkampf* ('Channel Battle') – and, by extension, the *Luftschlacht um England* – began.

This initial phase of the Battle of Britain has long been characterized by the British as not commencing until 10 July 1940. This date was set by Air Chief Marshal Sir Hugh Dowding, the commander-in-chief of Fighter Command, due to intensified Luftwaffe attacks along the Channel coast.[20] More cynical observers, on the other hand, have insinuated that Dowding, who by his own admission had 'somewhat arbitrarily'[21] chosen this start date, plumped for this because the loss ratio from that day on reflected well on Fighter Command.

Between 10 and 13 July 1940, forty German aircraft were shot down in return for twenty-nine British fighters, of which ten were able to be repaired.[22] Churchill received Dowding at Chequers shortly afterwards, animatedly remarking that 'the previous four days had been the most glorious in the history of the RAF'.[23] It was undeniably favourable that the commander-in-chief opted for these dates as the start of hostilities. For the Germans, however, this was merely the renewed harassment of Allied shipping and coastal targets that had swirled around the English Channel since late June 1940. Thus, to properly analyse the German mindset during the Battle of Britain, it is imperative that we adopt their unique chronology of the campaign.

In the 1–2 July 1940 issue of the *Baruther Anzeiger*, the main headline on the front cover was 'Bombs on British Airfields',[24] reporting on a concerted effort to attack the British mainland the day beforehand. For a more 'official' beginning to the Channel Battle, Göring ordered Richthofen's VIII. Air Corps to start attacking 'Channel sea traffic' on 2 July 1940.[25] One of the key early attacks took place on 4 July, when twenty-six Ju 87s of III. *Gruppe* from the *Sturzkampfgeschwader* (dive-bomber wing) StG 1* swarmed over a convoy, OA 178, just off Portland and around Portland's naval base. With HMS *Foylebank* sinking a day later, and 176 Royal Navy sailors killed, the Stukas were scoring deadly hits against British seaborne forces.[26]

That same day, the Luftwaffe's fighter pilots claimed their first victims of Fighter Command in the Channel Battle: one Hurricane was shot down and four Spitfires damaged, with no loss to the victorious Bf 109s.[27] What marked a particularly significant escalation in the Luftwaffe's air war during this

---

* This was redesignated as II./StG 1 two days later. See A. Saunders, *Stuka Attack!: The Dive-Bombing Assault on England During the Battle of Britain* (London: Grub Street, 2019), 18.

period, however, was the concentrated shift towards maritime targets on the British coastline. Fink claimed that 'Previously, we [the bombers] had not been allowed to fly over England: we were NOT to attack anything, even if most favourable, on land. Because Hitler was still hoping it wouldn't come to war with England.'[28] He further asserted that 'We could NOT attack even ships in ports, though we saw them there.'

So this was a distinctive new phase for the Germans, a development that was marked when Fink, the *Kommodore* of the bomber wing KG 2 was bestowed with the title of '*Kanalkampfführer*' ('Channel Battle Leader') – or *Kanakafü** for short.[29] An assiduous commander with a lived-in face and steadfast temperament, the forty-five-year-old Fink had been freshly decorated with the Knight's Cross of the Iron Cross, as reported by the *Baruther Anzeiger* newspaper on 2 July 1940:

> He played an outstanding role in defeating the enemy air force in the breakthrough across the Meuse and in later shielding the southern flank. Combined attacks by the wing under his personal leadership on aerodromes and troop camps had resounding success. In the most difficult weather conditions and during night operations, *Oberst* Fink led his wing in repeated missions and successfully completed all sorties with a minimum of losses.[30]

What this report also demonstrates is that the Luftwaffe had barely stopped since those triumphant campaigns. For some personnel, it was only through witnessing newsreels of the recent fighting that the enormity of what they had just accomplished sunk in. On 4 July 1940, a Luftwaffe *Soldat* based with an Airbase Company at Quedlinburg in central Germany wrote:

---

* This title does not fully make sense in German, reading as 'Kana' instead of 'Kanal' – but it is the abbreviated title cited by Stephen Bungay and Robert Chaussois, so it has been repeated here. It is possible that the 'L' in 'Kanal' was dropped to make the title easier to pronounce, and it is consistent with the German penchant for using abbreviations that contain the first letters of a word.

'Yesterday, before the beer night, we went to a cinema and watched the latest newsreel. The film was overwhelming. When you think about everything, it's hard to believe...'[31] The *Soldat* confidently assured his letter's recipient that 'By the autumn, everything will be done...'[32]

Thus, some Luftwaffe personnel remained optimistic that the newest aerial campaign over Britain was more of a formality than a stand-off. But to others, who were now tasked with flying over the English Channel, this new Channel Battle hardly seemed to be the beginning of the end. Just after Case Red, *Oberleutnant* Kurt Scheffel – now a Ju 87 Stuka dive-bomber pilot with I./StG 77 – was transferred back to Celle, where he initially enjoyed rejuvenating swims and restorative walks across its quaint landscape. Then, on 6 July 1940, Scheffel flew to Kitzingen in Bavaria, and had dinner with his former *Staffelkapitän*'s wife, *Frau* Sayler, and the *Kommandeur*, *Hauptmann* Friedrich-Karl von Dalwigk zu Lichtenfels, as well as Dalwigk's wife.[33] It was a welcome lull after the frenetic campaign in France.

The fledgling Channel Battle appeared to be going well: on 8 July 1940, the *Baruther Anzeiger* newspaper crowed about the Luftwaffe's 'Hard Blows Against England'.[34] But, when flight operations recommenced for Scheffel that same day, something was wrong. As he prepared to take off from the airbase and fly west, he noticed Dalwigk's wife flinging her arms despairingly around the *Kommandeur*'s neck. With tears streaming down her flushed cheeks, and with a choking gulp in her goodbye, *Frau* von Dalwigk planted mournful salty kisses across her husband's face. Scheffel later mused: 'Should she have suspected that it was a farewell for forever?'

The men from the I. Gruppe started their journey towards the II. *Gruppe*, located in Flers, Normandy. They subsequently reunited with their own *Gruppe* at La Ferrière-aux-Étangs, preparing for operational readiness at a moment's notice. On the very next day, the *Gruppe* transferred its staff and the 1. *Staffel* to Théville; Allied ships had been spotted nearby in the English

Channel. Finally, their orders came through in the afternoon to immediately attack a cluster of ships located between Portland and Torquay off southern England. The Stukas took off and roared towards Portland, spearheaded by the *Kommandeur*. He was flanked by the highly capable pilot *Oberleutnant* Karl Henze to his right and Scheffel to his left.

Closing in on the British peninsula, Scheffel recounts, 'We were flying at an altitude of about 2,000 metres and were still about 500 metres away from the *Kommandeur* when I suddenly saw English fighters coming from below from Portland.' An alarmed double-take rattled his bones; Scheffel observed 'ten to fifteen Spitfires coming towards us incredibly quickly. Neither the *Kommandeur*, nor Henze, nor the three on-board radio operators, had seen them'. Scheffel immediately reported the enemy sighting over the radio waves – but there was a deafening silence from the *Kommandeur* and Henze, who steamed ahead. Scheffel urgently repeated his warning, adding for good measure: 'Turn left into the clouds!' Nothing.

Unable to catch up to the *Kommandeur* – who had now pulled five minutes' flying time ahead – and because 'the Spitfires were already breathing down our necks', Scheffel veered off towards a protective cloud formation that constituted his only silver lining. He was momentarily distracted when he spotted Henze peeling off towards the south – and then, 'a great dance began'. Scheffel hurtled for the clouds, his radio operator stippling the sky with bullets from the rear machine-gun. The British fighters were so close that Scheffel could even make out the silhouettes of the pilots within their canopies. He took evasive manoeuvres by flipping the Stuka onto its back and then immediately turning it upright, careening the aircraft from left to right.

With five enemy bandits locked onto his tail, Scheffel 'had only one goal: the clouds, the clouds. I don't remember how long the dance lasted. My radio operator kept shouting: "From the left, from the right, from above, from below," and he kept cursing.' Such was the chaos of the engagement that the empty bullet casings from the rear machine-gun, which were meant to drop orderly into a nearby bag, malfunctioned; they now whizzed past the radio operator's ears, stamping him with painful bruises. Miraculously, the besieged Stuka crew made it to the clouds. Scheffel breathed a quick sigh of relief, but the air soon caught in his throat as his radio operator informed him: 'Aircraft to the left.'

Weaving in and out of the clouds once more, Scheffel's determination paid off: the guns eventually fell silent on both sides as he relied on blind flying with his instruments to get them back home. After confirming that his radio operator was fine apart from the bullet casing bruises, the two men headed towards the French coast. 'We carefully emerged from the clouds,' he noted in his diary. 'Two pairs of eyes scanned the airspace like lynxes, but there were no planes to be seen.' Finally, they managed to land back at base. As Scheffel went to park the Stuka, however, his radio operator stiffened and called out: 'Three hunters from the sea!' Scheffel's heart dropped like a stone – had the British followed them over the Channel?

Turning off the engine, Scheffel and his radio operator flung themselves onto the ground by the Stuka, protectively cupping the backs of their heads with both hands. But, as the deadly trio soared over the airfield, the Stuka crew peeked up and sagged with relief at the familiar shark-like outline: they were three Bf 109s returning from a mission. 'So, false alarm!' Scheffel recalled with amusement. 'We looked at each other and laughed.' Although the shaky laughter temporarily elevated Scheffel's mood, it did little to distract him from the long wait to see who would also make it back to base. With the German sea rescue service and Luftwaffe reconnaissance units informed of the engagement, Scheffel pulled up an empty bomb crate and sat down to wait for both the *Kommandeur* and Henze.

The anxious Stuka pilot began to beat himself up: 'I asked myself then, as I often did later, whether we should have flown onwards to the *Kommandeur* under all circumstances.' Scheffel tried to remind himself that they still had enough fuel to get home – but 'It was a tremendous feeling of emptiness and aloneness that I had back then. My eyes kept veering towards the direction of the Channel.' Just as all hope was lost, there was a faint rumbling in the distance. Scheffel's eyes roved over to the source of the sound: it was a Ju 87! Peering closer at the newest arrival, he spotted the green ring that denoted it as a staff aircraft – either Henze or the *Kommandeur*. As the dive-bomber curved inwards to land, the letter 'B' on the side of the fuselage confirmed that it was Henze.

After landing, Henze confirmed that the *Kommandeur* had not responded to the warning and then Henze became embroiled with the Spitfires, before fortuitously emerging over the recent German acquisition of Guernsey. So, if Henze had made it – perhaps the *Kommandeur* also found a way out? Two and a half hours passed, but there was still no sign of him. In the evening, with Henze waiting in Théville to hear of any updates, the German sea rescue explored the area – but they had to turn back after the guns of a British cruiser boomed threateningly in their direction. Finally, it became apparent that the *Kommandeur* had fallen in combat. '*Hauptmann* Freiherr Friedrich Karl von Dalwigk zu Lichtenfels,' Scheffel wrote mournfully in his diary. 'This was a hard blow for all of us.'[35]

Back in the Reich, the German public were starting to mourn their own losses. Despite focusing its attacks on industrial and maritime targets, the British navigational aids and bomb-aiming technology in this period were at an early stage of development, meaning that Bomber Command could not entirely avoid killing German civilians as well as workers. On 2 July 1940, the *Baruther Anzeiger* issued a public reminder to properly take cover during British air raids. After some workers who did not heed the warning siren were killed in the west of Germany during early July, the newspaper chastised that:

> These losses could definitely have been avoided if air protection regulations had been followed. Just as the soldier at the front

seeks cover during air raids, every German at home also has the obligation to protect themselves by going to the air raid shelter. This warning is all the more appropriate as it has been shown time and again that English pilots do not shy away from attacks on the civilian population.'[36]

Four days later, a vexed resident of Wesseln near Kiel wrote that 'Tommy makes the area unsafe here every night – even during the day, he has been over Misburg a few times and has bombed the oil wells there. Thank God we don't live in the city, otherwise we'd have to rush down to the basement every night...'[37] One woman living in Paderborn scribbled nervously on 7 July 1940 that 'Tommy won't let us rest here either. There has been an air raid alarm every night for fourteen days now. A lot has already been done here. These days, they were there in the middle of the afternoon and wanted to bomb the main train station. Then a fierce dogfight ensued directly over Paderborn...'[38]

That same day, a citizen in the small Hanseatic town of Uelzen wrote about how the air raids had become so intense that 'We always have ten to fifteen firefighters in the basement who didn't have a place to stay, and now they usually sit down in the garden with every alarm.'[39] For the underfunded firefighters of Uelzen to lack their own accommodation – grimly clutching their equipment as they waited for the British bombs to fall – illustrated the stark inadequacies of German air defence in certain regions during the Channel Battle. The resident of Uelzen ultimately wrote in frustration: '*When* is England going to be dealt the decisive blow?'[40] Such was the public concern over when the hammer blow would come against Britain that Joseph Goebbels, the Reich minister of propaganda, wrote tensely in his diary that 'The people fear that we have missed the right moment.'[41]

This hesitation was particularly frustrating for the occupying German forces in France, who felt as if they could almost reach out and touch the defiant island. When flying over Calais on 10

July 1940, Scheffel described in his diary how 'You could see the chalk cliffs of Dover and the harbour very well. You could also see far into the country.'[42] The lack of closure with the British seemed even more puzzling considering that, as Scheffel observed of Dunkirk during his flight, the abandoned BEF equipment – 'weapons, ammunition, gas masks, boats, etc.' – remained desperately strewn across the beaches. As he swooped over the ghost town, he added, 'Everything still looked very desolate, destroyed and burnt. I smelt the smoke all the way up in my machine.'

With no reports on the invasion of Britain in the newspapers, the German public relied on letters from their serving loved ones for updates. Another member of the aforementioned family in Wesseln wrote on 14 July 1940: 'We have set up four weeks' harvest leave for Anton, but I don't think he's coming. Apparently, they are already on the way to England! And how long will it take then? Here you get the illusion that the war will be over in fourteen days. Do you believe in it?' The writer then sighed that 'As nice as it would be, I can't imagine it! It's already taken for granted that all the young people are gone. I think it will be quite unusual if you all come back here in one fell swoop! But when will the day come? Maybe later than we think!'[43]

In certain areas of the Reich, a number of Luftwaffe antiaircraft units perceived a considerable surge in British air raids during the early Channel Battle. A *Flieger* attached to the Luftwaffe Guard Command at Tessenow, Mecklenburg, wrote on 14 July 1940 that 'After an eventful night, I'll continue to write to you. I was on watch from 1 a.m. to 3 a.m. Suddenly, heavy anti-aircraft fire broke out. Tommy was back.' The *Flieger* continued that:

> I have never experienced anything like it. At a distance of 50 km the sky was bright as day, so strong was the defensive fire. The tracer ammunition was also eerily beautiful, pure fireworks. The spectacle lasted almost an hour. All you could hear

was a thunderous roll and roar. But I didn't really pay much attention to it. The feelings and thoughts are indescribable.[44]

Despite these early British air raids in 1940 being characterized by some German eyewitnesses and post-war historians as largely ineffective, we should not discount the distraction that they caused to Luftwaffe flak personnel who were stationed away from home while protecting other German cities. As one *Unteroffizier* with Luftwaffe Flak Regiment 49 wrote on 13 July 1940: 'How is it with you in Mannheim? I've heard that no planes have come lately. Is it true that the Mannheimers shot down an Englishman?'[45]

It is important to remember, too, that being the undisputed conquerors of the Continent did not render some of the Luftwaffe's men immune to a gnawing feeling of homesickness and isolation. Billeted in unfamiliar territories and loathed by some of the subjugated locals, the letters from certain personnel expressed a deep hankering to go home to the Reich. Luftwaffe doctors and medics were especially worn down by the endless violence: after all, even the most dazzling military victories still came with mangled limbs to amputate, weeping burns to sterilize and bad news to deliver. A Luftwaffe junior doctor with the III. *Gruppe* of the bomber wing KG 2 wrote to his wife on 6 July 1940:

> We lonely people at the front are gripped by a burning longing for our homeland and for our dear women. But we are also aware that we must shield and protect this, our greatest happiness on earth, for all eternity through our deprivations and our struggle. For all of us, there is only this single, deepest and most sacred task. Many of our bravest young comrades had to give their lives, many graves line the streets and paths. When we come back to you, we are no longer the old versions of ourselves – we have become men of iron, in loyal comradeship, we have found our idealism again and have grasped like

no other what our dear wives gave us at home and what the word 'home' means.⁴⁶

Yet, the doctor recognized that his focus was now on supporting his bomber unit through the next campaign: 'The war with France is over, we are facing England, I hope everything will end well soon. When the war is over, the first way will be to each other...'⁴⁷ These frazzled accounts demonstrate, then, that the strains of war were already beginning to show among the Germans in the opening weeks of the Channel Battle. The Luftwaffe may have enjoyed free rein over the beaches from which an historic Allied retreat had been carried out – but, at the same time, they had simply pushed the RAF further into their fortress. Now, it was time to lure them out back out again.

8

The Overture

The face of the victorious German *Heer* turns towards Great Britain across the Channel. The hour is here that forces one to make the final decision and where the life-and-death struggle begins. Europe will not find peace until England is brought to her knees. The standard-bearers of a new era, Germany and Italy, are now turning their weapons and their hearts with all their might against the English Empire. The dying generation of those plutocrats who so often poured out their excessively arrogant buckets of slander on Greater Germany and its *Führer* in the Houses of Parliament, working hand in hand with those stock market-jobbers not far from their headquarters, are now in their 'fortress'.[1]

Taken from an issue of the *Deutsche Luftwacht* ('*German Air Watch*') magazine on 15 July 1940, this article not only intended to assuage the concerns of the German public at the unforeseen extension of hostilities, but also to deflect blame from the Nazi war leadership onto the 'warmongering' British. By the middle of July 1940, many citizens of the Reich were calling for a resolution with the British. The Nazi leadership could not afford to risk public sentiment souring, so German propaganda began to promote the impression that the Wehrmacht was poised to invade the island nation.

The German press held the British 'plutocrats' responsible for the protracted war, while also taking a jab at the 'stock market jobbers' that they were seemingly in cahoots with. This was linked to the antisemitic conspiracy that Jews controlled high finance for their own insidious gain: since the seventeenth century, Jews had been used as scapegoats for any unfavourable fluctuations in European stock exchanges.[2] In addition, as stockjobbers earned their income through buying and selling shares on their own behalf, the term soon became conflated in antisemitic rhetoric with the 'self-serving' Jews in finance.[3] This was then extended to the British, who were characterized by the Nazi press as being financially greedy due to the British Empire, whose rapacious nature would be its downfall.

Various historians have busted the myth that the island nation stood alone during that 'Spitfire Summer' of 1940 – in reality, of course, a fifth of Fighter Command was made up of Commonwealth personnel and the airmen of Nazi-occupied countries who were hellbent on avenging their fallen homelands.* Little has been made of the fact, however, that German propagandists also created their own 'stand-alone' myth: presenting Britain as being isolated, unpopular, vulnerable and ready to be gutted by the Luftwaffe. The *Deutsche Luftwacht* article from 15 July 1940 bore the headline 'England's Isolation Brings the End' and added that:

> From day to day, from hour to hour, the vice grips tighter around the island that is no longer an island. Norway, the great armoured flank against England, the coasts of Holland,

---

* In addition to Polish, Czech, Slovakian, French and Belgian air force personnel who had retreated to the 'island of the last hope', the airmen from the Commonwealth who participated in the Battle of Britain included Australians, New Zealanders, Canadians, South Africans, Rhodesians, a Jamaican, a Barbadian and a Newfoundlander. There were also a handful of Americans and an Austrian.

Belgium and the whole French coast have taken front across the Channel to the island kingdom. We shook hands with our Spanish friends. The Channel Islands of Jersey and Guernsey, fortified by the British but recently evacuated, have been occupied by a bold coup. The steel ring from which there is no escape becomes ever tighter. For the first time on her long, predatory road to world power, England stands alone. The term 'Western Powers' belongs to the past.[4]

The idea of there being 'no more islands', as Hitler had boasted after the fall of Poland in October 1939, was often reinforced among the Luftwaffe aircrews in order help them bypass the psychological barrier of flying over the English Channel.[5] As one pilot with the bomber wing KG 26 noted, 'When we get into the aircraft to start the enemy flight, we read a passage that is like a routine for us pilots: "there are no more islands. Adolf Hitler." Our commander had that typed out on a typewriter and placed at the entrance, where every man in the crew sees it and must read it before each flight.'[6]

Following the preliminary contemplation of an amphibious invasion, Hitler met *Großadmiral* Erich Raeder – commander-in-chief of the German navy – on 11 July 1940 to further discuss whether Sealion was feasible. That same day, Air Fleets 2, 3 and 5 were ordered to commence 'intensive air warfare against England',[7] with Jeschonnek including both the British armament industry and the RAF as official targets.[8] Nevertheless, on 12 July 1940, a further survey entitled 'First Deliberations Regarding a Landing in England' acknowledged that 'the landing will be difficult'.[9]

Recognition of the British Royal Navy's prowess, as well as the losses sustained by the *Kriegsmarine* in Operation Weser Exercise North, can be seen in the document's admission that 'England is in command of the seas – therefore, a landing will only be possible on the South coast of the Channel where we can substitute our lack of sea supremacy by air supremacy and where the sea crossing is short.'[10] This was in direct contrast to Nazi propaganda, with *Deutsche Luftwacht* magazine crowing three days later that 'It can no longer be proven by the British today that the English Channel, which was previously called English, is their water – it has now become the precursor to the final dramatic prelude and is undisputedly ours.'[11]

Despite Raeder's grievous misgivings about whether the *Kriegsmarine* was prepared for a cross-Channel operation, Hitler pressed on with planning for Sealion.[12] On 16 July 1940, he issued Directive No. 16, which confirmed that 'I have decided to prepare and, if necessary, carry out a landing operation against England. The purpose of this operation is to eliminate the English motherland as a base for continuing the war against Germany and, if necessary, to fully occupy it.'[13] The next day, Air Fleets 2, 3 and 5 were ordered to achieve full operational readiness for this endeavour.[14] A flurry of conferences regarding Sealion was further organized from mid-July, including one joint service conference at which Jeschonnek represented Göring, while the latter hosted another top-level conference at his country home, Carinhall near Berlin.[15]

This turned up some disturbing findings for the Luftwaffe on the true readiness of the *Heer* and *Kriegsmarine* regarding Sealion, but Hitler's belief during this stage of the Channel Battle was, as he explained to Raeder, 'If the effect of the air attacks is such that the enemy air force, harbours and naval forces, etc. are heavily damaged, *Unternehmen "Seelöwe"* will be carried out in 1940.'[16] However, it would be far preferable for the Germans if the Wehrmacht did not have to fight this perilous campaign at all. So, on 19 July 1940, Hitler delivered his infamous 'Last Appeal to Reason' speech to the Reichstag:

> A great world empire will be destroyed. A world empire which I never had the ambition to destroy or as much as harm. Alas, I am fully aware that the continuation of this war will end only in the complete shattering of one of the two warring parties. Mr Churchill may believe this to be Germany. I know it to

be England. In this hour I feel compelled, standing before my conscience, to direct yet another appeal to reason in England. I believe I can do this as I am not asking for something as the vanquished, but rather, as the victor, I am speaking in the name of reason. I see no compelling reason which could force the continuation of this war.[17]

To express his gratitude for Case Yellow and Case Red, the *Führer* now elevated Luftwaffe commanders Erhard Milch, Hugo Sperrle, and Albert Kesselring to the rank of *Generalfeldmarschall* ('Field Marshal'). Most notably, Hitler bestowed Göring with the unprecedented rank of *Reichsmarschall* ('Marshal of the Reich') in recognition of the 'Wonder of 1940'. Hitler declared of the Luftwaffe that 'The manner of their deployment in the operation in general, as well as their adjustment to the tactical demands of the moment, was exceptional.' But one of the most revealing aspects of his speech was his reiteration of what the Luftwaffe's key objectives had been in the campaign:

> One – to destroy the enemy air forces, i.e. to remove these from the skies.
> Two – to support directly or indirectly the fighting troops by uninterrupted attacks.
> Three – to destroy the enemy's means of command and movement.
> Four – to wear down and break the enemy's morale and will to resist.
> Five – to land parachute troops as advance units.[18]

This passage was intended to show that since the Wehrmacht had achieved all this before, it could theoretically do it again to any other country that dared to stand in its path. Hitler's speech also touched upon the mounting concerns among the German public that the ideal moment to strike at the British had passed. He assured the Reichstag – and, by extension, the Reich – that 'If I do not speak for some time, or nothing much happens, then this does not mean that I am not doing anything.' Even more revealingly, Hitler alluded to the nascent plans for Sealion:

> Besides, hens would be ill-advised to cry out to the world every egg just laid. It would be all the more ill-considered of statesmen to announce projects barely beyond the planning stage, in nervous chatter, to the surrounding world, so as to inform it in a timely manner.[19]

Thus, the *Führer* appeared to publicly confirm that a military offensive was in the works and that, if the British refused to sue for peace, it would be left to them to find out what the full terrifying repercussions would be. His speech also demonstrates that the Nazi regime had cleverly positioned itself to either accept a peace offer or prepare for more fighting from what it called the 'warmongering' island nation, which it presented as choosing hate over harmony. The Luftwaffe then dropped leaflets containing Hitler's 'Last Appeal to Reason' over Britain's mainland; the next time, it would be bombs.

One day after Hitler's speech, the British bombers that buzzed the Ruhr Valley – the industrial *Waffenschmiede des Reiches* ('armoury of the Reich') which had propelled the Nazi war machine towards its prolific military victories – demonstrated that the threat of Bomber Command had not subsided by the mid-Channel Battle. In June and July 1940, the RAF's 'Striking Force' had been deployed against the Ruhr and the north German coast in the belief that its attacks would be detrimental to the Reich's economic base, wearing down the Germans through 'carefully planned bombardments of vital objectives'.[20] In response, a dedicated *Nachtjagdgeschwader* ('night-fighter wing') – NJG 1 – had been created in anticipation of renewed Bomber Command attacks on Germany. Now, in the early hours of 20 July 1940, this experimental flying arm would be truly put to the test.

Shortly after 2 a.m., a British bomber formation passed over the Ruhr, which was half-dozing under a protective quilt of clouds. At 6,000 feet, *Oberleutnant* Werner Streib blazed through the moon-soaked clouds towards the oncoming bomber wave in his Bf 110, flying brazenly towards the German flak zone. The searchlights that quickly swivelled to track the aerial pursuit gave Streib the necessary illumination to spot his unfortunate prey: a lumbering Armstrong Whitworth Whitley bomber. Seized by the thrill of the hunt, Streib closed in rapidly on the stricken Whitley, which now loomed on his gunsight as a massive black shadow.[21] Such was his excitement that he initially shot too wide; he had to go again.

Readjusting the Bf 110 until he was directly on the Whitley's tail, Streib regained his head. This time, he unleashed a concentrated fusillade right into the bomber's vulnerable fuel tank. It caught fire instantly and exploded, engulfing the Whitley in flames and sending it into an irrevocable tailspin. It marked the very first time that a night-fighter shot down an opponent in the Second World War – and it demonstrates that, although the Channel Battle is often thought of as a daylight campaign, the fighting sometimes raged long after sunset. The bombing of the Ruhr was most embarrassing for the newly minted *Reichsmarschall*, who had famously given an impassioned speech in Essen on the perceived impregnability of German air raid defences shortly before the outbreak of war:

> As Reich aviation minister, I am personally satisfied with the measures that were taken to protect the Ruhr area. In the future I will take care of every battery myself, because we will not allow a single enemy bomb to fall on the Ruhr area… if a single enemy bomber ever gets that far, then I don't want to be called Göring anymore – then you can call me Meier![22]

Zorner quotes the surname as 'Meier', a surname of Latin and Middle High German origin which is the third most popular surname in Germany[23] – equivalent to a 'Williams' in Britain or the USA.[24] As such, it is possible that Göring was challenging the crowd to call him an 'average Joe' and not the 'great' *Reichsmarschall*, if he were proved wrong about the Luftwaffe air raid defences. However, this quotation often spells the name as 'Meyer', which comes from the Hebrew surname of Meir.[25] As such, the most common interpretation of this quotation is that Göring was goading his listeners to publicly humiliate him by addressing him with a Jewish surname if he was wrong – a theory which easily fits with his antisemitic rhetoric. Whatever his original intention, Bomber Command had soon made *Herr* Meier eat his words.

At a Carinhall conference that ran from 18 to 21 July 1940, Göring confirmed his rather overambitious plan that the Luftwaffe would subdue the RAF within two to four weeks.[26] During the concurrent 'Last Appeal to Reason', the Reich hung on to the *Führer*'s every word. Rather like the way Churchill tried to galvanize the British with his speeches, certain key addresses by Hitler resonated strongly with many of the Reich's citizens. This was not just because the *Führer* was a powerful orator, rapidly delivering his words like an endless belt of ammunition: it was also because they helped to update the population on the war situation and their forces' morale.

In the 'Last Appeal to Reason', for instance, Hitler clearly stated that his speech was intended 'to give our *Volk* insight into the historic uniqueness of the events we have lived through'. He then progressed to criticising the recent smattering of Bomber Command air raids so that he could justify any future intensification of the German air war as a necessary evil to counter Britain's seemingly faltering ethics. The *Führer* claimed that Churchill:

> ...Launched this war in an arena in which he apparently believes he is quite strong: namely, in the air war against the civilian population, albeit beneath the deceptive slogan of a so-called war against military objectives [...] These objectives have turned out to be open cities, markets, villages, residential housing,

hospitals, schools, kindergartens and whatever else happens to be hit.

In the face of this apparent British immorality, some Germans began to see the prospect of concentrated Luftwaffe attacks over Britain as a matter of divine adjudication. Hitler often cultivated the image that his 'crusades' were backed by God, preaching in his 'Last Appeal to Reason' speech that 'It is almost painful to me to have been chosen by Providence to give a shove to what these men have brought to the point of falling. It was not my ambition to wage wars, but to build up a new social state of the highest culture. And every year of war takes me away from my work.'[27] A civilian living in Görlitz, east Germany, wrote on the same day as Hitler's 'Last Appeal' that:

> Now we're just curious to see when the strike against England will start. It will be more than just a mere defeat. This is simply going to be an eerie judgement of God. And with all of that, as a Christian, [...] Dr W. Schmidt, inventor of the superheated steam locomotive and 2,000 other inventions, said in 1919: 'England is dying of its horrible dishonesty and the resulting impenitence. God will make it a burnt crater among the nations unless it repents.' And now we're going to witness that judgement of God.[28]

On 18 July 1940, the *Baruther Anzeiger* newspaper boasted that 'the incessant attacks by German planes on England have apparently caused a great deal of concern there and have considerably increased nervousness' – apparently forcing the British War Cabinet 'to give lengthy radio speeches to the English'.[29] The newspaper denigrated Churchill's attempts to 'instil confidence in the people of victory' and asserted that the British had 'claimed against their better judgment that the British air force was superior to the German one'. A few days later, the same newspaper disturbingly reported that a 'Jewish-plutocratic clique bears responsibility' for the protracted hostilities.[30]

Under the headline 'England Remains Incorrigible', the newspaper claimed that:

> The whole world is under the influence of the great *Führer*'s speech in the *Reichstag*, which was broadcast by almost 1,000 broadcasters in around thirty languages. It was understood that the *Führer* had spoken his last word and, once again, made a victorious appeal to English reason. In this respect, it now depends on what London has to say in response. This answer will decide England's fate.[31]

The German press now watched the UK Parliament like a hawk, with the *Baruther Anzeiger* article claiming that 'The British government will not give an official response, but Churchill will perhaps only comment on the *Führer*'s speech in the House of Commons on Tuesday in response to a question from MPs.' The newspaper report attempted to portray Hitler as a benevolent humanitarian after his 'Last Appeal', asserting that 'The neutral countries understand that England has been given a last chance, but they agree that reason, which could hold down the curtain on the final scene in the European war, will not have a say.' The article further alleged that a Stockholm newspaper had reported how 'Everything points to the possibility that the world would soon experience a victory on a completely different level than ever before.'

Whilst it is reasonable to assert that Britain – engirdled by the powerful Royal Navy – was never truly in danger of being subdued by Sealion, the press in neutral countries generally presented a rather bleak outlook for the island during the Channel Battle. On the same day that Hitler gave his 'Last Appeal to Reason' speech, the Swiss *Zürcher Illustrierte* magazine published an intriguing report entitled 'Threatened England':

> Today, the whole world is concerned with the question of

whether the German *Heer* will attempt to land on the English coast in order to bring the war into the country. The English government is already taking defensive measures on a grand scale, so it seems to be expecting the possibility. The course of England's history so far shows that enemy armies have actually successfully landed in England several times. Even if no conclusion can be drawn from these undertakings with regards to present times, in which the defensive options have been considerably improved in relation to the offensive measures, an overview of invasions of England by enemy armies will be of interest right now.[32]

The *Zürcher Illustrierte* then documented past invasions of Britain by the Romans, Anglo-Saxons, Vikings and Normans, demonstrating that the Swiss press considered a German occupation of the British Isles to be quite feasible. Indeed, some of the British anti-invasion preparations that the article referred to were being openly mocked in the Nazi press. On 12 July 1940, the German *Baruther Anzeiger* newspaper scoffed at the hurried formation of the Local Defence Volunteers (LDV) – later known as the Home Guard – in case of enemy invasion, claiming that 'the English measures to defend against German attacks are becoming increasingly nervous and hasty'.[33]

The article further sneered that 'procurement of steel helmets seems to be particularly difficult for the private shooting clubs that go by the high-sounding name "Local Defence Volunteers"'.[34] This defensive activity, however, illustrated that the answer Britain had to the 'Last Appeal' was to rally its forces in the face of the invasion threat. Mines laced the island's winding beaches; barbed wire spiralled around its key strongholds; and, beneath the flying arms of the RAF and the Royal Navy's Fleet Air Arm, the RAF's Balloon Command, the doughty LDV and the British Army's Anti-Aircraft (AA) Command, largely staffed by the Territorial Army, primed to protect their homeland.

The German public were divided in their response as they

waited to see whether the silent island would accept Hitler's 'olive branch'. A housewife from Constance in southern Germany wrote on 23 July 1940 that 'Unfortunately, the Englishman reacted badly to our *Führer*'s well-intentioned peace offer. Well, what will happen next? It's a shame that the English people don't want to accept reason when they see that they will never be masters against us...'[35]

That same day, however, a female resident of Tübingen wrote animatedly: 'One is really excited to see how it's going to be done against England until they've been forced to their knees. You will also have heard the great *Führer*'s speech! Wasn't that fine? It was foreseeable that England would not give in. And they say it's going to be devastating! Well, they don't deserve it any other way.'[36]

After no official response was forthcoming from Britain by the last week of July, the ominous headline of the *Baruther Anzeiger* read: 'England Has Chosen.'[37]

With the Luftwaffe's bombers and dive-bombers heavily reliant on their fighter escorts, a number of fighter units were now transferred to the French coastline. Rall noted that 'While bombs are already falling on British shipping escorts in the English Channel, we are moving back to the Reich. We practise the air war against England in Werneuchen for three weeks.'[38] Eventually, on 22 July 1940, Rall's unit was relocated to Coquelles on the Pas-de-Calais – much to the bitter protest of a local French farmer, who was forced to prematurely mow his prized rye field to make space for their new runway! Other Luftwaffe personnel were shifted around German-held territory to fill in some of the gaps in the aerial defences, with a *Gefreiter* in Flak Regiment 231 penning two days later:

> I am now here in Belgium near Ghent, as we have been withdrawn from France again. However, a lot of English planes have come here. A few days ago, we shot down another English plane

while it was burning. Before we came here to Belgium, our position was in Calais, just on the coast of the Channel. On a clear day, we could see the English coast.[39]

Rall made a similar comment on being able to see the white cliffs of Dover from his new airbase at Coquelles, pointing out how close Britain seemed to be to the Luftwaffe personnel poised across the Channel. Yet, in Rall's case, he quickly spotted the potential for disarray at his new site: 'The *Staffeln* are spread out around the levelled field, which is slightly elevated in the middle – just enough so that the opposite edge of the runway cannot be seen from any point on its edge.' The peril this posed, as Rall explained, is that it 'makes large-scale scrambling a tricky affair. You never know who you'll meet in the middle of the pitch during a full starting run'.[40]

Historians rightly point out the 'home advantage' wielded by the RAF's Fighter Command pilots during the Battle of Britain – from not having to start all of their aerial engagements with a cross-Channel flight, to baling out over friendly territory. To add to this, the Luftwaffe had to operate units from makeshift airfields and runways in France that, under less pressurized circumstances, might have been rejected as unsuitable. Amid the ongoing Channel Battle, however, precious time could not be squandered on finding alternative bases. Rall's *Staffel* of Bf 109s was given just one day to settle in and prepare before their first real sortie of the campaign: escorting a group of Ju 87 Stukas to attack British maritime targets at Dover.

'This is about as difficult as trying to get a family of hedgehogs unharmed across a motorway,' Rall mused unhappily. He recalled that 'The Stukas reach a cruising speed of around 250 km/h [155 mph] in horizontal flight with full set-up weight – not much more than the speed at which a fully fuelled and ammunitioned Me 109 can just barely stay in the air.' The discrepancy in speed between the Stukas and their fighter escorts meant that the Bf 109s had no choice but to defend the Stukas by peeling off and securing wide swathes of the airspace surrounding them. Or, as

Rall evocatively put it, 'You can only save a family of hedgehogs on a motorway by blocking traffic from a considerable distance away and not by flanking them left and right.'

On 24 July 1940, Rall and his comrades took off from Coquelles at 1.10 p.m. and headed towards Dover. They were ably led by *Hauptmann* Wolf-Heinrich von Houwald of the III. *Gruppe* from the fighter wing JG 52, a seasoned *Gruppenkommandeur* who had become a fighter ace during the Spanish Civil War. Houwald seemed the perfect choice to lead Rall and his confrères into their first serious aerial battle over Britain. Despite still being relatively wet behind the ears, Rall was already developing the strong gut hunches of a fully-fledged fighter pilot. As the Bf 109s met up with the Stukas, he later recalled how 'Our armada sluggishly cruises northward from Boulogne across the Channel in bad weather; Cap Gris-Nez can only be seen dimly to our right. This is one of those missions that you fly with a certain trepidation from the first minute – it almost smells like disaster.'

Rall ignored the nauseating intuition gnawing at his stomach and pressed on through the squally British weather: 'Suddenly, a pack of Spitfires pounce in between us like hawks in a pile of chickens. I see [our *Staffelkapitän*] Lothar Ehrlich, who is leading the first swarm in front of me, turning towards the opponent, and then I'm already busy defending myself. My swarm burst apart in a split second, and everyone is fighting for themselves in the wild scuffle. There is no longer any thought of protecting the Stukas.' Managing to survive the sudden onslaught, Rall somehow landed in one piece at Coquelles at around 2 p.m. He soon joined the returning fighter pilots standing 'in small groups of excited discussion around the apron, greedily inhaling their first cigarettes' – but he also recalled their consternation that 'The Tommies caught us just as we feared, like leaden ducks.'

Once they had earnestly recounted their own tussles in the dogfight, the young pilots shifted their gaze to the runway, waiting tensely to see who else had made it through the mêlée.

Solo and grouped Messerschmitts continued to land at sporadic intervals. Sometimes, the pilots clambered out with good news: Jupp Zwernemann of the 7. *Staffel*, for instance, had scored his first aerial victory near Margate against a Spitfire, with his wingman Edmund Roßmann having also survived and bagged a kill in the fraught exchange. But, as the flood of returning fighter pilots dwindled to a trickle, it became clear that fate had its favourites. 'The bodies of Houwald and [Erich] Frank were found washed up near Dunkirk a few days later,' Rall narrated sombrely. In addition, Ehrlich and the *Kapitän* of the 7. *Staffel*, Herbert Fermer, failed to return that night.

Soon after the Dover raid, Rall was unexpectedly appointed to replace Ehrlich as leader of JG 52's 8. *Staffel*. 'I had turned twenty-two in March,' he related incredulously, 'and, with a single sentence, I was given responsibility for twelve aircraft pilots and a good eighty ground personnel. Most men are significantly older than me.' Despite the loss of war-grizzled pilots like Houwald and Fermer, the relative seniority of the men that the fresh-faced Rall now commanded demonstrated how the fighter units still had plenty of *alte Hasen* ('Old Hares') to call upon during the Channel Battle. With scarcely a second to process his new appointment, though, Rall noted that:

> The very next day, in the early afternoon, there was another escort mission to the port of Dover, and less than two hours after our return, we were in the air again to provide cover for Stukas during an attack on a convoy. My first twenty-four hours as a *Staffelführer* end with a third mission – again to protect Stukas off Dover, from which we return to Coquelles at 2005 hours.[41]

Although the Channel Battle is often neglected by historians in favour of studying the *Adlerangriff* ('Eagle Attack') which took place from mid-August, the accelerated strain that the Luftwaffe was put under during July 1940 should not be dismissed. It flew

1,300 sorties across the first five weeks of the Channel Battle – an average of 260 a week, or just under forty per day – showing that this was hardly a sporadic operational phase.[42] Among some of the Stuka crews, the magnitude of the campaign started to hit home. A Luftwaffe *Soldat* attached to an airfield operations company with the dive-bomber wing StG 2 wrote on 20 July 1940 that:

> It's raining right now, and our planes are in the forest. So, I lay down on the floor to write to you. We have a big victory behind us, but the biggest fight is still ahead of us. We don't have earth beneath us, but water – that's more dangerous. We hope for the best and wish to return to our homes as victorious soldiers.[43]

Nevertheless, the Channel Battle gave the Luftwaffe enough breathing space to regain its strength in other areas. It recovered from losing nearly 40 per cent of its twin-engine bombers in the preceding campaign on the Continent, and by 20 July 1940, the frontline bomber force was slightly bigger than the force that the Luftwaffe had wielded in the Low Countries and France.[44] A further success was seen in the fact that the Luftwaffe's *Seeflieger* ('naval aviators') had sunk twenty-five steamships totalling 50,000 tonnes in just over a month of fighting in the Channel Battle.[45] Indeed, from 19 July 1940 the British Admiralty had to temporarily suspend all merchant shipping passing through the Straits of Dover in daylight hours.[46]

As the Luftwaffe's wrath spread over the Channel and toward the mainland of Britain, some of its personnel felt increased confidence in a final German victory. On 20 July 1940, a serviceman with a Luftwaffe station command wrote, 'I've been in Belgium for a week, near the French border... now England will be overthrown, then the war will soon be over, definitely before the end of this year.'[47] Witnessing the subjugation of the Belgians likely bolstered his confidence, with the writer noting that, 'I'm fine here, the Belgians don't want anything to do with the war anymore. Ooh, this poor country has suffered! Two great wars in a short time. We can be thankful that our fatherland was spared such misery...'[48]

Back in the Reich, some German civilians shrewdly recognized the dangers which lurked ahead, but many remained convinced by the Nazi propaganda which emphatically declared that karma was about to befall the 'greedy' Brits. A resident of the small municipality of Glashütten wrote angrily on 23 July 1940 that:

> England has now decided with its megalomania. It won't be long now either, and the war can take its unstoppable course. It is sad that a few such moneybags can have so much power over a people and bring them to ruin. Unfortunately, we also have to bite the bullet. You can't get out of the fear of death. If only you were more protected! Now there will probably be the heaviest battle that has ever taken place...[49]

This further demonstrates how some Germans anticipated that the existing aerial campaign against Britain would metamorphose into the most difficult military campaign to date. Yet, it was the frontline Luftwaffe units who would be forced to deliver this aerial offensive – and, for some of them, their nerves were already beginning to disintegrate by the end of July 1940.

In a letter from 26 July, the junior doctor's mounting psychological trauma at treating Luftwaffe bomber crews during the Channel Battle was evident. 'We want to finish off England as soon as possible,' he wrote urgently, 'and then we have only one wish: to go home to our dear wives. We can't spend the winter here; our nerves can't take it anymore. The horrors of war on the front line and also the pain of parting from you will not be bearable for me in the long run.'[50] He lambasted the lack of an effective rest rotation system in the Luftwaffe, heatedly declaring that:

> I also consider it an injustice to leave the same men at the front all the time. The many slackers at home should also be seen for once, filling their pockets while we are here. We recently had to bury another brave young comrade who was shot down by enemy fighters. Despite all the celebrations of victory in the Reich, it is sad here...

Contrasting his characterization of the later Channel Battle with Rall's sanguine appraisal of its beginning demonstrates how the first seeds of doubt were being sown in the Luftwaffe during the *Luftschlacht um England*. By the end of July 1940, the Nazi leadership fervently hoped that the Channel Battle had put enough pressure on Britain that the afflicted island would throw in the towel. But, like a groggy boxer grimly resolved for the next round, Britain staggered once more onto her weary feet – much to the palpable shock of her opponent.

With no sign of an impending British surrender during the Channel Battle, Hitler knew that he could not afford for the troublesome island to recover its strength and threaten his Western conquests.[51] Thus, his earlier declaration on 16 July that 'I have decided to prepare, and if necessary to carry out, a landing operation against [Britain]' was pursued with renewed vigour.[52] For the *Kriegsmarine* to have the best chance of defeating the Royal Navy and landing German troops unhindered onto the island, the Luftwaffe needed to definitively suppress the RAF in a more concentrated aerial campaign than the Channel Battle. So, Hitler now ordered that the Luftwaffe would:

> [...] prevent all enemy air attacks, and will destroy coastal defences covering the landing points, break the initial resistance of the enemy land forces, and annihilate reserves behind the front. The accomplishment of these tasks will require the closest cooperation between all individual units of the air force and the invading army units.[53]

By 30 July 1940, the *Führer* 'had ordered that preparations for the major battle of the German Luftwaffe against England should be made immediately and with the greatest speed so that it could begin twelve hours after the *Führer* issued the order'.[54] The imminent campaign, Eagle Attack, would see 'flying units', 'ground installations' (including radar), 'supply organizations' and the British aircraft industry come under intense attack. Five weeks – spanning from 8 August to 15 September – had been allocated for the Luftwaffe to overpower 'the English air force with all the forces at its command, in the shortest possible time'.[55] Thus, as Rall put it, 'The overture to the Battle of Britain ends – and its first act begins.'[56]

9

A Very Bad Piece of Work

Here in France, we are all slowly going crazy. If we don't get relieved in the autumn, I think almost all of our men will be nervous and exhausted. What has been asked of us for years is too much. God grant us that the last stroke of violence succeeds. As a doctor, I see how things are with us. People gather their last strength to win. The last few days have brought us some terrible experiences. We have lost many brave and loyal comrades and are immensely sad. In addition, we have a lot of work, we can't think of vacation for the time being, because we think the war with England has taken form. We all still want to do our best to win, but it's impossible to spend the winter in France; they have to relieve us then, otherwise we'll all be crazy. None of us want a vacation, we want a quick win, and then all the way home to you!...[1]

Written on 13 August 1940, this anguished letter from the junior doctor of the bomber wing KG 2 to his wife demonstrated the Luftwaffe's rising fatigue by *Adlertag* ('Eagle Day'). This was the fateful moment that the Eagle Attack – the next phase of the *Luftschlacht um England* – would be put into action. With the Channel Battle having failed to force the British to sue for peace, apprehension began to grip the overtired Luftwaffe. The doctor added in his letter that:

> Our nerves are shot, but we have to persevere to protect the homeland. The England mission has slowly begun. Apparently, they want to achieve peace before winter after all. I also believe that this is the only possibility for our victory. Right from the start, most of us are under constant psychological strain and gather the last of our strength. I myself have been working hard lately, this business can't be endured much longer, then we're at the end...[2]

Admittedly, the doctor perhaps had a more macabre view of the *Luftschlacht um England* than the average Luftwaffe serviceman, as he tended to the men who had been broken by the Channel Battle. He bore witness to the raw, unfiltered reality of the daily campaign in a way that the Nazi regime and the *Oberbefehlshaber der Luftwaffe** ('Supreme Commander of the Luftwaffe', Ob.d.L.) were sheltered from.

The doctor closed his letter with a prayer: 'May the *Führer* never forget our heaviest sacrifices of nervous strength! Here in France, we are living the hardest days of our lives...'[3] His observation of combat fatigue in their bomber wing shows that the Luftwaffe was potentially in a far worse psychological state by Eagle Day than is often acknowledged. Some personnel were clearly already suffering with great exhaustion and agitation by the onset of Eagle Day.

---

* It is often written that top orders in this period were issued by the *Oberkommando der Luftwaffe* ('Luftwaffe High Command', OKL). During 1940, however, the power lay with Göring's command as the *Oberbefehlshaber der Luftwaffe* and the *Reichsluftfahrtministerium* ('Reich Air Ministry', RLM). In 1944, the Reich Air Ministry was restructured to formally establish the *Oberkommando der Luftwaffe*. See L. Patterson, *Eagles Over the Sea, 1943– 45: A History of Luftwaffe Maritime Operations* (Barnsley: Pen & Sword, 2020), xi; P. Caddick-Adams, *Monte Cassino: Ten Armies in Hell* (Oxford: Oxford University Press, 2013), 299.

Even for personnel stationed in the Reich, the relentless galvanisation of the Luftwaffe was evident to its personnel by early August 1940. An *Unteroffizier* of a Luftwaffe armoured train unit wrote from Oranienburg in northeast Germany on 1 August 1940 that 'Hopefully it will continue soon so that we can all get home soon. No one has been released from us yet, we are all needed...'[4]

The Luftwaffe ground crews protecting Germany's skies, meanwhile, found themselves utterly frustrated by the British aircraft roaming around the Reich. By Eagle Day itself, British air raids over the Reich were perceived as increasingly criminal and indiscriminate against German civilians by the Luftwaffe flak crews and watchkeepers. The *Soldat* based at Tessenow wrote on 13 August 1940:

> Today at night, exactly 12.55 a.m., I was on watch. Tommy came and circled over us. That was the first time we took cover. We ran wherever we could run. Then we experienced the war and realized that it's not a summer holiday. We didn't have an alarm. It was a good thing that, as always, we were lucky. Low clouds obscured all visibility, even though he was flying very low. High-explosive and incendiary bombs were thrown in one spot. A barn caught fire. When they wanted to start extinguishing the fire and the whole village was on its feet, the planes flew low and covered the place with machine-gun fire.[5]

Tessenow is a tiny German village which held little military significance during the war, although its state of Mecklenburg did contain a number of pertinent military targets.* Consequently, such reports of British aircraft strafing German towns added

---

* The state's capital of Schwerin – a patchwork quilt of agriculture and some industrial production – had already been attacked by the British in July 1940, while the state also contained the famous experimental weapons testing station at Peenemünde, of which British intelligence had been aware since 1939.

to the public yearning for the Luftwaffe to subdue the British for good. The same day that the letter was written, Jodl signed a report which proclaimed that 'The landing operation must not fail under any circumstances. Failure could have political consequences that may stretch far beyond the military ones.'[6] The role of the Luftwaffe in the Eagle Attack was 'the counteraction of the English Air Force until it is – or can be – virtually eliminated':

> One should not undertake operations for war aims, but fight for victory. England's will to resist must be broken by spring. If you can't do it with one landing, you can do it with all other means. Everything else has to take a back seat to this most important task. We are now entering the decisive battle against England. This means that the general principles of war must also come into force within our coalition, namely to concentrate all our forces at the critical place – that is, the air and *U-Boot* ['submarine'] war against the English motherland.[7]

Nevertheless, his report emphasized that Eagle Attack might not herald the launch of Operation Sealion, noting that 'the next eight days will clarify this' and that 'there are other ways to bring England to its knees'. It was this chronic indecision over an amphibious invasion that would ultimately affect the Luftwaffe's operational performance in the Battle of Britain. Indeed, nearly two weeks before Eagle Day, Hitler's hesitation over Sealion had been evident in an OKW report that detailed his provisional decisions on the invasion:

> The preparations for 'Sealion' must be continued in the *Heer* and Luftwaffe and also scheduled for 15 September. After eight, or at the latest fourteen days, counted from the start of the major air battle against England – which can begin around 5 August – the *Führer* will decide, depending on the outcome of this battle, whether or not the 'Sealion' operation will take place this year. If the decision is made against carrying out

'Sealion' in September, all preparations must still be continued, but in a form that excludes serious damage to the economy through the paralysis of inland shipping.[8]

Here, we start to see the incongruence that dogged German decision-making leading up to Eagle Day. On the one hand, the Luftwaffe was expected to fight as if a decisive victory was on the horizon; on the other, the OKW's uncertainty about whether an amphibious invasion would even take place in 1940 signified that a lack of confidence underpinned the overall endeavour. It was hardly surprising, though, that the German leadership was concerned about extending an aerial campaign that was significantly different from the Luftwaffe's winning formula on the Continent.

Rather than dutifully assisting the other Wehrmacht branches, the onus was now placed on the German air force to construct its own war-winning strategy. *Oberst* Paul Deichmann – the chief of staff for the II. Air Corps – claimed after the war that 'There was no army attacking, no navy attacking.'[9] Admittedly, the plan for Eagle Attack shared many similarities with the tasks that the Luftwaffe had succeeded at before: striking hard at the opposing air force and swiftly taking it out of the equation. This time, however, there were some unavoidable differences.

Firstly, Britain could not be shaken by *Blitzkrieg* in the same way as the countries who shared a land border with the Germans. Secondly, the sheer breadth of the English Channel posed a new obstacle for the Luftwaffe that the narrow fjords of Norway had not adequately prepared them for. Around two-thirds of the Bf 109's fuel capacity was used up just getting to and from Britain: once they arrived, they had only around twenty or thirty minutes of flying time in the active combat zone.[10]

Thirdly, and most crucially, Britain wielded a better equipped and larger air force than the Luftwaffe's defeated adversaries, along with possessing the jewel of radar in the crown of a sophisticated integrated air defence system.

Nevertheless, the essence of the Luftwaffe's previous *Blitzkrieg* campaigns remained the same on Eagle Day: shatter the enemy air force in the shortest time possible; cripple its military–economic infrastructure to an unrecoverable level; and exert pressure on the enemy government to capitulate. In other European countries the Luftwaffe had a successful track record of giving the *Heer* and *Kriegsmarine* the necessary breathing space to undertake their own unhindered manoeuvres. So, it was perhaps not yet fully evident that the German air force was ill-suited for the task in this particular situation.

Indeed, Rieckhoff was of the opinion that 'The balance of power in August 1940 was exceptionally favourable for Germany. England was the only air enemy in Europe.'[11] He further explained that 'The base gained from Holland to Brittany through the Western offensive made it possible to deploy in a wide arc opposite the British defensive ring and to attack it concentrically.' Moreover, he pointed out that 'The extensive nature of the German ground organization made British countermeasures difficult. The German fighter base made it possible to flank English air raids against the mainland and to intercept returning units before landing, or to smash them on the ground shortly after landing.'

Where the Luftwaffe was put in an especially difficult situation, however, was the fact that it was ordered to lay the foundations for an operation that the *Heer* and *Kriegsmarine* were not even sure they could deliver. The *Kriegsmarine* did not wish to progress with Sealion until the ports at Dover, Plymouth, Portland and Portsmouth were more heavily bombed and mined by the Luftwaffe.[12] The *Heer*, meanwhile, felt that it required at least forty divisions to be transported by the *Kriegsmarine* if Sealion had any change at success – but the *Kriegsmarine* would only commit to being responsible for getting ten divisions into England.[13]

Consequently, although Fighter Command lacked the Luftwaffe's advantage of choosing when, where and how to strike, it can be argued that the RAF had a key defensive advantage in being able to call on its fellow armed forces for direct support. For Britain, the overarching goal was clear in the Battle of Britain: repel the Germans in the air and, if necessary, at sea and on land. The British Army assisted Fighter Command in this endeavour by manning the anti-aircraft units, guarding prisoners of war, and passing on visual information.

At the same time, around half of the Royal Navy's destroyers were recalled to protect the Channel from invasion, while constant minelaying and minesweeping operations were conducted to protect Britain's harbours.[14] The German air force, on the other hand, had to contend with multiple opinions by the *Heer* and *Kriegsmarine* about how their campaign should be fought. In addition, the Luftwaffe ran the risk of losing its airmen for nothing if its sister Wehrmacht forces were not ready to pounce upon Britain in Operation Sealion.

As the OKW situation report noted about Eagle Day, if the *Heer* and *Kriegsmarine* did not adequately prepare for Sealion, 'an invasion of Britain would be an act of desperation by no means justified in the present situation'.[15] Yet, all eyes were on the Luftwaffe to secure the aerial conditions required to carry out an invasion. This disconnected vision among the German leadership was hardly conducive to creating a consistent synergy for the preparation and potential execution of Sealion. Thus, it is perhaps not entirely unreasonable that Deichmann felt as though the Luftwaffe had to shoulder the full burden of the campaign.

The depth of intelligence that the Luftwaffe compiled on Britain prior to Eagle Day, however, further blows out of the water his assertion that 'The famous Sealion was only an idea', even if it was lacking in full commitment. Luftwaffe intelligence during the Battle of Britain has long been presented as woefully inadequate, with much of this being attributed to the particularly roseate intelligence reports that came from *Major* Josef 'Beppo' Schmid, the head of the 5th Department of the Luftwaffe air intelligence service. The professional rivalry between Schmid's unit and the 3rd Department of the Signals Service – headed by the Chief of Luftwaffe Signals, General Wolfgang Martini – often resulted in inaccurate assessments being passed on.[16]

Additional weaknesses within Luftwaffe intelligence included its delayed grasp of British radar and poor signal security, which had long supplied the RAF with a fairly astute picture of Luftwaffe numbers, units and air traffic.[17] This also helped Fighter Command to pinpoint departures and anticipate incoming German aircraft prior to an air raid.[18] However, the Germans surveyed the British mainland itself in great detail leading up to the launch of the *Adlerangriff*. A Luftwaffe situation report from the West reported at the beginning of August 1940 that:

> Signposts, town signs and milestones have been removed throughout the country to make orientation more difficult for invading German troops. The defence against paratroopers and airborne troops by state riflemen, homeland and local defence forces is prepared in the most remote parts of the British Isles. Particularly important points, e.g. bridges, are constantly guarded.[19]

The report added that 'The ringing of church bells is considered an alarm signal when paratroopers and airborne troops land,' and detailed that 'The [British] population is warned urgently by the press and radio not to talk about military matters. In the event of a landing by German troops, the population should remain in place, hide food and bicycles, and make motor vehicles unusable so that they do not fall into German hands. There are signs that the population will take an active part in the struggle.'[20]

For Luftwaffe intelligence to look so deeply into the British anti-invasion measures, then, shows how seriously it was sizing up the island for an amphibious landing – even if it was not yet convinced of its feasibility. In addition, the German war leaders were successful enough in cultivating a widespread belief that the invasion would go ahead that it even influenced the decisions of some young German men to choose the Luftwaffe as their professional service. Kurt Gerhard Raynor recalled how, on the same day as Keitel's memorandum, he had voluntarily reported to the training depot in Bitterfeld near Leipzig and applied for the Luftwaffe:

> Ah, you say, one of those Nazis fighting for their fatherland. Not true! The only reason for volunteering was, I am sure in most cases, that the prospective recruit could choose between Army (Wehrmacht*) [sic], Air Force (Luftwaffe) and Navy (Marine). I chose the Luftwaffe for one overriding reason; I had no hankering to foot-slog it with heavy equipment on my back while conquering Britain![21]

Whilst Raynor's post-war defensiveness regarding his choice to voluntarily fight for a criminal political regime is evident, his comment nevertheless demonstrates just how convinced some Luftwaffe recruits became that Britain would be invaded. But the Luftwaffe was not finding it easy to exhaust the RAF. Just before the Battle of Britain, the decorated Luftwaffe fighter ace Werner Mölders had flown a captured Spitfire and Hurricane at the Rechlin test centre. He reported that 'Both types are very simple to fly compared with our aircraft, and childishly easy to take off and land. The Hurricane is very good-natured and turns well, but its performance is decidedly inferior to that of the Bf 109. It has heavy stick forces and is "lazy" on the ailerons.'[22] Mölders added that:

> The Spitfire is one class better. It handles well, is light on the controls, faultless in the turn, and has a performance approaching that of the Bf 109. As a fighting aircraft, however, it is

---

* It is a common misconception that the 'Wehrmacht' referred only to the German *Heer* ('army'). It actually means 'armed forces' in German, encompassing all three forces – *Heer* (army), *Kriegsmarine* (navy), and *Luftwaffe* (air force).

miserable. A sudden push forward on the stick will cause the motor to cut; and because the propeller has only two pitch settings (take-off and cruise), in a rapidly changing air combat situation, the motor is either over-speeding or else is not being used to the full.[23]

Crucially, however, Mölders had flown versions of the British fighters that were now outdated; Fighter Command had been working hard to rectify some of the Hurricane and Spitfire's glaring weaknesses compared to the Luftwaffe's fighter planes. During June and July 1940, both fighters were being converted to constant-speed de Havilland/Rotol propellers – meaning that a Spitfire Mk I's climb rate suddenly jumped by 730 feet per minute.[24] In addition, the widespread switch to higher 100-octane fuel in time for the Battle of Britain powered the new constant-speed propellers more effectively, further boosting the British fighters' rate of climb and maximum speed.[25]

The important edge that this gave the British could be seen in how it took until late August 1940 for the Germans to find out about the fuel, which only happened when they analysed a force-landed Spitfire that had 'green' fuel of a much higher octane than the standard 'blue' 87 that German fighters used.[26] Despite some of these key modifications, however, the Luftwaffe fighter pilots looked pityingly upon the Hurricane, which flew between 10 and 30 mph slower than the Bf 109.[27]

Galland once claimed that 'The Hurricane was hopeless – a nice aeroplane to shoot down. But the Spitfire was dangerous – its armament, climb, manoeuvrability.'[28] Johannes Steinhoff agreed that 'the Hurricane was a big disadvantage to [the British]; the rate of roll was bad – we were lucky to meet Hurricanes.'[29] Thus, despite the Hurricane going on to secure over 60 per cent of Fighter Command's kills in the Battle of Britain, it was the Spitfire that tended to be revered by the Luftwaffe.[30] Admittedly, more Hurricanes were deployed in the campaign than Spitfires: by early August 1940, for instance, 55 per cent of the RAF's operational fighter aircraft were Hurricanes and only 31 per cent were Spitfires.[31]

In addition, there was some targeted use later into the campaign where Hurricanes were despatched to take on the more vulnerable and more numerous bombers, while Spitfires were directed to take on the fighter escorts.[32] So, a fighter deemed technologically inferior by the Luftwaffe was still able to be statistically effective in the Battle of Britain. However, the German pilots also derided the Hurricane because of a phenomenon known as 'Spitfire snobbery' – the frequent practice of downed Luftwaffe pilots saving face by insisting that they had been shot down by the more technically refined Spitfire. This often led to the Germans assigning some of the exceptionally skilful victories achieved by Hurricane pilots to Spitfires.[33]

Moreover, for the first eight months of the war the Luftwaffe's fighter pilots had come up against the Hurricane before the British fighter was improved. Galland noted of the Phoney War, for instance, that 'We did not see much of the English in those days. Occasionally we met a few Blenheims. The Belgians for the most part flew antiquated Hurricanes, in which even more experienced pilots could have done little against our new Me 109E. We outstripped them in speed, in rate of climb, in armament and, above all, in flying experience and training.'[34] The Spitfire, on the other hand, had only really made its first real mark at Dunkirk, and it subsequently maintained its esteemed reputation with additional modifications in the summer of 1940.

Finally, the growing circulation of 'de Wilde' incendiary bullets proved incredibly helpful to Fighter Command because, as the New Zealander fighter ace Alan Deere once testified, the luminous flash they made on impact meant that 'For the first time, pilots were able to confirm their aim; a most important consideration in the conservation of ammunition and the delivery of the *coup de grâce*.'[35] Deere also noted that there had been greater standardization of a trick that Spitfire pilots had realized after Dunkirk: harmonizing their Browning machine-guns to converge with one concentrated burst at 250 yards ahead of the wings proved to be far more effective than converging at the typical 400 yards.

These vital innovations utilized during the Battle of Britain were crucial towards keeping Fighter Command's momentum going – and the German pilots were beginning to feel their effects. By early August 1940, Rall had already identified of his fighter wing that 'The combat effectiveness of both the *Staffel* and the *Gruppe* has fallen so much that we have to be withdrawn from the front. Lack of operational experience is certainly one of the reasons that led to this decline.'[36] However dangerous these new strides by Fighter Command were for the German fighter pilots, though, it did not compare to the acute peril that now enveloped the vulnerable bombers and dive-bombers they had to escort.

As first light flooded the cockpit of his Ju 87 Stuka on 9 August 1940, Helmut Mahlke was en route to Théville with the aim of finishing off a limping British convoy that had just passed Dover. 'With a strong fighter escort,' he later wrote, 'we set out across the Channel. The convoy was spotted some ten nautical miles to the south-west of the Isle of Wight. As the vessels had by this time closed up again, I decided that we would attack in *Ketten* ['chains'].'[37] The Stukas circled the ships like famished vultures and began their attack, ready to turn them into contorted lumps of metal on the seabed. Yet, before they could land the final blow, British fighters sneaked in at low-level and suddenly pounced upon the Stukas from behind.

Within moments, as Mahlke recounted tensely, 'We were embroiled in another furious mêlée. I had the feeling that our opponents outnumbered us this time. Whether this was actually true or not, I couldn't say. But I do vividly recall at one point seeing no fewer than three enemy fighters bearing down on me from three different directions! I could only turn towards one of them.' Mahkle possessed great faith in his responsive Stuka to get him out of this predicament, but he quickly realized that the situation was spiralling out of control:

It was absolutely impossible to keep track of them all. Time and again we turned in towards them, hoping to get in a few shots with our two forward machine-guns, but it was no use. Our opponents were much too fast for us. They could choose where and when to launch each assault. They dictated the terms and the course of the fight. And if we did find ourselves in a position to meet them head-on, they always sheered away. They obviously didn't like confronting us face-to-face, even though they each had eight machine-guns to our two.[38]

Eventually, their Luftwaffe fighter escort managed to repel the British fighters; veering back towards the French coast, the Stuka crews were lucky to be alive. Some, however, teetered more on the edge of death than others. Mahlke recalled how '*Oberleutnant* Klaus Ostmann of the 8. *Staffel* had been seriously wounded: a bullet had shattered his right knee when his machine was attacked by a fighter.' As Ostmann drifted in and out of consciousness from blood loss, his *Staffel* desperately tried everything to keep him awake long enough to land.

Suddenly, they hit upon the only way to revive the fading *Oberleutnant*: pulling up abruptly into his line of sight to startle him enough to react instinctively and keep flying! 'The shock of seeing another machine suddenly filling his windscreen seemed to put new life into him,' Mahlke noted wryly. Back at base, the Stuka crews immediately dissected this close shave. 'Later that same evening I called my *Staffelkapitäne* together to discuss the day's events,' Mahlke further recalled. 'We went over our shared experiences and tried to work out what conclusions could be drawn from them. How could we improve upon our attack procedures and, above all, how could we prevent – or at least reduce – future losses?'

The prevailing conclusion they reached was that some of the less experienced crews had been afflicted with what Mahlke claimed happened to every new Stuka pilot: they have been 'gripped, almost mesmerized, by scenes and events that we should have totally ignored'. It was essential, then, that if Eagle Attack was going to succeed, the Luftwaffe's more excitable pilots needed to keep a cool head and a discerning eye.

The plans for Eagle Attack had been gradually refined. During the operation, the Luftwaffe aimed to eliminate aerial resistance from the RAF's Fighter Command by bombarding its airfields, infrastructure and installations on a semicircle to the south of London, stretching up to a radius of 93 miles.[39] With concentrated, heavy attacks over the course of five days, they would gradually advance in three phases towards London. In early August, the jagged radar towers and accompanying stations dotted around Britain's coastline were also identified as a key target for the Luftwaffe.[40]

Operation Eagle Attack would also include striking RAF airfields which could pose a serious hazard to an amphibious German invasion; seizing any favourable opportunities to attack Royal Navy targets, in port and at sea; and paralysing British defences through attacking aircraft factories, communication hubs and ports.[41] On 12 August 1940, the Luftwaffe moved towards attacking RAF Manston, Lympne and Hawkinge, as documented by a situation report the following day:

> In the early afternoon hours, there was heavy air fighting in the south-east and south of England, with a focus on Dover and Portsmouth, in the course of which heavy losses were inflicted on the enemy. The fighter defence in the attack areas was more relaxed than before and therefore no longer particularly effective. The enemy fighters avoided the fight at individual positions. To what extent this is due to planned reticence or a lack of willingness to attack cannot yet be assessed.[42]

This day gave no reason for alarm among the Luftwaffe forces, with one *Soldat* serving in the bomber wing KG 51 claiming on 12 August 1940 that 'Today our *Staffel*, or rather our bomber wing, made a joint mission against England. Eighty-nine of ninety machines came back; one had to make an emergency landing on the water 20 km from land.' He added that 'Our planes were able to carry out the attack plan in full, which was even announced by a special report...'[43] Otto-Wolfgang Bechtle – a Do 17 pilot with the 9. *Staffel* of the bomber wing KG 2 – noted, such attacks on 12 August 'were comparable in scope and intensity to all previous air battles in the British War'.[44] The tempestuous British weather, however, was not yet favourable for the all-out assault of Eagle Day.

For the morning of 13 August, the Luftwaffe Central Weather Service Group declared 'Southern and central England largely clear, slightly hazy, mostly good visibility.'[45] Thus, as Bechtle recalled, 'The codeword "Eagle Day" was issued for 13 August'. But, as any Briton can testify, rarely does the island's weather match the forecast. At 6.30 a.m., Eagle Day opened with a thick, damp fog that draped over Kesselring's airfields, while a citadel of clouds arched over southern England and held on tight.[46] Fink, as the commander of KG 2, went through the typical rigmarole of making the last checks in his Do 17 bomber. Satisfied that everything was in place, he started up his 'Flying Pencil' and signalled to the Luftwaffe ground crew that he was ready to roar over the island:

> We took off, flew to the assembly point, found the fighters there in the air, and, to my surprise, they kept coming up and diving down in the most peculiar way. I thought this was their way of saying they were ready. So I went on, and found, to my surprise, that the fighters didn't follow. But I didn't think much of that; hardly noticed it. And as I had arranged for many more machine-guns to be built into my Do.17, I didn't worry much. We'll manage it somehow.[47]

The Bf 110 fighters that were wiggling frantically in front of Fink were in fact trying to get his attention: *Adlertag* had been postponed yet again by Göring due to bad weather, but the long-wave radio set in Fink's bomber had broken down and he had not received the cancellation order.[48] In addition, a handful of Luftwaffe bomber units based around Arras and Cambrai had also heard no word of this cancellation, as the crystals in their radios had not been upgraded to match the fighters' communication system.[49] These uninformed bomber units had taken off between 0450 and 0510 hours and were now heading into the lion's den without a whip. Despite the miscommunication and inconsistent deployment, however, the Luftwaffe went on to conduct 1,485 sorties on Eagle Day.[50]

In the afternoon, Coastal Command airbases in the south bore the brunt of Eagle Day. At Detling airfield, seventy-eight personnel were killed and twenty-two Avro Anson reconnaissance aircraft were destroyed on the ground.[51] Over at Eastchurch, twelve people were killed and forty injured, though the Luftwaffe had mistaken the ten Blenheims destroyed on the ground for Spitfires.[52] Despite the mixed results, the Nazi press unsurprisingly went into a frenzy when reporting on *Adlertag*. On 14 August 1940, the front page of the *Baruther Anzeiger* reported on the 'New air raids on England's south and southeast coast.'[53]

It boasted of the 'ceaseless stranglehold' that the Luftwaffe held over Britain: 'The English, who had babbled in their lying report that the Germans would not be able to repeat their attacks, are now seeing the continuation of these attacks with a tenacity and constant increase that astonishes the whole world.' The article concluded that 'What is particularly noteworthy about these air battles is the high number of kills achieved, which amounts to 236 English aircraft destroyed over three days of fighting.' While overclaiming aerial victories was rife on both sides, this particularly ludicrous 'statistic' put British losses on *Adlertag* at being over fifteen times higher than they actually were.

By the end of the day, the Luftwaffe had in fact lost forty-two aircraft to Fighter Command's fifteen* – a loss rate constituting seven per cent of the striking force.⁵⁴ Nevertheless, the *Baruther Anzeiger* further gloated that 'The clear superiority of the German Luftwaffe is thus proven and can no longer be ignored by the British side, no matter how big their distortions and lies are. This is also the clear verdict of foreign military writers, who are fully aware that Germany's victory is beyond question.'⁵⁵ Yet, the article was careful to provide the caveat that:

> With this clear statement of German superiority, the extraordinary severity of these air battles must in no way be underestimated, which would mean criminally belittling the outstanding achievements of our pilots, who are dealing with an extremely tough opponent who can be assessed in a completely different way to, for example, the Pole, who collapses immediately after the third blow.⁵⁶

The obvious irony of this statement is the fact that some of Fighter Command's fiercest pilots were Polish. However, presenting the British as an unusually formidable opponent bought the Nazi propagandists more time by explaining that the *Luftschlacht um England* might not be a swift victory like the earlier *Blitzkrieg* campaigns. Furthermore, it presented the Luftwaffe's successes as being even more glittering since they were achieved against a more 'difficult' foe. The article continued that 'It is the hardest and most difficult war work that has to be done every day in these furious air battles. The utmost is made of man and machine, and the huge successes can only be explained by the death-defying and unconditional commitment of our brave pilots.'

---

* Holland, *The Battle of Britain*, 459. Christer Bergström points out that although the general figure for RAF fighter losses on 13 August is thirteen, the loss reports of the participating fighter squadrons indicate that a further three aircraft were lost on this day. See Bergström, *The Battle of Britain: An Epic Conflict Revisited*, 111.

As Bechtle later noted, however, 'The "*Adlertag*" did not have the hoped-for success. Due to the unfavourable development of the weather situation, only a portion of the planned attacks could be carried out. For the same reason, the required fighter protection for the bomber units was very patchy. The losses that day were heavy.'[57] Indeed, Fink would likely have screwed up the *Baruther Anzeiger*'s coverage of *Adlertag* in disgust had he read it. After somehow managing to survive the calamitous attack without fighter escorts, he recalled that 'After I'd landed, in this over-excited condition after an operation, I went straight to the phone, got on to Kesselring, and shouted down the line exactly what I thought about it. I told Kesselring bluntly that his staff had done a very bad piece of work indeed.'[58]

PART III

CONSOLIDATION

10

The Hour of Judgement

The air battles over England since Thursday last week have found no other official response in England, as have the military events in Poland, Norway, Holland, Belgium and France. It is the well-known lying tactic of Churchill to withhold the truth from the English people and to lie until the truth can no longer be withheld from them, even if the beams are caving in. The hundreds of destroyed Spitfires and Hurricanes, the roaring flares of the tethered balloons, and the sunken ship wreckage in the Channel make the English people themselves witnesses to the *Luftschlacht um England*.[1]

Published by the German *Rheinsberger Zeitung* newspaper on 16 August 1940, this article declared that 'The hour of judgement is at hand!'
Claiming that the British had lost 143 aircraft, the newspaper painted 15 August 1940 as a day of national reckoning. In reality, however, on that day, which became known as *Schwarzer Donnerstag* ('Black Thursday'), the Germans had suffered their heaviest losses of the entire campaign so far – seventy-five

aircraft.* This was twice the number lost by Fighter Command, and 15 August was later dubbed the 'Greatest Day' for the RAF.² It was particularly humiliating for the Nazis given the immense Luftwaffe resources deployed on 'Black Thursday', with the Germans launching 2,000 sorties to counteract the unfavourable results of Eagle Day two days earlier.³

Indeed, the fact that the Luftwaffe was still reeling from Eagle Day could be seen in the fact that Göring held a staff conference the same day as Black Thursday to dissect the German air force's previous mistakes on 13 August. Although the RAF also claimed far more victories than actually happened – citing around 180 German aircraft shot down – the lopsided dynamic of the dogfights in the RAF's favour that day was undeniable.⁴ With the Luftwaffe shedding another forty-five aircraft on 16 August 1940, it was evident that Eagle Attack was not proceeding as planned.⁵ Trudging towards the first week of this new phase, 297 German aircraft were lost on combat operations – a humongous rise of nearly 600 per cent compared to the average weekly loss during the entire six-week campaign of the Channel Battle.⁶

Then, on 18 August 1940, both sides would take their collective heaviest losses so far in what became known as the 'Hardest Day'. The Luftwaffe succeeded in shooting down thirty-four RAF aircraft and damaging a further thirty-nine in the air, before destroying twenty-nine aircraft and damaging twenty-three on the ground – but this was in exchange for sixty-nine Luftwaffe aircraft destroyed and thirty-one damaged.⁷ As the Ju 87 pilot Scheffel wrote of the 'Hardest Day':

> It started out so harmlessly. On this day, the 1. *Staffel* organized a sports competition on the sports fields of Caen. The commander had gone there early. Henze, Platzer and I had taken

---

* Fighter Command's 13 Group exacted particularly heavy losses on *Luftflotte* 5, with the Germans losing between fifteen and twenty Luftwaffe aircraft over north-east England at no cost to the defenders. 'Black Thursday' marked the first – and last – mass raid that *Luftflotte* 5 would launch against Britain in daylight.

some time after the long night and were just about to drive to the field in Maltot when a call came from the *Staffel*: operational readiness, immediately relocated to Tonneville. We then drove straight to the sports fields and passed the order on to the commander. The sports matches were immediately stopped, and we went to the airfield in Maltot.[8]

Little did they know that the worst Stuka losses of the entire *Luftschlacht um England* were just on the horizon. As Scheffel noted, it would become 'a day that no member of the I. *Gruppe* will ever forget in their life'.[9] His unit was scheduled to attack RAF Thorney Island – a predominantly Coastal Command station – with one fighter wing to perform a preliminary free hunt ahead of the bombers. Meanwhile, additional German fighters would accompany the Stuka *Gruppe* as close fighter escort.

So far on his outward flight, everything seemed routine to Scheffel: 'Cherbourg lay below us in the sunshine, the surf was clearly visible on the coast. The sky had a slightly blue tint and there was a light haze over the Channel.' Then, out of nowhere, Scheffel spotted 'English fighters pushing steeply towards us from the top right.' As his rear gunner, *Gefreiter* Otto Binner, blasted away, Scheffel began to defensively rock their Stuka from side to side. Their Luftwaffe fighter escort become embroiled in a tense dogfight with the British as the Stukas continued anxiously yet determinedly towards their target.

Suddenly, a hail of bullets rattled the cockpit of their Stuka, embedding themselves into flesh and metal alike. Scheffel felt a hard blow to his left shoulder, causing him to sag over to that side, while a roaming, 3-cm-long shard of metal split open his right thumb that had been firmly gripping the control column. With the throbbing pain making his head swim, Scheffel then heard the most unsettling noise that a Stuka pilot can hear: a tortured cry like a wounded animal rung out from his rear gunner, and the unsettled Scheffel watched his colleague slump over his machine-gun in the rear-view mirror.

Helpless to assist his gurgling rear gunner, but also unable to abort the sortie, Scheffel persevered towards his target. As he saw the Stukas ahead of him fan out over the target, he threw himself after them and deployed the dive-brakes, a crimson river of blood now spurting from his right thumb. Friendly and enemy fighters continued to swirl around one another as if sucked into a tornado, but Scheffel homed in on the British airfield below. Dropping his bombs and heading back to the French coast, he called out repeatedly to his rear gunner over the intercom. The deathly silence confirmed to Scheffel that he was truly on his own as a British Boulton Paul Defiant fighter closed in hungrily towards the Stuka; the predator had now become the prey.

But, somehow, Scheffel was able to manoeuvre himself away from the domed turret of the Defiant and sneak off, joining up with a few straggling Stukas on the way back. As he headed towards Cherbourg, Scheffel finally got a chance to take a mental inventory of the Stuka's damage: 'The left windows were shattered, and the wind whistled through them. Almost no instrument on the instrument panel was still intact. Holes and hanging parts everywhere. Lots of holes in the areas outside. Hopefully the tanks would remain sealed so that the fuel would allow us to reach the coast.' He assessed himself, too, and the results were about the same: battered and bruised, but somehow hanging on.

'I couldn't move much,' he later wrote. 'I sat slumped to the left. The right thumb with the splinter was still bleeding, the entire control stick was soaked red. I wanted to pull out the splinter with my left hand, but I didn't have the strength to do so. In the mirror I saw that my face was also bleeding – it was from the shards of the reflector sight.' Upon landing at Maltot in Normandy, Scheffel recalled that 'After the experiences of the last few hours, I probably was pretty much at the end of my strength. It was likely also the exhilarating feeling that you've done it, now nothing else can happen.' His dead rear gunner was carefully lifted out from the back of the Stuka, while Scheffel's injuries were finally attended to by the medical team.

Notwithstanding Scheffel's quick reflexes and situational awareness, he was exceedingly lucky to be alive: his Stuka had been riddled with eighty-four bullet holes, fifty of which were concentrated in the fuselage. Many of the Stuka crews that were deployed on that day, however, did not come back. Having lost 20 per cent of the Stuka force over the last ten days, the 'Hardest Day' was the final straw for the Stukas: they were largely withdrawn from the *Luftschlacht um England* thereafter.[10] As the slowest operational German aircraft in the Battle of Britain – and kitted out with insufficient armour against the British fighters – the Ju 87 had become a major Achilles heel for the Luftwaffe during the campaign.

However, in contrast to the entrenched notion that it had been withdrawn due to the unsustainable losses, the Stuka pilot Paul-Werner Hozzell claimed that 'We were consequently "withdrawn from service" for the time being, so as to allow our heavily decimated forces to rehabilitate in preparation for new operations.'[11] Indeed, only fifty-two Stukas had been completely destroyed by 19 August 1940,[12] while the worst affected Stuka wing on the 'Hardest Day' – the dive-bomber wing StG 77 – fully replaced its seventeen aircraft lost[13] and their respective crews within a week.[14] Nevertheless, it is undeniable that the Stuka's deadly reputation was severely dented by its disappointing performance in the *Luftschlacht um England*. As Bechtle noted in a subsequent report on the campaign:

> The English fighter defence inflicted unusually high losses on the dive-bombing units, despite strong German fighter protection, so that after a few attacks the Ju 87 had to be abandoned again during the day against targets on the English mainland.[15]

Bechtle also noted that 'The Bf 110 also did not perform as expected' in the Battle of Britain.[16] In recent years, however, there have been attempts by some historians to rehabilitate the image of the Bf 110 as a fighter escort. These ideas highlight that the

Bf 110 was impressively armed, with two 20-mm MG FF cannons and four 7.92-mm MG 17 machine-guns slotted into the nose, with its automatic cannons wielding three times as many shells as a Bf 109.[17] In addition, some observers have downplayed its inability to outturn Hurricanes and Spitfires because a fighter's speed, rate of climb and diving performance were considered just as important as its manoeuvrability in the Second World War.[18]

Certainly, Galland often criticized some of the elder Luftwaffe generals for being 'stuck on the idea that manoeuvrability in banking was primarily the determining factor in air combat'.[19] He explained that they 'could not or simply would not see that for modern fighter aircraft the tight turn as a form of aerial combat represented the exception, and further, that it was quite possible to see, shoot and fight from an enclosed cockpit'. This held true in any 'free hunt' engagement where the Bf 110 had the advantage of surprise and initiation – but, of course, aircraft were often shot down when put on the defensive, where the Bf 110's bulkiness counted against it.

As *Hauptmann* Wolfgang Falck of the *Zerstörergeschwader* ('Destroyer Wing') ZG 1 recalled, when coming up against Hurricanes in France, 'The British Hurricanes were an opponent to be taken seriously, as we provided them with a significantly large target due to the size of our aircraft, and they were naturally more manoeuvrable than us. Apart from that, the British were excellent pilots and hard fighters!'[20] But, despite this widespread criticism, the long-held anecdote from Luftwaffe commanders that the Bf 110 required its own fighter protection does not seem to be corroborated by official Luftwaffe documentation of the time. Such claims often seemed to be stated by German pilots and bomber crews who looked for something to blame for the Luftwaffe's shortcomings in the *Luftschlacht um England*.

Rieckhoff, for instance, recalled witheringly that 'The so-called Bf 110 "*Zerstörer*" ['Destroyers'], originally intended for combatting enemy fighters and as fighter escorts for bomber crews, were conspicuously inferior. When they were used, they themselves required fighter protection and were therefore worthless as escorts for the combat units.'[21] The Luftwaffe bomber pilot Werner Baumbach damningly claimed 'The German Me 110 destroyers, which had been planned as heavy fighters to escort the bombers, failed completely. They were so slow and so inferior in flying that they themselves needed protection from hunters.'[22]

Again, Deichmann brought up the same myth: 'Our twin-engined fighters were no fighters, they were as vulnerable as the bombers, and needed their own fighter protection. This restriction made it impossible to gain air superiority.'[23] Nevertheless, the Bf 110's heavy armament meant it was highly favoured by some of its pilots, particularly against slower Hurricanes. *Leutnant* Jochen Schröder of the destroyer wing ZG 76 recalled:

> We felt at ease in the Me 110 after only a few initial flights. Naturally, we soon realized that she was much heavier and clumsier in combat than the Me 109 C and E, but despite this we had a lot of faith in the impressive armament: two 2-cm cannons and four machine-guns with a high rate of fire. When practising shooting at ground targets, we established that the Me 110 provided great accuracy and was also a very stable firing platform in the air. We believed that this machine was very suitable for future missions as a long-range fighter to accompany bombers like the He 111, Ju 88, Dornier 17 and succeeding types.[24]

The Bf 110 also demonstrated a better rate of climb than the Hurricane and was particularly suited to diving. Its fall from grace during the Battle of Britain seemed more dramatic because the destroyer concept had been placed in great faith by Göring,[25] as reflected by the production ratio between the Bf 109 and Bf 110 in 1939 (2.9:1) having almost levelled out to 1.6:1 by 1940.[26] However, the considerable Bf 110 losses by the end of

August 1940 – fifty-three aircraft, constituting around 20 per cent of the force – demonstrated its liability as a fighter escort in the campaign.[27] No matter how their individual aircraft fared, however, they were of little use if they were intercepted before they could successfully make their attack.

On 'Black Thursday', Göring had claimed that 'It is doubtful whether there is any point in continuing attacks on radar sites, in view of the fact that not one of those attacked so far has been put out of action.'[28] The *Reichsmarschall*'s characteristic impatience was placing his airmen in unimaginable danger. 'Chain Home' (CH) – a protective wreath of receiver and transmitter radar towers dotted around the Isle of Wight and the south coast of England – was capable of detecting enemy aircraft when they were still twenty minutes away from Britain's coastline.[29] In order to fill any potential gaps in the CH system, 'Chain Home Low' (CHL) had also been developed to detect aircraft flying at lower altitudes.

By the summer of 1940, there were twenty-two CH stations and thirty CHL stations established, albeit not all operational, to scan the skies for German aircraft heading towards Britain.[30] Radar, initially referred to as RDF (Radio Direction Finding) by the British, was incorporated into the Dowding System. This pioneering integrated air defence system was named after the commander-in-chief of Fighter Command because of his involvement in its development as Air Member for Research and Development from 1935.[31]

With the radar findings being funnelled via landline to a filter room – namely at Fighter Command Headquarters, RAF Bentley Priory – the Operations Rooms were subsequently provided with critical real-time information on the number, height, location and estimated destination of Luftwaffe formations during the Battle of Britain.[32] This intricate communications web had many vital cogs, including two-way radio and telephone links between radar operators. Their findings were supplemented by reports from the Observer Corps, a civil defence organization that visually tracked incoming aircraft over land. In addition, Women's Auxiliary Air Force (WAAF) map plotters worked tirelessly in the Filter Room, Group and Sector Operations Rooms and radar units to keep track of incoming Luftwaffe aircraft and the outgoing Fighter Command squadrons that were scrambled to intercept them.

It was long claimed that the Germans did not recognize the importance of radar, but that is not strictly true. Firstly, they acknowledged its importance as a discipline to the extent that German scientists actually pulled *ahead* of their British counterparts in certain respects during the inter-war period. *Freya* – the main radar scanner designed to cover around 160 kilometres (99 miles)[33] – operated with a frequency of 1,000 Hz and a 1.2-metre (47-inch) wavelength, a mere tenth of the standard British CH.[34] *Würzburg*, which detected aircraft within a shorter distance, operated at over 500 Hz with a tiny wavelength of just 53 centimetres (21 inches); this allowed for far more sensitive detection across 70 kilometres (43.5 miles).[35]

As shown by Milch's provocative comment during his visit to Britain in October 1937, the Germans had long recognized the possibility of Britain wielding a similar technology. Their key failing, however, was that they felt *Freya* had to constitute the upper limit of radar. So, when the airship *Graf Zeppelin* electronically probed Britain in 1939 – and did not immediately detect Chain Home, which operated at 50 Hz and on a 12-metre (39-foot) wavelength[36] – they were initially confident that the British had not developed anything of note in the field of radio detection.[37]

The British system, however, leant more towards *how* their radar was deployed, rather than the more sensitive but localized calibration of German radar. It was only discovered by Luftwaffe signals units in late July 1940, when they realized that their aircraft were being picked up by the coastal towers, with the RAF fighters being guided via radio to intercept the attackers.[38] After the war, Galland declared of British radar and the Dowding System that 'the success was outstanding', partly because the Luftwaffe lacked such a sophisticated ground-to-air communication system:

> Each of our movements was projected almost faultlessly on the screens in the British fighter control centres, and as a result Fighter Command was able to direct their forces to the most favourable position at the most propitious time. In battle we had to rely on our own human eyes. The British fighter pilots could depend on the radar eye, which was far more reliable and reached many times further. When we made contact with the enemy our briefings were already three hours old, the British only as many seconds old – the time it took to assess the latest position by means of radar to the transmission of attacking orders from Fighter Control to the already-airborne force.[39]

Galland's assessment of British radar was arguably too gushing, given that the system did not come without technical glitches, delays in relaying information and atmospheric interference. Nevertheless, by August 1940 Luftwaffe fliers were at a significant psychological disadvantage, knowing that they could come up as a blip on Britain's radar at any moment. What also weighed heavily on their minds, of course, was the fact that if an RAF airman baled out, he would land over friendly territory and, if uninjured, would be up in the air once more as soon as possible.

Not only did the Luftwaffe have further to fly – and sometimes needed to send its damaged planes all the way home from occupied France to Germany for repair – but, as Bob notes, 'We envied the British pilots in so far as they were fighting over their home territory and could save themselves on their own territory by parachute or forced landing in an emergency. We, on the contrary, had to put up with captivity or risk a forced landing in the Channel, i.e. in the open sea, which meant certain death.'[40] In some ways, though, the more severe consequences of baling out perhaps sharpened the Luftwaffe's focus and innovation, with Bob once improvising a rather unusual gliding technique to stay out of British hands:

> The Me 109 had in case of motor failure a gliding angle of 1:13, for example at a height of 1 kilometre you could, in the best possible way, cover a distance of 13 kilometres. One day I was shot at over England, approx. 80 kilometres from the French coast, at a height of 4,000 metres, in such a way that the cooling system of the motor failed. Usually, we had to put the air screw to gliding position and try to glide, as mentioned 1:13, homeward. According to my consideration, this was normally not possible from a height of 4,000 metres. Therefore, I let the motor, in neutral gear, become cooler first.[41]

After cooling the engine, Bob continued, 'I switched the ignition on again, opened the throttle and went as high as possible until the motor was overheated again.' Miraculously, after repeating this several times, Bob just managed to force an emergency landing on the French coast. 'According to my experience report,' Bob concluded, 'this method was later on called "to Bob" (baab) across the Channel.'

Yet, the fact that the Luftwaffe could not commit as unflinchingly as Fighter Command often forced them to veer away first. Flight Lieutenant Dennis L. Armitage of No. 266 Squadron recounted how:

> Big bomber formations had to be broken up before they could be attacked with reasonable safety and head-on attacks were usually the best way. Fortunately, the enemy always seemed to break away first. I do not think this was any reflection on the nerves of the German pilots but rather because we were, after all, playing the match on our home ground.[42]

Indeed, the heightened stakes of repelling the Germans over the skies of Blighty stoked a stronger fire in the bellies of Fighter

Command, with Pilot Officer Frederick Desmond Hughes – who flew Boulton Paul Defiants with No. 264 Squadron at Duxford – recalling that 'The most vivid memory was the first sight I had of a German aircraft. First, some specks away to the south, their progress punctuated by bursts of 3.7-inch flak, accurate for height but behind.' As he flew hell for leather towards the shadowy apparitions, 'the specks grew into long pencil-slim silhouettes', and then –

> Suddenly, there were the black crosses, insolently challenging us in our own backyard! John Banham had positioned us perfectly and, as the range closed, the Defiants' turrets began hammering away. Fred Gash took the second Dornier in the leading flight and his de Wilde incendiaries twinkled all over it. It began to fall out of the formation, a hatch was jettisoned, two baled out and the stricken aircraft went down increasingly steeply with its starboard engine well alight.[43]

During Eagle Attack, then, the Luftwaffe faced a number of challenges which would ensure that this more concentrated phase of the *Luftschlacht um England* would not be fought with ease. Nevertheless, in order to paint an accurate picture of both sides in the middle of August 1940, it is also important to document the advantages that the Luftwaffe *did* have. It did, for instance, retain the psychological advantage of initiating the attacks, which kept the island in a permanent state of tension. In addition, while the Luftwaffe had long been the Wehrmacht's gilded eagles, the RAF was fighting to redeem itself in the eyes of the nation after the fall of Dunkirk – particularly in the eyes of the British Army and the Territorial Army.

As Eagle Attack opened, H. V. Cossons – a newly signed-up soldier in a Home Defence battalion – soon found himself at the heart of the Luftwaffe's wrath. While he was digging in the lawns of Reinden House, located near Hawkinge airfield in Kent, 'Dive-bombers screamed down on the airfield and there followed the most violent explosions which shook the very trenches we were in. More diving, more blood-curdling screaming, then more terrible explosions and what sounded like metal being ripped apart.'[44] Cossons recalled after the raid that:

> There was bad feeling between us and the R.A.F. – it had existed since Dunkirk – and we delighted in trying to be clever with the airmen and they returned to camp at night and were challenged by us in no uncertain way. This led to a lot of sauce from them, we in turn threatening to shoot them with bullets if need be! It really came close to that at times.[45]

Thus, the RAF had more to prove to its military comrades than the Luftwaffe – but Cossons did concede that 'As the situation developed, we realized that we were all in it together, all on the same side, and we began to appreciate that we all had a job to do.' Now, the nation began to swing fully behind Fighter Command as it faced the fight of its life.

By the end of Eagle Attack's first week in August 1940, 12.5 per cent of the Luftwaffe's engaged force over Britain was lost; then, across the full month, 18.5 per cent of the Luftwaffe's deployed units were lost.[46] However, German intelligence officers were distorting the operational picture of the campaign by reporting grossly inaccurate combat victories, despite the Luftwaffe's verification of kills actually being more rigorous than that of Fighter Command.*

One German intelligence report claimed that the Luftwaffe had lost 174 German aircraft in exchange for 624 British aircraft between 12 and 19 August 1940.[47] The actual statistic during

---

* Luftwaffe aerial victories claims had to be independently verified by another witness, either on the ground or in the air, and were then passed through to the supervisory officer, who confirmed or rejected the claim. From there on, the claim went to *Geschwader* level and on to the Reich Air Ministry for the final confirmation. See U. Feist and T. McGuirl, *Luftwaffe War Diary: Pilots & Aces, Uniform & Equipment* (Mechanicsburg: Stackpole Books, 2014), 51.

13–18 August, albeit from a slightly shorter length of time, was 131 British fighters compared to 247 German aircraft lost.[48] It proved greatly puzzling to the Germans, then, when Fighter Command appeared in better health than the Luftwaffe's claimed kills would suggest.[49] This discrepancy was partially down to Schmid's ignorance of the Civilian Repair Organisation (CRO), which worked around the clock to patch up RAF fighters and send them skywards once more. Such was the stellar work of the CRO that it succeeded in putting 1,872 repaired fighter aircraft back into the air between June and September 1940.[50]

In addition, although the Germans managed to produce 3,106 Bf 109 and Bf 110 fighters in 1940, the British managed to churn out 4,283 fighters.[51] Indeed, Fighter Command's available fighters consistently increased in numbers and never dipped below 700 throughout August, September and October 1940.[52] The British aircraft industry became galvanized, as can be seen by the fact that it outproduced the Germans from 1940 until 1944, despite employing fewer workers.[53] The Luftwaffe, on the other hand, was struggling to replace the gradual haemorrhaging of its bomber crews; this was compounded by the fact that the Germans had no meaningful reserves, which gave the frontline bomber and fighter units little respite.[54]

In such circumstances, the German airmen found it particularly disrespectful when the newly promoted *Reichsmarschall* berated them for apparently lacking courage. Since Göring also took frequent vacations and missed staff meetings, many Luftwaffe commanders were losing any respect they had for *der dicke Hermann* ('fat Hermann'). Initially, he had shown willing by attempting to inject young blood such as Galland and Steinhoff into the Luftwaffe's *Kommodoren* ('commodores'),* with all of the new appointments in July and August going to men under

---

* '*Geschwaderkommodore*' is often translated as 'Wing Commander', but this is not quite accurate – it was more of a position than a rank that technically translates to a 'Wing Commodore'.

thirty years of age.⁵⁵ In September 1940, a captured Luftwaffe *Oberleutnant* was overheard praising this practice by British intelligence:

> It all depends upon the leadership. With the young leadership we have at the present moment, even the highest do not spare them themselves but knuckle down to it in an emergency; they will mix with the common soldiers in the thick of it and set them a brilliant example. In this especially, we are ahead of these old men in the English government and thereby we have been able to save ourselves from many a difficult situation.⁵⁶

As Rall noted, 'Göring had also made a very good choice in the men who were given high levels of responsibility here: they had all gone through careful pre-war training and had gradually been able to grow into such positions.'⁵⁷ In general, though, Göring was a rambunctious liability, leading the Luftwaffe with a pomp and circumstance that remained ignorant of the new air war. Günther Lützow, the *Geschwaderkommodore* of the fighter wing JG 3, once allegedly snapped at him: '*Herr Reichsmarschall*, could you stop talking for fifty minutes and listen to what is really going on, otherwise this meeting will be meaningless?'⁵⁸

A famous early incidence of retaliatory insubordination came when Göring demanded to know what else Galland would want from him and Galland retorted that 'I should like an outfit of Spitfires for my *Staffel*!'⁵⁹ As a reflection of how the Luftwaffe's role in the Battle of Britain has often been used to bolster the British side of the story, this quotation has been reported within British popular culture as suggesting that Galland preferred the Spitfire over the Bf 109. As Galland clarified in his 1954 autobiography *The First and the Last*:

> It was not really meant that way. Of course, fundamentally I preferred our Me 109 to the Spitfire. But I was unbelievably vexed at the lack of understanding and the stubbornness with

which the command gave us orders we could not execute – or only incompletely – as a result of many shortcomings for which we were not to blame.[60]

This fraught exchange demonstrated that, two months into the *Luftschlacht um England*, tempers were fraying around the infuriating Nazi leadership. Göring's fluctuating moods, which careered from energized and enraged to lazy and despondent, often led to the Luftwaffe feeding him highly optimistic statistics on German kills and losses. Early into the Eagle Attack, for instance, he was informed that the RAF was down to just its last 300 serviceable fighters; the true figure was actually 706 by 18 August 1940.[61]

Although Göring did appoint some highly capable young commodores, the upper echelons of the Reich Air Ministry and the Luftwaffe remained studded with the old guard of First World War veterans, some of whom had only learnt to fly in middle age or could not fly at all. Politics and favouritism, too, starved the Luftwaffe and Reich Air Ministry of key leadership talent. That the hapless Ernst Udet, an old fighter ace comrade of Göring's in the First World War, was elevated to positions of critical decision-making importance illustrated the dangers that Party politics posed to the German air force's operational development. The appointment of Nazi yes-man Hans Jeschonnek as the Luftwaffe's Chief of the General Staff on 1 February 1939 was another keen example of how political ideology robbed the Luftwaffe of vital command experience.[62]

Rather than douse the burning rivalries that ignited between Milch, Jeschonnek and Udet, Göring often elected to fan the flames – creating a volatile melting pot in which politics and personality often counted for more than competence and experience.[63] The lack of a proper Luftwaffe general staff added to the tumultuous nature of its command style.[64] In any case, the leadership's war was a million miles away from that experienced by the Luftwaffe personnel on active duty who continued to take unimaginable risks for *Führer* and Fatherland. Sometimes flying up to five sorties a day in the Battle of Britain, many Luftwaffe airmen collapsed into a crippling malaise when they landed.[65]

As with their counterparts in the RAF's Fighter Command, some German pilots would vomit with nerves during their briefing: every sortie they flew, they knew that Death was sharpening his scythe.[66] Another terrifying possibility, of course, was returning with life-changing injuries. The considerable rise in sorties naturally led to an increased chance of fatigued blunders, crashes and debilitating injuries. Throughout the Battle of Britain, the most common injuries sustained by Luftwaffe airmen were soft tissue injuries (16 per cent), full body shattering (just under 13 per cent), broken legs (11 per cent), fatal skull fractures (10 per cent), broken arms (6 per cent), bruises on the torso (5.5 per cent) and fatal burns (5 per cent).[67]

Luftwaffe airmen who had broken their spines were twice as likely to die than to survive, while broken legs were nearly twice as common as broken arms – presumably due to botched landings when baling out or their leg position when crash-landing.[68] Fewer than 1 per cent of fatalities between July and December 1940 were due to drowning or freezing to death, although naturally there were many aircrews whose bodies were never recovered.[69] Mundane accidents, too, sometimes claimed lives. On 22 August 1940, the commanding general of Air District Command Western France, issued a Daily Order which declared that:

> The number of motor vehicle accidents has increased dramatically in recent times. I attribute this to a lack of supervision of drivers by their superiors and inadequate measures to punish accidents. I repeatedly ask the commanders to personally commit themselves to impeccable driving discipline within their area of command. In the future, in the event of motor vehicle accidents, it will always be made immediately possible to

determine whether there is a lack of supervision on the part of the responsible superiors.[70]

A conversation between captured Luftwaffe airmen overheard by British intelligence on 19 September 1940 sheds more light on the causes of accidents: 'In Paris I took myself in hand, otherwise I'd have gone to pieces altogether. Others played fast and loose. We took no notice of traffic signals at all, whether they were green or red. We simply rushed through with the military cards. And one-way streets did not exist for us, we enjoyed driving up those in the wrong direction.'[71] In some cases, then, the folly of youth and the intoxication of flouting the rules in an occupied territory delivered karma on swift wheels.

What could also prove dangerous, of course, was the aircraft they flew, with the demanding Ju 88 '*Schnellbomber*' ('fast bomber') proving especially hazardous for new pilots. Although the Ju 88 was by far the most impressive German bomber in the Battle of Britain, it was not as dependable as the He 111 or Do 17. The bomber wing KG 4 'General Wever' issued a warning on 14 August 1940 that 'Repeated operating errors during take-off and landing with the Ju 88, which are due to a lack of understanding of the flying characteristics of this aircraft model on the part of the young pilots, have led to serious fractures and accidents.'[72] As Peter W. Stahl wrote of the Ju 88:

> The bird is something of a diva. He seems to know that he is beautiful and interesting and behaves accordingly. He is able to do things suddenly and without warning that you are not prepared for. It makes a noise, especially when rolling while starting. You can keep the direction clean with the rudder until the point where the tail lifts off the ground and the fuselage comes into horizontal position. The diva doesn't seem to like this movement.[73]

1 (*below*) German airmen planning a sortie during the Battle of Britain, 1940.

2. (*right*) A Nazi propaganda cartoon from the *Luftschlacht um England*, 1940. It shows Britain – anthropomorphised as Britannia – being uncomfortably harassed by the Luftwaffe, depicted as Death. Underneath, German bombs are being dropped all over the English countryside.

3. (*below*) German fighter pilots on standby during the Battle of Britain, 1940. This was presumably taken on a hot summer's day due to their lighter state of dress!

4. (*left*) A shot of a Luftwaffe pilot just before take-off on a flight to England, 1940. He is wearing a compass on his right arm and a wristwatch on his left.

5. (*below*) German shot of the White Cliffs of Dover from the French side of the English Channel, 1940. British barrage balloons and Chain Home radar receiver towers can be seen on the coastline, while the smoke out in the distance is from a recent Luftwaffe air raid over Canterbury.

6. (*right*) Messerschmitt Bf 109s from the fighter wing Jagdgeschwader 3 fly over the English Channel during the Battle of Britain, 1940. The White Cliffs of Dover can be seen in the background. The central pilot is Leutnant Detlef Rohwer, who was a Technical Officer (TO) with the Geschwader.

7. (*left*) Luftwaffe personnel chatting during the Battle of Britain, 1940. One German pilot recounts a dogfight upon his return from a recent sortie. In the background is a Bf 109.

8. (*left*) A Luftwaffe aircrew discuss their mission during the Battle of Britain, 1940. Behind them is a Messerschmitt Bf 110 *Zerstörer* ('destroyer').

9. (*below*) A German bomber crew practice inflating their rescue dinghy during the *Luftschlacht um England*. Here, the bellows are being connected.

10. (*left*) German personnel inspect the remains of a British aircraft shot down over the English Channel, 1940.

11. (*below*) During the late Eagle Attack, British workers are seen carrying fuselage sections of a downed German aircraft to an existing pile of crashed aircraft debris, 31 August 1940.

12. (*right*) Halfway through the Eagle Attack, Luftwaffe airmen listen intently to a demonstration of flight manoeuvres against British maritime targets, 24 August 1940.

13. (*below*) A Junkers Ju 88 bomber crew sing along with an accordion during their downtime on a French airfield, 6 September 1940. The nose art of the bomber unit's emblem has been retouched.

14. (*above*) While on standby at a French airfield, a visibly stressed *Leutnant* tries to distract himself with a game of chess during the *Luftschlacht um England*, 1940/41.

15. (*below*) Bf 109 fighters prowl around the White Cliffs of Dover, 7 September 1940.

16. (*right*) Bruno Loerzer, the Luftwaffe commander-in-chief Hermann Göring and Hugo Sperrle visit an air base on the Channel coast. From here, Luftwaffe aircraft will be taking off for attacks on England during the first day of the Blitz, 7 September 1940.

17. (*below*) A Nazi propaganda cartoon illustrates the threat of Luftwaffe bombs to London, shown by a deathly spectre holding a bomb and a sword, almost like a violin, above Big Ben and the Houses of Parliament. This is presumably from the Blitz phase of the *Luftschlacht um England*.

18. (*above*) *Reichsmarschall* Hermann Göring (front), commander-in-chief of the Luftwaffe, and his Chief of Staff, General der Flieger Hans Jeschonnek, at a briefing session in France about an air offensive against England, October 1940.

19. (*left*) Luftwaffe ground crew load a German aircraft with bombs for an imminent attack on London, 23 September 1940.

20. (*right*) German schoolboys take part in a History lesson based on a Wehrmacht combat report during the Battle of Britain, 12 October 1940. They are following the fighting on a map of Europe, specifically focussing on the Luftwaffe's bombing raids against England. The original caption states: 'today, History in school is not dry material from past centuries: no, it is the living present, because the Wehrmacht reports give the boys a much more vivid picture of the course of historical events than anything else. After all, they are now experiencing the great historical upheavals for themselves.'

21. (*left*) A German cameraman films a Luftwaffe sortie during the Battle of Britain, 22 October 1940. Early on in the war, accommodating to a journalist on board meant a vital crew member getting left behind. To overcome this, Luftwaffe war correspondents were then trained as rear gunners and were not allowed on war flights before they had completed this training.

22. (*left*) Following a heavy Luftwaffe air raid on London, a torn Union Flag sticks out from the resultant bomb damage, 7 October 1940.

23. (*above*) Roll call of equipment at a Stuka unit in France; the German airmen hold up their life jackets, October 1940.

24. (*right*) A downed Luftwaffe airman is escorted by his British captors to a train that is waiting to take him to a prisoner of war camp, November 1940.

Stahl added that the Ju 88 'then makes an almost instantaneous turn to the left without any noticeable warning, which can no longer be stopped if you react a moment too late with the rudder'. As Stahl further outlined, a Ju 88 required modification of the bombardier's thinking as well, because 'In addition to his task of supporting the pilot in navigation and operating the fuse-setting devices and bomb-dropping devices, he observes the engine monitoring devices and, on the instructions of the pilot, flicks a number of switches on the pumping system that are more convenient for him to control than for the pilot.'

Studded with four machine-guns and able to carry a payload of 4,000–5,500 lbs of bombs over 1,900 miles,[74] the Ju 88 was a swifter but wilder beast to tame compared to the He 111 and Do 17, both of which flew 50 mph slower.[75] As well as flying much faster and thus not holding the fighter escort up as badly, the Ju 88 could carry a payload 1,100 lbs heavier than the Do 17.[76] In addition, its combat radius (360 miles[77]) was greater than that of both the He 111 (300 miles[78]) and the Do 17 (210 miles[79]). Thus, to address the debate as to whether the Luftwaffe truly had the capacity to strike at the heart of the RAF during the Battle of Britain, it arguably had the necessary tool in the Ju 88. But, by early 1941, the Ju 88's temperamentality contributed to a shortage of pilots who knew how to pacify the 'diva'.[80]

Here, the Luftwaffe demonstrated a lack of foresight by not accurately determining its training needs throughout and beyond the *Luftschlacht um England*. In August 1940, however, all the Luftwaffe could think of was the here and now, Operation Eagle Attack. But, as the fighter ace Johannes Steinhoff later admitted, 'Those four months – that is, from the beginning of August to the beginning of November – were the most trying, both psychologically and physically, I was to experience throughout the entire war.'[81]

11

Don't Talk so Loud!

Early into the Eagle Attack, a lone German bomber was hurtling uncontrollably towards the ground in the north-east of England; the aircraft whined in protest at the sharp hiding it had just taken from a Spitfire. As the English countryside became an undiscernible blur, the Luftwaffe bomber crew realized they had no choice but to do the unthinkable: bale out over enemy territory. One by one, the five airmen slipped out of their stricken aircraft, pulled their parachute cords and watched helplessly as their one ticket back to Germany was shredded before their eyes. The swaying aviators landed twenty-five yards away from a humble farmhouse, but their eyes soon boggled at two enormous double-barrelled shotguns pointing at them.

As two British farmers jerkily gestured towards the farmhouse with their guns, the resigned German aircrew bowed their heads and raised their hands. For them, the *Luftschlacht um England* – and, indeed, the Second World War – was over. As they were being ushered into the farmhouse, the language barrier between the captors and the captives became painfully apparent. The two farmers, who were brothers, began to whisper furiously to one another about what to do next. The oblivious German aircrew, with hardly a lick of English between them, were unsure how to convey that they were not a threat. Catching sight of the farmers' wives and their young children, they knew that something had to be done to break the tension between the two sides.

Eventually, they decided that smiling, clicking their heels together, outstretching their arms in a Nazi salute and crying out '*Heil*!' would endear them to their British hosts.[1] Bemused by this awkward encounter, the brother's wives nevertheless led them through to the kitchen table and pulled chairs out for them to sit on, plying the German airmen with the trustiest of all British libations: a steaming cup of tea. One of the wives recalled on 16 August 1940 that 'We gave them biscuits and all the cigarettes we had, and they tried by gestures to thank us.' While the Luftwaffe bomber crew gratefully downed this proffered drink, two eleven-year-old girls – Wendy Anderton and her cousin, Cathie Jones – peered curiously around the door at the peculiar sight.

Their initial shyness soon gave way to doe-eyed innocence. They decided to befriend their unexpected visitors. 'The two little girls had a grand time,' Cathie's mother added in amusement. 'For over an hour, they entertained the airmen' as they waited for local authorities to pick the Germans up. Having been up in the air not even half an hour ago, dicing with death over Britain, most of the burly German aircrew were now creasing with laughter as the two little English girls joked, twirled, pranced and danced. 'It was almost like a circus at one time,' Cathie's mother recounted, shaking her head in bemusement.

She added that 'The children were trying to explain to one young airman that we used the double-barrelled gun for rabbit shooting, but although they imitated rabbits, he could not grasp what they meant and simply roared with laughter, repeating, in broken English, "antique!"' The young German airman that had particularly taken to the young girls tried to gift them the only items he had left on his person: 'a metal penknife, a piece of chocolate (part of the crew's iron ration), a German–French phrase book and his name and address in Germany'. All of the German aircrew laughed at the little girls' chatter except for one, who rocked back and forth 'with perspiration breaking out on his brow!'

This farcical exchange demonstrates the strange transformative effect that an encounter with the enemy could have during the Battle of Britain – but it also shows the awkwardness experienced by each side as fate unceremoniously thrust them into the roles of captor and captive. Before documenting the rest of the campaign, it is important to not neglect the combat mentality, experiences and morale of the captured German airmen whose *Luftschlacht um England* was cut unexpectedly short. Sometimes, as Bartels recalled while flying over London's docks during the Channel Battle, this could be in truly unpleasant circumstances:

> We were heavily outnumbered by the Spitfires we met in the air. It was obvious that they were out to get Galland. Before I knew it, I had a Spitfire on my tail, and my leg was all but shot off. A whole group of muscles was just gone. One hand was also shot to pieces. In addition to all of this, I had been hit in the head. I felt for my crotch. If there was nothing there, I saw no reason to land. But my 'family jewels' were okay, so I decided to try somehow to crash-land the plane.[2]

Ducking under some telegraph wires and eventually bellylanding his Bf 109 in a English field, Bartels collapsed in a heap of cabbages and potatoes. He came to when receiving an urgent blood transfusion from his captors, his leg suspended in painful traction and his arm hoisted above him like a broken marionette. The diligent medical care shown to the injured Luftwaffe airmen was deeply appreciated and often indebted them to their British captors. Walter Knappe, the Ju 88 navigator who had previously been shot down over Dunkirk, had been unluckily downed again in the *Luftschlacht um England*:

I woke up several days later in a British hospital. I had a brain concussion, a fracture at the base of the skull, and arm and leg fractures. My three crew members had died on the way to the hospital. As I was told later, the doctors had worked hard to save me since I had shown signs of life. They succeeded in pulling me out of a coma that lasted eight to ten days. I had to be fed intravenously.[3]

He added that 'In the hospital, I felt quite a sense of *esprit de corps* with the British there. All of the soldiers had come from Dunkirk. Although I was their prisoner, we didn't live together as enemies, but as soldier comrades. The experience of war brought us together.' In Knappe's case, he claimed the incident had given him a wholly different appreciation of both his own life and those of his 'enemies'. He recalled how 'A German Jewish woman living in England was sent to help me learn to speak again. She gave me the feeling that it was important to speak. I felt as if I were learning to talk all over again.'[4]

Other newly downed Luftwaffe airmen, however, displayed a haughty and surly attitude towards their new hosts. Major Guy Hadley recalled that 'The Battalion Headquarters of the 7th Devons was a collecting centre for prisoners and there were usually about half a dozen new arrivals daily. Most of them, especially among the bomber crews, were sullen young thugs who, in other circumstances, would have been sent to a reformatory.' He also reported a story he had heard: 'A particularly truculent Nazi spat in the face of a Somersets Major. The Major wiped his face, ordered the German's boots and socks to be removed and marched him briskly round a field of stubble for ten minutes.'

Interestingly, however, Hadley noted that 'The fighter pilots seemed to be of slightly better quality, and I even met one pleasant young man who could talk like a normal human being.' The British soldier confidently informed the German fighter pilot, 'This aircraft was one of 175 German planes brought down that day by the R.A.F., whereupon he shook his head with a smile and said that he "really didn't expect me to believe *Doktor* Goebbels and I mustn't expect him to swallow the B.B.C."[5] That the fighter pilots appeared to be more amiable, refined and intelligent than the bomber crews was also reported by the British press. On 28 September 1940, J. L. Hodson wrote in the *Aberdeen Press and Journal* that:

> During the past two days I have talked with various officers who have probably had 200 captured Germans through their hands. They find a marked difference between fighter pilots and bomber pilots and crews. The first are a better type, as a rule, more chivalrous and better mannered. The bombers are both more impudent, cruder and more brutal. Their treatment on capture varies according to their behaviour. One German who started throwing things about on a trawler came ashore stripped and wearing a blanket. At an Army headquarters, another German sat on the floor and said, 'English pig.' On the other hand, a fighter pilot captured to-day was described as a charming youth by the major who had charge of him.[6]

It is possible that the classic 'knights of the sky' aura promoted by the Luftwaffe fighter pilots may have left an impression on their British captors, as it certainly did in the post-war period. Yet, there were also multiple stories of Luftwaffe bomber crews, as shown at the beginning of this chapter, who had acted politely and sometimes warmly to their captors. So, it is possible that more variable behaviour was observed among the bomber crews simply because they were more numerous, and therefore perhaps offered more of a cross section of behaviours and personalities in a way that fighter pilots did not. It is also possible that the group dynamics of a bomber crew – the 'strength in numbers' – may have emboldened them to act more defiantly against their captors in a way that was simply not possible for a lone fighter pilot.

But there were naturally some unpleasant interactions among individuals as well. William David, the commander of No. 213

Squadron, alleged that during the Battle of Britain, when 'The German Stukas had just dive-bombed Tangmere with disastrous effect and killed lots of WAAFs [...] one of the [captured] Germans was seen to smirk. And the RAF commander hit him very hard to stop him laughing.'[7]

Depending on the transgression, however, some RAF personnel took revenge on their odious prisoners in a more light-hearted manner.

Sergeant Arthur G. Ethridge, an RAF cook and butcher, recalled how a recently captured German airman gave them merry hell. 'Every time the bombs dropped,' he recalled, 'the MPs [military police] had their work cut out to hold him down. He could hand it out, but not take it. He was taken from the guardroom past the butcher's shop for his meals at the Sergeants' Mess, and how arrogant he was doing the goose-step and Heil Hitler all the way.'[8] Ethridge and his men's solution to this insufferable German airman exemplifies why you should never bite the hand that feeds you: 'A few of us got together to talk about this, and ways to stop it, we decided we would enlist the aid of the cooks to keep him quiet, we knew he always had a plate of soup without fail for his lunch.'

He added that 'On our visits to the village shop, we bought among other things a packet of Feen-a-mints [a strong laxative], but instead of putting one in the soup, we, with the help of the cook, put in a whole packet. Needless to say, that airmen did not surface for quite a long time – he was, we were told, spending his time moving between the toilet and his cell.' Although hardly approved of by the Geneva Convention, this trick certainly relieved the annoyance of the RAF cooks as swiftly as it did the bowels of the German airman!

These personal testimonies illustrate the diverse range of interactions which downed German airmen experienced with the British population during the *Luftschlacht um England*. They provide an insight into how Luftwaffe personnel – when at their most raw and vulnerable – perceived their enemy, and vice versa.

The first moment of their capture was also the point when the German airmen were at their most guarded – both about what they knew and how they felt about the air campaign. If there is one universal truth about aviators, though, it is that they cannot keep their escapades secret from their peers for long. A *Gefreiter* who had been a radio operator on bombers issued an astute warning to his fellow prisoners of war to refrain from loosening their lips:

> Once, we were advised by an old fox who knew all about it as to how things might be in captivity, and he advised us to boast as little as possible. There's no point in it. In this way you only jockey yourself into an awkward position and they'll notice it and your treatment may suffer. Besides, should the evidence at any later date fall into German hands, it might somehow be interpreted as treason and for that reason it is better not to talk at all. [...] You must always imagine that it will come to German official quarters one day. Everything we say is put down on paper. That I assume without a doubt.*

His hunch that all of their interrogations were recorded by the British was utterly correct. What he did not know, however, was that this seemingly private conversation was being recorded and transcribed as well. On 26 October 1939, the British Directorate of Military Intelligence unit MI1 (H) was redesignated as the Combined Services Detailed Interrogation Centre (CSDIC UK).[9] It was a specialist interrogation centre that served as a second line of processing for prisoners of war who merited further investigation and questioning after being

---

* AIR 40/3071. S.R.A. 801, 'A 565 – Gefreiter (Bomber-W/T Operator), A 602 – Gefreiter (Bomber-W/T Operator) & A 596 – Obergefreiter (Bomber – W/T Operator), 22 October 1940. The original document says 'image' instead of 'imagine', but as this does not quite make sense in English, 'imagine' has been used to enhance its comprehensibility. It is possible that the original German transcript was translated more literally than figuratively by the translator.

passed on from the prisoner of war 'cages' managed by the Prisoner of War Interrogation Section (Home), (P.W.I.S.(H)).[10]

Originally based in the Tower of London until December 1939, CSDIC UK was then transferred to the Trent Park estate ('Cockfosters Camp') in Barnet, Hertfordshire.[11] During the Battle of Britain, Trent Park was used as a temporary holding facility for Luftwaffe prisoners of war across all ranks, before later being used as a dedicated prisoner of war camp for high-ranking generals and staff officers.[12] In total, 685 German airmen were brought to Trent Park and over a thousand reports were compiled on their interrogations and conversations up until the end of December 1940.[13]

When examining these accounts, it is important to remember that we are dealing with hearsay and anecdotal testimonies, and the Allied translators did, of course, choose what they personally felt was most pertinent to report. Some German units had also been better briefed than others about not sharing information, with both Stuka personnel and the men of the bomber wing KG 77 proving to be exceptionally silent.[14]

Finally, as prisoners of war, their comprehension of the ongoing air campaign was frozen on the day that they were taken prisoner. These caveats notwithstanding, the CSDIC transcripts provide an unusually revealing insight into German mentality and morale during the *Luftschlacht um England*. Sifting through British intelligence files allows us to pinpoint what Luftwaffe prisoners of war thought about the Battle of Britain, if they felt an invasion would ever be launched, and – most importantly of all – if they still believed the Germans would prevail.

During the first phase of the Eagle Attack, the number of German prisoners of war being retrieved increased by nearly 30 per cent.[15] It was initially reported on 21 August 1940 that the morale of the Luftwaffe prisoners remained high, with no tangible effects from loss of materiel and personnel.[16] Nevertheless, it was revealed that some Luftwaffe flying arms felt that they were appreciated less than others. In a further report from Wing

Commander S. D. Felkin, it was added that 'Morale and "*Esprit de Corps*" are high throughout the Coastal Reconnaissance Units, but promotion is slow, and they feel rather slighted that so few decorations are awarded; they have to do eighty hours' war flying before even getting to the E.K.II [*Eisernes Kreuz II* or Iron Cross 2nd Class].'[17]

It did not help, either, that the German reconnaissance units had suffered heavy losses at the hands of the British: 166 reconnaissance aircraft of all types were lost during the *Luftschlacht um England*.[18] It was also hard for them to secure their war flights, too, as half of their units were often withdrawn at any time to regroup or to train further aircrews.[19] The stinging disappointment of Luftwaffe prisoners of war who were unable to obtain an Iron Cross was evident, with a report noting that 'One man captured on his first War Flight had missed his Iron Cross, and hoped rather wistfully that perhaps he would get "A Prisoner of War Medal"'![20]

In a reflection of how their hunger for medals and victories could sometimes get the better of them, a captured *Leutnant* and *Oberleutnant* discussed how 'We underrated the English fighters. Recently everyone who wanted a few easy additions to his bag flew over, but they got a surprise.'[21] Others, however, focused more on what the war had cost them rather than what it could have given them. Later into the campaign, two *Oberleutnants* were overheard sharing their grief at the loss of their men, with one lamenting that:

> [There was] another loss yesterday: Grabbow. Who's left now in the '*Staffel*'? Hocker, the only officer; the old *Oberfeldwebel* Wilpert, who's a tired old man and can't do much more; *Oberfeldwebel* Oswald. The only one who got back from the 'dust-up' is *Feldwebel* Volpert. He is another of the old brigade. The only one. Four men of the old gang. All gone! All gone! No one is left![22]

The *Oberleutnants* dismissively concluded that 'The great aces have moved into senior positions and what is coming up from below is worthless.'[23] Although this could be a sign of their bitterness at not being able to prevent themselves from being shot down, frustration at the *Luftschlacht um England*'s development is also evident in a conversation between an *Unteroffizier* bomber radio operator and an air gunner who noted that:

> The loss in aircraft is on average three times as high as it was in France and then pilots, who are shot down here, are taken prisoner. At the beginning, in France, we had fifteen aircraft in my '*Staffel*', of these, eight aircraft were completely shot to pieces in three weeks, but the pilots are not dead. Some of them were posted as missing two or three times, but always turned up again.[24]

Indeed, a number of pilots wistfully spoke of the campaign over France compared to the more challenging *Luftschlacht um England*, as recounted by a captured *Oberleutnant* in a British intelligence report from September 1940: 'It was child's play in those days, we flew in the sun at 8,000 m, I mean during the attack on France. Now things are different, we get a "*Staffel*" of eight or nine, etc., and as soon as they see half of them it is all over.'[25]

It is not surprising that the Luftwaffe prisoners of war recounted their experiences in previous aerial campaigns in an effort to comfort and convince themselves that the tide would eventually turn in Nazi Germany's favour. In doing so, however, a number of German airmen incriminated both themselves and their organization by revealing some of the immoral behaviour they had witnessed or perpetrated in earlier campaigns. Some truly relished the death and destruction they were causing. For example, an *Oberleutnant* was overheard claiming on 17 July 1940 that 'Throwing bombs has become a passion with me. One itches for it; it is a lovely feeling. It is as lovely as shooting someone down.'[26]

Another Luftwaffe prisoner was reported to have discussed bombing hospitals in Britain: 'Well, the people are half dead anyway, and if the hospitals are bombed, no fresh casualties can be admitted.' Another admitted he had tried to bomb Hyde Park in central London, because he knew there were a lot of air raid shelters there.[27] Such unsettling revelations were not just contained to the British campaign, but also stretched way back to the beginning of the war, with a small number of Luftwaffe prisoners of war gleefully documenting the violence wreaked against the populations in Nazi-occupied territories – and asserting that such heinous measures would be replicated with British civilians if they were conquered.

A bugged conversation between three *Oberleutnant* fighter pilots was recorded in September 1940 in which one claimed that 'England is invincible. That is the mentality of the people here, and if every civilian defends his home stubbornly, then it's a difficult matter.'[28] Another of the men replied, 'They'll simply be shot.' The first *Oberleutnant* said, 'That is not allowed,' to which he received the reply: 'The whole question of what is allowed or not is finally a question a power. If you have the power, everything is permissible. In spite of that, our troops should not massacre civilians who do not shoot (at them).'

One of the other men adds that 'Terrible things happened in France', to which the reply was, 'It did happen that a man fired a shot out of a house, and they were unable to catch him. Then all the men in the village were shot.' Other Luftwaffe prisoners were in full agreement that the British should be handled roughly in the event of an invasion. When discussing how British resistance would be put down if Operation Sealion were to be launched, a *Hauptmann* bomber observer informed a *Leutnant* bomber pilot:

> Do you know how they work in Poland? If only one shot is fired there is trouble. Then the procedure is as follows: from whatever town or district of a town shots have been fired, all

the men are called out. For every shot fired during the following night, in fact during the following period, one man is executed.[29]

That his companion replied 'Splendid!' to this disturbing revelation should be kept in mind when considering Steinhoff's generalization that the Luftwaffe had been 'fighting sportsmanlike' leading up to the Battle of Britain. However, since some of the German airmen approved of this murderous approach being applied in Britain, too, it is clear that no civilian population – whether the Nazis considered them to be racially 'subhuman' (as they described the Polish people) or not – was immune from Luftwaffe-approved violence. Indeed, one German airman was overheard telling a particularly horrific story from France:

> Near Dinard, a drunken *Unteroffizier* – a married man with two children – entered a house, seized a fifteen-year-old girl by the hair, laid out the mother and sister who tried to intervene, dragged the girl outside and violated her. As he was getting onto his motor bicycle to ride away, the father appeared and attacked him. The German hit the father so hard that he died. He was condemned to death, not because he had killed the father, but because the girl was under sixteen. This the prisoner of war seemed to find amusing.[30]

His fellow prisoner of war, on the other hand, commented that 'Such incidents were severely punished,' and that he 'feared that such incidents would increase during the coming winter when the troops had little to do'. This shows, then, that other Luftwaffe prisoners of war did not always share a sadistic attitude and could be horrified by the stories that were told. Indeed, some expressed warmth towards their British hosts and wished them no harm, with one fighter pilot noting that 'The English were extraordinarily kind to me.'[31]

However, some of their praise for the 'English' came from

the antisemitic belief that the English people's hands were being tied by a wider Jewish world conspiracy driving the wheels of war. One Luftwaffe fighter pilot, for instance, was surprisingly complimentary of Churchill – but his praise for the British prime minister was wrapped in a poisonous jab at how it was '80 per cent Jewish capitalism' behind the war that was holding up peace talks:

> Churchill is playing a role. I'm sure of it. The fellow is tough. He's a good man, Churchill, himself. My conviction is that of *Generalfeldmarschall* Milch. When [Milch] said goodbye to our lot at Döblitz he remarked, 'That fellow is tough. He's a good man.' Our press naturally abuses him, doesn't it? And the English abuse Hitler. But if they knew him as a man, they would not pull him to pieces so much.[32]

This passage demonstrates how the Luftwaffe prisoners could hold multifaceted opinions on the propaganda they were exposed to: rejecting the Nazi newspapers' demonization of Churchill, but accepting the antisemitic view that it was Jewish bankers who were prolonging hostilities. Antisemitism was a common motif among the captured German airmen, with a British interrogation report stating that 'All Jews now have to work for a living, which [*Feldwebel* Schmidt] described "as a change for them".'[33]

Schmidt added during his interrogation that 'It was interesting to see numbers of formerly opulent Jews, sometimes still in their fine raiment, taken to the public squares or aerodromes to shovel snow away in winter.' There was also disturbing comments from the Swiss-born fighter ace Franz von Werra, who was captured by the British on 5 September 1940 after being shot down over Kent. Werra later became notorious for being the Luftwaffe's Houdini – he was the only Axis prisoner to successfully escape from Canada and return to the Reich during the war.[34] Werra was said to be 'the most optimistic of the believers in Hitler's victory' and was considered by MI9 to be 'a completely successful product of Nazi education'.[35]

When listening in to his conversations shortly after his capture, British intelligence discovered that Werra 'approves pogroms and the beating of prisoners in concentration camps* ... von Werra's opinions are quoted at length for their psychological interest. His supreme confidence is exceptional.'[36] However, some Luftwaffe prisoners of war were more concerned about international opinion of the Nazi persecution of Jews than the violent acts themselves, with a *Hauptmann* bomber observer claiming that:

> I could not understand the object of that plundering. It if were not desired that Jews should own property, it could have been confiscated. The only thing I was not sorry about in the Jewish pogroms was that the synagogues were burnt. The plundering damaged our reputation abroad to a considerable extent. We have to defend these actions to foreigners, but in our hearts, we must be ashamed of them.[37]

Indeed, as the intelligence report noted, 'Many think that Hitler made his first big mistake in throwing them out, not because Germany thereby lost many outstanding scientists and artists, but because Jewish propaganda has raised the democracies, and especially America, against Germany.' Such testimonies display the complex ways in which antisemitism manifested among the captured Luftwaffe prisoners: some in full approval of antisemitic persecution, others embarrassed at the international reaction to it while remaining apathetic to Jewish suffering.

On the other hand, some of the German airmen took advantage of their captivity to vent their disapproval of the Nazi leadership. It was noted by British intelligence that 'For

---

* Nazi concentration camps were not yet designed specifically as death camps in this period, but mistreatment and murder were already typical in such facilities.

some time, there is known to have been considerable conflict and rivalry between Senior Air Force Officers and people like Goering, who are looked upon more as Political Leaders than as Officers.'[38] A British intelligence report from 19 September 1940 noted how a captured *Oberleutnant* fighter pilot claimed to his companion, a bomber pilot, that 'The *Führer*'s speeches are a form of mass suggestion. At the "*Reichsparteitag*" [Nuremberg Rallies], I was never able to join in the cheers of the crowd.' His companion replied 'nor could I. When the *Führer* passes and the crowd cheer and shout, I was rather annoyed with the people to tell the truth.'[39]

In addition, one critical *Oberleutnant* fighter pilot was overheard complaining that 'People who used to buy obscene reading matter now read *Der Stürmer*.* You can see it in them at once, as with lustful eyes they put the paper in their power. I know that the *Stürmer* appeals to a certain class of reader, and that it is perhaps the right thing for these people. Therefore, it is a disgrace to culture.'[40] A surprising number of Luftwaffe personnel listened to BBC propaganda broadcasts despite the severe penalties associated with doing so. The report details an interesting means by which some of the German airmen tuned in to international news:

> His brother-in-law has a radio business, and Schmidt, who is a good mechanic, put in a bit of time lending a hand in the workshop. One of the more interesting 'repairs' was fitting blank dials in place of the normal printed station dials. ('War-time frequencies are always being changed anyway, so it is no use having them written out'.) 'Of course, if I hear English or French spoken, I switch off at once – but if I hear news in German, how can I tell that it is from a foreign station until the announcer says so?' Not a bad story to have on tap if the Gestapo calls.[41]

---

* *Der Stürmer* ('The Stormer') was one of the most prominent national tabloids in Nazi Germany – characterized by its virulent antisemitism and obscene caricatures.

Indeed, another British intelligence report on captured Luftwaffe airmen in 1940 noted that 'Out of eleven fighter pilots recently interrogated by this Section [...] five admitted that they sometimes listened to the B.B.C. German broadcasts, and that these five all belonged to different units. In one case the pilot said that he was instructed to listen-in in order to obtain names of PoWs [prisoners of war]. In view of the communal life led by GAF [German Air Force] airmen in France, listening-in to British broadcasts must be done openly.'[42]

The report added, though, that 'The attitude to the broadcasts seems to be that they are just so much propaganda. One PoW said that he believed the English figures of German losses, while another said our exaggerations made him laugh.'[43] Thus, although the Luftwaffe airmen had been forged by a totalitarian regime, some of them retained a degree of individual agency and autonomy in their opinions towards the war – even going so far as to risk being disciplined for listening to foreign broadcasts in order to make sense of the war from the Allied side.

Amid all of these diverse influences on mindset and morale among the captured Luftwaffe prisoners, there was one predominant factor which determined how optimistic they remained about the war: their belief in the likelihood of being liberated by an imminent German invasion. Some German prisoners of war had the sobering epiphany that the onset of autumn rendered an invasion more unlikely. 'We have missed the best moment for the invasion,' an *Oberleutnant* said to another man of the same rank on 14 September 1940.[44]

Moreover, in a British intelligence report from 11 September 1940, a conversation had been recorded between a *Hauptmann* and an *Oberleutnant* in which the latter complained that 'I simply cannot see how we are to win the war. To starve them out is fantastic, impossible, just as they could also not starve us out.'[45] Others, however, held so firm to their conviction that the German invasion was imminent that they were overheard discussing the possibility of escaping in conjunction with the invasion.

A British intelligence report noted on 12 September 1940 that an *Oberleutnant* was overheard discussing an escape: 'We must go towards the West, as planned,' he informed two other *Oberleutnants*. 'Don't talk so loud!' one of them hissed at him, but the first *Oberleutnant* was adamant that 'I'm going, in any case, even if the rest of you stay here...'[46] He added that when 'The invasion starts, we must absolutely get out of this place. Otherwise, you'll go to Canada... you must not forget, they [the Germans] could cut off the whole of the southern part of England, including London. I think that, tactically, that would be right.'

The captured airmen who had the most to lose back home, however, often learnt to become content with their lot in captivity. One *Unteroffizier* on bombers was overheard claiming that 'previously, I always thought I'd rather be dead than a prisoner of war. But when one sits here now and thinks of wife and children, one gets quite different views. As a matter of fact, we should not be of use to the country if we got shot here in an attempt to escape. As far as my character is concerned, I think that I will go home a better man than I came. One has so much time here for thinking things over.'[47]

12

Like a Thunderbolt He Falls

The world is under the spell of the great final European battle. A new phase in the liberation of Europe from the gold-greedy plutocratic rule of England has begun, the last one that is now taking place in the spotlight of the entire world. The attacks of the German air force roll over the island kingdom without a break; death flies in the thunder of the German combat units over England's skies; daily and hourly our fighter *Staffeln* break through the enemy's protective barriers in order to expose the island fortress, praised in arrogant tones as insurmountable, so that our bomber *Staffeln* can land their deadly blows.[1]

This was how, on 1 September 1940, *Deutsche Luftwacht* magazine fiercely vocalized its support for the renewed concentration of Luftwaffe attacks during the second phase of the Eagle Attack.

By declaring that 'The air offensive against England has begun!' the article was papering over the cracks in the Channel Battle and first half of Eagle Attack, treating them as a mere apéritif before the main offensive in September. Downplaying the earlier stages of the Battle of Britain, however, did not change the fact that Nazi propagandists had previously shouted about their importance. On 16 August 1940, the *Briesetal-Bote* newspaper had borrowed the American journalist Alexander Seversky's turn of phrase to dub that phase of the war as a *'Trafalgar des Luftkrieges'* ('Trafalgar of the Air War').[2] 'Should England lose this mighty air battle,' the article cited, 'it will have lost the war.'[3]

As would often happen during the *Luftschlacht um England*, the Nazi propagandists were altering the timelines of the aerial campaign in order to portray the German air force in the very best light. Given the Luftwaffe's variable performance during the early days of the Eagle Attack, it is not surprising that Nazi propagandists felt the need to cover up such disappointment. Yet, in the second half of the Eagle Attack – stretching from 24 August until 6 September 1940, when the Luftwaffe renewed its focus on RAF sector stations[4] – things were genuinely starting to look up for the Germans.

By 31 August, the Luftwaffe was levelling out the disaster of the early Eagle Attack by closing in on a loss ratio of around 1:1 during its best days.[5] Sinking its talons into the RAF's airfields with unfettered ferocity, the Luftwaffe inflicted an especially high British loss of 137 Spitfires and Hurricanes between 31 August and 6 September.[6] True, the Germans had lost 200 aircraft themselves during the same period – but they continued to use their offensive advantage to run Fighter Command utterly ragged.[7]

While the British aircraft industry continued to work double-time to replace the fighter losses, the RAF's pilot reserves, though greater in number than the Luftwaffe's, remained under constant strain. By 7 September, the British were losing around 120 pilots a week.[8] Furthermore, although Fighter Command's frontline strength rarely dipped under 650 fighters during the Battle of Britain, its reserves had been whittled down from 518 in the first week of June 1940 to just 292 at the end of the first week in September.[9]

Thus, although the RAF was not yet at breaking point, this stage was the closest that the German air force would get to defeating Fighter Command in the Battle of Britain.

Most concernedly, the ongoing summer expansion of the Luftwaffe's paratrooper units had now reached its peak by the end of August 1940. As an original report on the scheme detailed, 'All units in the enlarged paratrooper section were set up. By this point they had at least reached the level of conditional operational readiness. Through hard training work, it improved quickly from week to week.'[10] Amid this more successful half of the Eagle Attack, the German *Heer* and Luftwaffe planned for the invasion of Britain with renewed vigour.

In order to truly assess their preparations, it is important to consider what kind of information the Luftwaffe was feeding to its sister Wehrmacht branches about the ongoing aerial campaign and conditions for invasion. A Luftwaffe *Koluft* (*Kommandeure der Luftwaffe*) was an air force liaison officer that was attached to every army headquarters and sometimes corps headquarters. They commanded the Luftwaffe anti-aircraft units and tactical reconnaissance squadrons attached to the *Heer*,[11] advised army commanders on aviation matters, and ultimately coordinated all Luftwaffe logistical support that was afforded to the *Heer*.[12]

However, the *Koluft* officers had no authority over combat aircraft, which the German air force did not desire to relinquish to ground commanders.[13] Their surviving documents counter the common belief that the three Wehrmacht branches were pulling away from one another like a pack of rambunctious scent hounds when it came to Sealion. Instead, the *Heer* planned the invasion in accordance with the guidance and requirements that trickled down from both the Luftwaffe and the OKW. This was demonstrated clearly in a survey circulated around Infantry Regiment 135 on 3 September 1940 that was entitled 'Study on the Use of Army Aviation Units and Anti-Aircraft Artillery in the "*Seelöwe*" Case'.[14]

It included a Luftwaffe report from the air liaison officer of AOK (Army Higher Command) 16 dated 26 August 1940, which noted that 'The attached study on the use of the army aviation units and anti-aircraft artillery in the "Sealion" case is based on the simulation games to be prepared in accordance with A.O.K.16 Ia No. 283/40 g.Kdos and cooperation with reconnaissance *Staffeln* and anti-aircraft artillery is to be taken as a basis.'[15] A day beforehand, the 'Study on the Deployment of Flying Army Units and Flak Artillery in *Unternehmen Seelöwe*' had been issued.[16]

The *Heer* reconnaissance units would aid their counterparts in the Luftwaffe by conducting 'detailed monitoring of the course of the battle when fighting for the coastal strip and creating a bridgehead'. The report explained that 'The focus is on reconnaissance of the enemy's main resistance pockets and artillery, as well as the early detection of counterattacks, movement of reserves, and enemy armoured forces in readiness and on the march.' Then, 'after gaining a foothold on the coast, the timely reconnaissance of the enemy's defence lines' would be required.

Shortly afterwards, an AOK 16 report from 30 August 1940 confirmed that the army reconnaissance pilots of Army Group A would continue to receive training via their Luftwaffe liaison officer, noting that 'The younger and inexperienced crews and the new young personnel in the specialist *Gruppen* who have been transferred to the *Staffeln* primarily require special training.'[17] It further declared that 'The focus of the training and activities of the *Staffeln* is on the tasks required in connection with "*Unternehmen Seelöwe*"; the study received by the *Staffeln* on the use of army aviation units and anti-aircraft artillery in the "*Seelöwe*" case serves as a basis for this.'[18]

Moreover, AOK 16 noted that 'If other material can be obtained through *Koluft* 16 [as a combined Luftwaffe and *Heer* flak initiative], it will be continuously sent to the *Staffeln*. Particular emphasis should be placed on map reading by subcontractors and drivers (taking into account the expected conditions in England).' In addition to 'maps, aerial photographs [and] military geographical descriptions of England', the men were also instructed in 'climatic and maritime conditions in the Channel area' and 'actions during emergency landings over water'.

In order to improve the survival rate of these *Heer* reconnaissance *Staffeln* during the invasion of Britain, the report added that 'According to the experience of the operational Luftwaffe, it is advisable to equip the crew with flare pistols and flare ammunition. Crews floating in the water were able to shoot with these and that was the only way they were saved.' This detailed correspondence shows that Operation Sealion was being thoroughly explored throughout the Wehrmacht during the more successful half of the Eagle Attack.

A renewed wave of optimism rippled through the Luftwaffe aircrews, who were thrilled to see their newest exploits reported in the German press. *Oberleutnant zur See* Jan Klatte, as part of the Luftwaffe's *Seeflieger* ('naval aviators'), was then assigned to the bomber wing KG 54 and wrote to his mother on 27 August 1940: 'The physical and mental strain of the missions is bearable. But please don't ask any curious questions about missions, flights, etc. Perhaps you heard in the Wehrmacht report today that a radio station on the Isles of Scilly was destroyed on 26 August. That was my work! I was glad it was mentioned.'[19] Naturally, the German newspapers continued to wildly overestimate the kills claimed by the Luftwaffe, as an article from the *Briestal-Bote* paper demonstrated on 2 September 1940:

> If one more example were needed to make the German air superiority over England clear to the whole world, the numbers here speak a very clear and irrefutable language. The following is a summary of German air warfare successes in August, from which it is clear that Germany is the victor on the battlefield over England. British aircraft and barrage balloon losses in August are contrasted with our own losses. England subsequently lost 1,565 aircraft and 177 barrage balloons, while our own losses amounted to 406 aircraft. It must not be overlooked

that Germany is always the aggressor and that the air war is taking place over the British Isles. If, as Mr Churchill and his agents maintain, the English air fleet could have equalled that of the Germans, then this chart would be looking very different.[20]

During August, Fighter Command had in fact lost 389 aircraft – just 25 per cent of the article's highly inflated figure – and the Luftwaffe continued to overclaim the true British losses by three or four times during this period.[21] Moreover, although the general situation was improving for the Germans, the Luftwaffe was still sustaining losses so concerning that a conference was held at the Hague to address the problem in early September.[22] But overall statistics were just one way that the German newspapers were able to showcase progress in the *Luftschlacht um England*; individual heroism among the Luftwaffe's airmen, too, remained a powerful tool at the Nazi press's disposal.

In particular, the *Jagdflieger* – known as *Experten** ('experts') – were watched in awe by young German and Austrian boys, some of whom ached to answer the Luftwaffe's call and fight in the *Luftschlacht um England* themselves. Just as Oswald Boelcke and Max Immelmann had vied with each other in the Great War, now the professional rivalry between Galland and Mölders captivated the nation. Their escapades in this more successful Eagle Attack stage had a powerful effect on recruitment for the Luftwaffe, as recalled by Klaus Deumling, a sixteen-year-old schoolboy who later flew as a bomber pilot in the reformed bomber wing KG 100 'Wiking':

---

* Andrew, 'Strategic Culture in the Luftwaffe', 361. A fighter 'ace' was generally recognized internationally as having five kills during the Second World War. The '*Experten*' often had far more than this – although Philipp Ager, Leonardo Bursztyn, Lukas Leucht and Hans-Joachim Voth point out that the *Jagdflieger* had particularly high claims because they were not rotated as frequently as the Allied fighter pilots. See P. Ager et al., 'Killer Incentives: Rivalry, Performance and Risk-Taking among German Fighter Pilots, 1939–45', *Review of Economic Studies* (2022) Vol. 89, 2257–92.

I was fascinated by the successes of our Luftwaffe in the *Luftschlacht um England*. Brand new reports came in every day about the well-known fighter pilot idols Galland and Mölders, and new aviation heroes emerged all the time. In the autumn of 1940, Adolf Galland and Werner Mölders were the first two fighter pilot aces who attracted attention during the 'Battle of Britain' with their outstanding number of kills and were the first pilots to receive the Knight's Cross.[23]

However, it was not just the swashbuckling fighter aces who inspired the German youth. 'What captivated me even more', Deumling confessed, 'were the bombers flying in formation against England – which was shown again and again in the newsreels in the cinemas. We were also often shown footage of how the crews worked in the planes and how the pilots controlled their machines. All of this affected me as a young person to such an extent that I developed an almost irrepressible desire to be able to fly such a bomber one day. As the war continued with its ever-increasing German successes, my wish became a vision.'

For many, it just took one more advert – in Deumling's case, a snippet in his local Bielefeld newspaper 'seeking applicants for active officer careers and pilot training in the Luftwaffe' – to make them bite the bullet and join up. Deumling recalled that 'In my school class there were four young people who all had the same interest and the same desire.' Despite the arduous requirements of the ongoing Eagle Attack, new recruits and officer candidates in the German air force were still given a thorough education. As shown by Luftwaffe training guidelines that were issued on 24 August 1940, its bomber crews were put together very early on during their training as a means of fostering bonds with one another:

> Pilots and radio operators have already been put together to form part of the crew at the blind flight school. The training lasts three months; of this, around one and a half months of

individual training and one and a half months of crew training. The start of crew training is left to the training managers. A complete aircrew always leaves the *Gruppe* bomber crew school.[24]

In addition, 'the bombardiers and co-pilots are complemented by Luftwaffe and Army personnel',[25] demonstrating an inter-service fusion of knowledge and experience. Radio operators, meanwhile, were to 'receive their on-board radio operator training at the air communications schools: they then go to blind flying school for one month and are then transferred to the larger combat pilot schools for weapons training'.[26] Finally, the air gunners received a 'month of preliminary training at the air gunnery schools with a focus on weapons training and are then transferred to the larger combat pilot school.'[27]

It is often asserted that the Luftwaffe provided far superior gunnery training compared to Fighter Command during the Battle of Britain.[28] Indeed, the Luftwaffe training guidelines from 24 August 1940 highlighted an extended fighter pilot training course, in which 'The training lasts three months in summer and four months in winter, the training objectives being: 1) mastery of flying fighter aircraft; 2) mastery of combat operations with fighter aircraft in pairs and as a swarm; 3) completed training on the on-board weapons.'[29]

These guidelines stated that the fighter pilots needed to know the '... structure, strength and equipment of a fighter *Staffel*; the structure of a fighter *Gruppe*; the tasks and fighting style of the fighter pilot up to the *Staffel* unit; and knowledge of aircraft types (Germany and enemy states), with a comparison of combat value and armament.'[30] They were also briefed on what to do if attacked on the ground, how to destroy their aircraft if they landed on enemy territory, and were kept up to date 'about the latest war experiences as a basis for tactics lessons'.

In addition to providing close escort duties, the would-be pilots needed to be prepared to go on the *'Freie Jagd'* ('free chase' or 'hunt'): a roaming fighter sweep in which they were able to scout the airspace near the target ahead of the bombers they were escorting. Instead of being forced onto the defensive through being caught out by RAF fighters, the free-chase tactics taught the German fighter pilot to pounce on an unsuspecting British plane and therefore clear the path for the Luftwaffe bombers.

As Galland noted, the free chase gave '... the greatest relief and the best protection for the bomber force, although not perhaps a direct sense of security. A compromise between these two possibilities was the "extended protection," in which the fighters still flew in visible contact with the bomber force but were allowed to attack any enemy fighter which drew near to the main force.'[31] As the British Pilot Officer Charles G. Frizell of No. 257 Squadron recalled:

> Their tactic was to wait 'high on the perch' a few thousand feet above the bombers they were escorting, and then they would swoop down in pairs on an unsuspecting Hurricane or Spitfire, and within two seconds have either shot it down or climbed up again to wait for another attack. They reminded me of Tennyson's eagle who 'watches from his mountain walls, and like a thunderbolt he falls'. It was a form of back-stabbing. When we had the chance, we did the same thing.*

The Luftwaffe's burgeoning confidence during the later stages of the Eagle Attack also stemmed from the fact that some of the British airmen who occasionally fell into German hands expressed lowered morale and heightened anxiety at the way the Battle of Britain was unfolding. The information obtained via

---

* MISC 3404 244/2. C. G. Frizell, 'Letter to Messrs. Denis Richards & Richard Hough, Battle of Britain Historians', 30 March 1988. Imperial War Museum, London. The poem that Frizell is referring to is 'The Eagle' (1851) by the esteemed Lincolnshire poet Alfred Lord Tennyson. It reads: 'Ring'd with the azure world, he stands/the wrinkled sea beneath him crawls/He watches from his mountain walls/And like a thunderbolt he falls.'

the German interrogation of captured RAF airmen was sometimes fed into the more general Luftwaffe situation reports. On 30 August 1940, as the German momentum in the Eagle Attack was at its zenith, a Luftwaffe situation report detailed some of the British prisoners' statements regarding their combat fatigue during this phase of the campaign:

> The British fighter pilots complain of extraordinary overexertion as a result of numerous daily missions. The crew of the standby *Staffeln* must stay directly next to the aircraft day and night; they sleep next to them in their flying suits during the night. The recent deployment meant that a hot meal could only be served once a day. A newly arrived prisoner reports that in one of the attacks on Norwich a line of bombs fell on a busy street, leaving about 200 dead and 240 wounded.[32]

The report further noted that 'A British fighter pilot (Spitfire) shot down over the French coast was carrying an unsent letter containing the following paragraph: "I am now with the R.A.F. in Eastchurch* (Kent). We have just experienced a tremendous attack with heavy bombing on this airfield. The officers' mess, three halls, the commander's house, the railway station and a petrol tank were hit. All our ammunition blew up in one of the halls. The fires lasted about two hours, but strangely the gasoline didn't burn."'[33] The captured Spitfire pilot's letter concluded that 'We are now being moved to Hornchurch, where the commander does not allow us to attack the enemy bombers, saying we are intended to protect our own light bombers.'

The Luftwaffe, then, was particularly keen to spot wobbles in morale among the British to counter the growing pressure

---

* Although Eastchurch was predominantly used as an RAF Coastal Command station during the war, a handful of Spitfire squadrons sometimes operated from it during the Battle of Britain owing to its important location within Fighter Command's No 11 Group.

from the German public for a final victory. As the Luftwaffe bomber pilot Werner Baumbach wrote in his war diary during the *Luftschlacht um England*: 'If someone achieves something tremendous, the large crowd constantly expects something else tremendous from him and is perhaps a little disappointed when reality, when the given facts, to a certain extent, prevent this tremendous achievement.'[34] His mounting frustration at the impatience of the Reich's citizens for victory is palpable among his pages:

> Hasn't it often been the case in the past few months that some Germans at home have asked: 'Why hasn't our air force yet destroyed England, the arch-enemy on the island?' To anyone who asks that question, we as airmen who are on duty every day answer: 'We know that England is the hardest nut to crack in this war. We at the front know that the final decision against England can only be made through planned cooperation between all parts of the Wehrmacht, through the uncompromising implementation of the elementary principle of combining forces on strategic focuses. These focal points are not necessarily identical to the opponent's strongest points.'[35]

If the German public was getting antsy about the prolonged duration of the *Luftschlacht um England*, it was nothing compared to the twitchiness of the Luftwaffe aircrews. Despite the upswing in late August 1940, they were no longer able to hide their exhaustion: the chronic strain of the campaign became too much for some. The normally docile Hinnerk Waller, a Bf 109 fighter pilot, struck fear into his colleagues when he snapped one day. Running towards their tent, he seized his pistol and hurried off into the nearby woods – apparently with the grim intention of killing himself.[36]

'Some of us rose to go after him,' recalled *Oberleutnant* Ulrich Steinhilper, adding that *Staffelkapitän* Helmut Kühle 'told us to sit down. He was a shrewd judge of character, and as we all sat waiting for what we thought was the inevitable shot, nothing happened.' It turned out that Waller had also been suffering from stomach complaints on top of his psychological distress. Despite the intense episode, he was deemed fully fit to fly again four days later. Steinhilper explained that:

> What we were seeing, although we didn't realize it at the time, was our first case of '*Kanalkrankheit*' ['Channel Sickness']: a combination of chronic stress and acute fatigue. At first there were isolated cases but, as the battle dragged on, there were to be more and more cases of this evil disease. The symptoms were many and various, but usually surfaced as stomach cramps and vomiting, loss of appetite and consequently loss of weight, and acute irritability. Typically, the patient's consumption of alcohol and cigarettes would increase, and he would show more and more signs of exhaustion.[37]

Galland confirmed that 'Many pilots were scared of the water – the small, dirty water.' He added that the men agreed 'If there were a bridge, it would be better,' and 'All Channel Flyers felt this. I emphasize: the "Shite Kanal"* ['Shit Channel'] did as much damage to our morale as the English fighters.'[38] The former *Kommodore* of the fighter wing JG 3, *Oberstleutnant* Carl Viek (or Vieck), said that 'The German system was absolutely wrong – no rest for pilots.'[39]

Viek added that 'I tried to help out by keeping them on the ground in bad weather; if possible, rest a *Staffel* for a day, sent them for a swim, etc. Physically, the strain on pilots wasn't so bad – it was more psychological.' Given that the men were allocated just two weeks of leave a year as standard, they often relied on their commanding officers to grant them additional leave, which the Luftwaffe medical corps claimed was a rather

---

* This source has been translated into a mixture of German and English, so it is not quite accurate in either language – but it certainly conveys the sentiment!

frequent practice.⁴⁰ Because he tried to protect his men's well-being, Viek was branded 'soft' by the Luftwaffe HQ, illustrating the lack of sympathy that the impatient generals were starting to show to its forces.⁴¹ Dowding, on the other hand, tried to give his Fighter Command squadrons at least one day off a week, though this was often abandoned whenever the campaign became too frantic.⁴²

Yet, away from the upper echelons of the Luftwaffe, German airmen who were suffering from operational exhaustion were often afforded a sympathetic ear by its medical corps. Admittedly, some of this compassionate attitude was to keep the German air force as effective as possible: the typical practice was to 'replace a whole *Gruppe* when it is felt that it has been operational so long that its efficiency is impaired'.⁴³ Contrary to popular myth, however, the Luftwaffe was not entirely a 'fly until you die' organization. Once a German airman started displaying symptoms of being '*abgeflogen*' – having 'taken off' mentally – he was sent to a dedicated Luftwaffe '*Erholungsheim*' ('convalescent home').

Being '*abgeflogen*' was a German term for war neurosis which had its roots in the belief that the physical sensations of flying, such as the vibrations of the aircraft and exposure to anoxia, 'have a specific effect, particularly on the autonomic nervous system'.⁴⁴ As German physicians initially believed that 'a physical basis for these disorders was present, though not yet demonstrated', this is possibly why Luftwaffe psychologists were initially more understanding than their British counterparts, because they thought that the mental distress had a tangible, physical quality that was easier to comprehend.⁴⁵ It is estimated that only 40 per cent of Luftwaffe convalescent home residents with a diagnosis of war neurosis ever returned to flying duties.⁴⁶

Even the rare airmen who broke down after only one or two sorties were sent to a home for rest and recuperation, with more severe mental breakdowns leading to a stay of up to a year or so for full recovery.⁴⁷ Such statistics suggest that the Luftwaffe medical corps genuinely cared for the mental well-being of its aircrews, most likely spearheaded by the organization's early adoption of dedicated aviation psychologists in the inter-war period. Their convalescent homes, as an American interrogation report noted after the war, were 'deliberately set up so as not to emphasize the medical nature':

> Doctors are available, but in the background, and hospital features such as nurses in white, bare walls, etc. are carefully avoided. These are generally very luxurious establishments. The length of the stay at these rest homes varies from a minimum of one month to three months but the official Luftwaffe attitude is to take a lenient view towards the length of stay so that most of the stays are longer. Flyers who recovered underwent additional tests to prove their suitability, while those who did not were either re-posted to a Luftwaffe ground service role or into a munitions factory if they were not psychologically suited for the task.[48]

After the war, prisoners of war from the Luftwaffe medical services testified that 'During the Battle of Britain, overwork was the rule and fatigue was serious'.[49] The leading cause of mental distress in the Luftwaffe was '… repeated and prolonged standing-by, particularly in fighters. Neurosis was a constant source of worry in fighters, but not bombers.'[50] It is important to remember, then, that the campaign remained gruelling and arduous for the Luftwaffe, even during its most successful phase. Indeed, despite the general upsurge in the later Eagle Attack between 24 August and 6 September 1940, the Luftwaffe lost 380 aircraft to the RAF's 290.[51]

Overall, the general downturn in the psychological well-being of Luftwaffe personnel during the Battle of Britain demonstrated just how challenging the campaign was for the German airmen. Then, as summer wilted into autumn, the course of the *Luftschlacht um England* would be irrevocably altered by an unexpected twist and a hubristic fumble.

On the night of 24–25 August 1940, the Luftwaffe unintentionally dumped bombs over the London residential boroughs of East Ham and Bethnal Green. With Nazi Germany appearing to have raised the stakes of the air war, Churchill now considered the heart of Berlin to be fair game for Bomber Command. The following evening, Handley Page Hampdens and Vickers Wellingtons went sent to pound the German capital.

The dense clouds that shrouded Berlin largely shielded it from any major damage; little was hit of the city itself.[52] But not all Berliners nonchalantly brushed off the British air raids. This was, of course, a German public that had not yet seen the destruction of Hamburg or Dresden: all they saw now was that British bombers were roaming across their capital on an unprecedented level. A day after the retaliatory blow, one resident of the district of Zehlendorf in Berlin wrote of how:

> This time, they really explored Berlin. You could hear the planes whizzing, the whole sky was as bright as day, and each time you had the feeling that everything was over. The alarm lasted from midnight to 3:15 a.m., or three hours. During this time there was continuous shooting. Today everyone was still agitated from the night before. As I said, it's impossible to describe how it makes you feel. I don't yet know what damage has been done.[53]

Another Berliner, Kirsten Eckermann, described the thunderstruck reaction of the local population in the following days. 'Göring visited the areas where the bombs had fallen,' she recalled, 'and people gathered around his car to listen to what he had to say to them. We really did believe that Berlin could not be bombed, and now that it had happened, we began to feel insecure and afraid.'[54] Göring 'made all kinds of stupid excuses and gestures, and I think people believed in what he told them. We did not believe at that moment that the RAF would be allowed to bomb our city again, but they did.'[55] In some cases, German civilians soon became fatalistic regarding their chances of surviving an air raid.

On 26 August 1940, a woman living in Emmendingen near the Black Forest recalled how 'We had barely been in bed for an hour or two on Sunday when the damn air raid alarm came. We had no choice but to get up and go to the basement. After a while we went back upstairs and lay down in bed until we suddenly heard bombs falling. We immediately jumped out of bed to check on the planes. Soon after, the humming stopped, and we went back to our beds. Yesterday, Saturday to Sunday, the alarm went off again, but this time I didn't get up...'[56]

A few days later, a resident of the district of Wilmersdorf in Berlin outlined their mounting anxiety about when the RAF's bombers would darken their doorsteps once more:

> We will probably have visitors every day, or rather every night. Hildegard has already described Monday to you, she said it here. We were in the basement for a long time, but since there were always longer breaks in the shooting, we also went upstairs during a break; I then immediately went back to bed because I didn't feel very well, but Haki had to stay up until the all-clear came, because it started shooting again when we were upstairs...[57]

It was not just the drone of engines and falling bombs, but also the rattling of guns from both sides which severely disrupted their sleep and kept them rooted to the spot until the immediate danger abated. One German schoolboy, who would later serve as a teenage Luftwaffe anti-aircraft assistant during the war, was twelve years old during the *Luftschlacht um England*. He recalled how:

> Air raid alarms were already not uncommon in Berlin in 1940. British aircraft, mostly reconnaissance aircraft, repeatedly invaded German airspace. They flew very high. There

was anti-aircraft fire, but that was all. This changed when the 'Battle of Britain' began in August 1940, which lasted until May 1941, after the Western campaign had ended so quickly and successfully.[58]

National attention swivelled towards Germany's major cities, with a resident of Göttingen writing sympathetically on 1 September 1940 how 'We don't have much to endure at all. What are the people in Hanover or in the industrial area supposed to say who have to go to the basement week after week. At the moment they have their sights set on Berlin, where they have been for three nights in a row...'[59] Such was the concern for the capital that false rumours of British plans to use poison gas began to hysterically spread. The resident of Zehlendorf in Berlin wrote anxiously on 31 August 1940:

> You must have heard about the air raids over Berlin on the radio. This week we were lucky to be able to sleep two nights without the Tommies bothering us. From Sunday to Monday, three hours; Monday to Tuesday, thirty-five minutes; Wednesday – Thursday, three hours; Friday – Saturday, three hours in the basement [...] A lot has already been broken. You can't find out more details. Everything is always immediately blocked off. An alarm was raised last night threatening that they would soon be coming with gas. Last night it was crazy again...[60]

The ongoing British air raids stoked paranoia among the Berliners, who started to believe that no corner of the Reich was safe. On 1 September 1940, the resident from Wilmersdorf wrote concernedly to his contact in Salzburg, Austria: 'We have mixed feelings about the reconnaissance that now often begins in the evening: it most likely means nocturnal visitors. It's unpleasant for my dear wife, because she often has sciatica pain and has to get out of bed. At the moment we have real coffee to cheer you up for the day's work! Where is your air raid shelter? I hope you don't have to go in, but the damn English people want to visit Salzburg too!'[61]

Thus, although the later Eagle Attack had been firmly turning in the Luftwaffe's favour, it would be inaccurate to assert that Bomber Command had zero impact on German morale during this time. Indeed, the increasing number of British air raids was considerably impacting the psychological well-being of Luftwaffe personnel stationed away from the cities. Torn between doing their duty on the front line and the visceral need to protect their loved ones back home, their distress is perhaps most evocatively captured in a letter from a Luftwaffe air signalman stationed with Air Signals Regiment 35. On 30 August 1940, he wrote a letter to his wife from France:

> Well, what is particularly important to me? You, I'm so worried about you. There are now so many bomb attacks on Berlin. Only yesterday about 150 bombs dropped on residential areas... there were quite a few fires. You never find out which parts of the city are affected. But I don't have a moment of peace here anymore. They're bombing in Berlin and we can't catch the dogs. Anyway, I'm telling you, if our house were to be hit, every house in England that I can get my hands on will burn.[62]

This letter is particularly significant in demonstrating the air signalman's belief that Operation Sealion would soon be carried out. Yet, the Luftwaffe air signalman also implored his wife to be vigilant against the British ordnance sloughing down onto the capital in response: 'If there's an air raid alert, or if you notice enemy planes in any other way, immediately disappear into the nearest air raid shelter. Yes, don't stay on the street or in the apartment and be curious. All of the Berliners who were killed and injured were not in the air raid shelter. Also, don't touch a dud or loiter near one. The ones with timers are the most dangerous. You must always keep me informed of which parts of the city are affected.'[63]

The air signalman signed off his letter in exasperation: 'If only we could get these guys. But nothing happens here, and yet all hell is breaking loose at home.' Yet, by the early autumn of 1940, the exponential German momentum in the late Eagle Attack appeared to suggest that a seismic change was finally on the horizon. As one worker in the *Reichsarbeitsdienst* ('Reich Labour Service')* Battalion 8/173 – which was supporting Luftwaffe units during the middle of the Eagle Attack – wrote animatedly on 2 September 1940: 'You back home can be really proud of us. Because it is precisely the Luftwaffe that is waging war. We give all our strength to the arm. You hear about the air battles over the Channel in the Wehrmacht report. Again and again, German commitment and success...'[64]

So, it seemed that there was a common feeling in the Reich that it was high time for a conclusive decision to be forced. With the Luftwaffe's Eagle Attack seemingly throttling the life out of the RAF, the time now seemed ripe to go in for the final kill.

---

* The *Reichsarbeitsdienst* ('Reich Labour Service', RAD) acted as a civilian labour auxiliary that facilitated the Wehrmacht's operational deployment. Founded shortly after the unveiling of the Wehrmacht in mid-1935, its military association was evident. Some of the young men conscripted to the RAD between the ages of eighteen and twenty-five later found themselves in the Luftwaffe, especially those whose construction units were transferred to the German air force in 1939.

13

You *Dummkopf!*

As Berlin's dead bodies piled higher under British bombs, Helena Vogel decided that she had had enough. During the RAF's early raids on the capital, she heard whispers that the Luftwaffe's commander-in-chief himself was coming to their district to inspect the fallout from the latest Bomber Command air raid. 'I wanted to confront this man and ask him questions,' she confided angrily.[1] To her irritation, she found that the swelling crowd blocked her from the *Reichsmarschall*, who glided through the throng with smiles, handshakes and reassurances that everything was going to be alright. Soon, however, Göring was struggling to contend with the volatile crowd, mainly consisting of elderly people and frazzled mothers with their fractious children.

Vogel recalled that 'They were clearly scared, and some were babbling and starting to become emotional. I could see that Göring was not comfortable with this situation, the look on his face said it all.' Failing to assuage their concerns, he announced that he needed to depart immediately for another engagement, but that he would return to talk again soon about the brand-new Reich air defence measures designed to tackle the RAF bombing raids. Already simmering with resentment, Vogel was rendered incandescent with rage by Göring's cowardly exit:

> As his driver began to drive away, anger overtook my common sense and I shouted, 'You fucking idiot, you *Dummkopf!*' ['fat head'] as loud as I could. No sooner had I shouted, two men in the crowd grabbed me from behind and dragged me out of the crowd. I shouted at them, 'Let go of me, you are hurting me.' I was taken to the police station and the two men took me into a locked room and started to ask me questions. I told them I had recently lost my man who had been serving in the Luftwaffe, and that I was angry with Göring over this and wanted to know why this was happening.[2]

Upon interrogation, things went from bad to worse for Vogel: the two men informed her that they were part of the Gestapo, the Nazi Secret State Police. 'You are to cause no further trouble… or else!' they warned her sternly. 'Insulting such a high officer in the Reich is treason.' Eventually, after cautioning her that there would be no further conversations or warnings if it happened again, they let Vogel go. Her impassioned outburst at Göring demonstrated her uncontainable grief as a Luftwaffe widow – but it also highlighted that the Nazi leadership could not afford for the *Luftschlacht um England* to become a *Luftschlacht um Deutschland*.

Even in a totalitarian regime, where one's tongue often had to be firmly held, visceral human reactions of grief, pain, frustration and devastation at the British air raids could not always be suppressed by the local population. Indeed, the renewed British air raids over the Reich were of sufficient concern by early September that Hitler was prompted to address them publicly. He did this during a speech for the opening of the annual 'Winter Relief of the German People' charitable drive in Berlin on 4 September 1940. Mentioning that 'It is something wonderful to see our people here in the war, in all their discipline', he added:

We are experiencing this right now at a time when Mr Churchill is showing us his invention of night air raids. He doesn't do it because these air raids are particularly effective, but because his air force cannot cross German land during the day. While the German pilots and German planes are over English soil every day, an Englishman cannot cross the North Sea at all in daylight. So, they come at night and, as they know, throw their bombs randomly and without a plan on civilian residential areas, on farms and villages. Wherever they see any light, a bomb will be dropped on it.[3]

Hitler's speech attempted to stoke continuing public anger at the British 'air pirates' seemingly dumping their bombs indiscriminately across the Reich during the Battle of Britain. It also displayed how he felt compelled to present himself as a public source of comfort amid the mounting attacks on the capital. Most notably, Hitler vowed in the speech that 'If they declare that they will attack our cities on a massive scale – we will wipe out their cities!'[4] Thus, the blueprints for the Blitz appeared to have been publicly drawn up and justified by the *Führer* himself.

In reality, though, Hitler merely viewed any potential intensification of the air war over Britain as a means of laying the foundations for Operation Sealion.[5] The OKW remained ambivalent over the feasibility of the invasion by the opening of autumn. A memorandum initialled by Keitel and Jodl determined on 3 September 1940 that the earliest possible date for the transport fleets to commence Sealion would be 20 September, with the commencement of '*S-Tag*' ('Sealion Day') and the subsequent '*S-Zeit*' ('S-Time') slated for 21 September.[6] A follow-up comment was added, however, that 'All preparations must remain liable to cancellation twenty-four hours before "zero hour".'[7]

Admittedly, this had been a standard procedure in other campaigns: Case Yellow, the invasion of France and the Low

Countries, had been postponed around twenty times.* Yet, for the OKW to still need to be able to slam the breaks on Sealion by the early autumn of 1940 demonstrated that their confidence in the undertaking had not been bolstered by the Luftwaffe's upswing during the Eagle Attack of late August. Likely irritated by the vacillating OKW dialogue regarding Sealion, Göring became enticed by the prospect of unleashing an all-out assault against London that he believed would finally bring the British to the negotiating table.

On 5 September 1940, a resident of Wotrum in Mecklenburg, north-east Germany, asked despondently, 'Will there be anything with England this year? I have soon given up all hope...'[8] Just two days later, Göring launched his first thunderous attack on London – 348 bombers, deftly escorted by 617 Bf 109s and Bf 110s[9] – which, in his own words, 'delivered a stroke right into the enemy's heart'.[10] In addition, Southampton, Portland, Brighton, Eastbourne, Canterbury, Yarmouth and Norwich were also attacked. Accompanied by over 1,000 fighter sorties, the first day of the Blitz killed 448 civilians and resulted in 1,337 casualties.[11] Thus, the newest and most protracted stage of the *Luftschlacht um England* – the Blitz – began.

It is often said that Göring's decision to switch to the Blitz was somewhat hare-brained and impulsive: a monumental error which determined the outcome of the Battle of Britain by giving the battered Fighter Command space to recuperate and hit back harder than ever. While this is true to some extent, it can be argued that Göring's tactical pivot was perhaps not entirely the illogical move that some historians claim. As Vogel's grieving tirade demonstrates, the *Reichsmarschall* was under mounting

---

* Richard Carswell claims that *Fall Gelb* was postponed seventeen times in R. Carswell, *The Fall of France in the Second World War: History and Memory* (Cham: Palgrave Macmillan, 2019), 51. Ryan K. Noppen, meanwhile, asserts that it was postponed twenty times between 12 November 1939 and 10 May 1940 in R. K. Noppen, *Holland 1940: The Luftwaffe's first setback in the West* (Bloomsbury, 2021) [Kindle Edition], 5.

pressure to retaliate against the increasingly frequent Bomber Command air raids.

To argue that the beginning of the Blitz was solely a consequence of the *Reichsmarschall*'s puerile temper, then, does not give the full picture. Nevertheless, it should be acknowledged that Göring's compulsive need to deflect blame from himself in any situation drove his particularly heavy-handed approach towards subduing the British for good. In addition, he knew that successfully retaliating against the RAF air raids would supply him with the public adoration that he ravenously craved. As a woman living in Hamburg exclaimed on 9 September 1940:

> In Winterhude, Barmbeck, Einsbüttel, Reichshof, Volkssorge and the harbour, residential buildings, streets, etc. were bombed. It was terrible! If you think about what it must be like in England, they have to sit in the cellar for ten hours. So, we slept until 10 a.m. today and were tired all day. I'm sending you the two reports – well, our *Reichsmarschall* will make sure that the bombardment doesn't go unpunished.[12]

Among the German public, then, the Blitz did not immediately provoke concern – and, perhaps more surprisingly, the same can also be said for certain sections of the Luftwaffe. Göring famously paid little attention to Deichmann's concerns that the escorting Bf 109s would be operating at the very limit of their range to facilitate the bombing of London, while Kesselring favoured attacking Gibraltar to strengthen the Luftwaffe's hand in seeking a British surrender.[13]

Even the over-optimistic Schmid had warned the *Reichsmarschall* that 'the general areas around London'[14] would be the most fiercely protected, although he claimed the British could not offer much resistance beyond the capital. Yet, many Luftwaffe servicemen among the lower ranks of the organization welcomed the newest escalation of the air war over Britain. One *Soldat* stationed with the Luftwaffe Station Command 16/XI wrote confidently the day after Göring's concentrated blow against London that:

> I have a lot of work these days. There is no longer any 'closing time' at all. Unfortunately, I cannot write to you about our special tasks. But you will have already heard on the radio in detail about the extensive destruction in England and especially in London caused by our bombers. We'll soften up the English. Our planes all returned from combat missions yesterday...[15]

A *Soldat* serving in an aircraft repair unit for the bomber wing KG 51 similarly documented on 9 September 1940: 'The last few days we have always been busy. It wasn't until late that work ended and the machines had to be prepared again for the evening. The English will probably not be happy about the free delivery of metal, which is now causing them a lot of clearing work.'[16]

The *Soldat* was now hopeful that 'they will destroy the whole gang in London soon so that there can finally be peace'. His perspective demonstrated that other Luftwaffe personnel aligned with Göring's view that the Blitz alone could potentially be sufficient to knock the British out of the war. But, for other Luftwaffe personnel who had long witnessed earlier German preparations for the amphibious landing with their own eyes, it was strongly believed that the renewed efforts over Britain in the Blitz finally heralded the beginnings of Sealion. As early as the fall of France, Bob recalled that:

> By the time the town of Dunkirk was under German control, we flew exploration and free hunts over that area, the Channel and the British southern coast. The British fighters were rarely seen. At that time, we still had a considerable number of fighter planes which ensured a respectable air superiority. On the occasion of such flights, I discovered around August in French harbours and along the coasts smaller boats and barges of inland invasion. Guesses and rumours developed into the firm knowledge

that a landing in England was planned. At that point of time, we considered an invasion of England possible because the British armies had become totally disbanded, and there were also only a few British fighter planes in the sky.[17]

Throughout the Battle of Britain, then, the growing armada on the French side of the English Channel gave further credence to the idea that Sealion would shortly commence, spearheaded by the Luftwaffe. Some of the German airmen in British captivity had even been overheard discussing aerial simulation exercises for Sealion that had taken place in September 1940. One prisoner of war from the bomber wing KG 1, for instance, 'said that the invasion of this country had been timed to take place on 20 September along the coastline between Worthing and Dungeness'.[18] He had further noted that 'The advancing boats would have been covered by large numbers of aircraft laying smoke screens.'

Nevertheless, 'On a trial during misty weather, over a short course, owing to bad seamanship, an operation time to take place in ten minutes in fact lasted half an hour.' Another Luftwaffe prisoner from the bomber wing KG 2 had spoken of carrying out practice runs '... with an He 111 carrying 32 v 50-kg. Smoke Bombs. These are dropped in a long stick and produce a smokescreen one mile long which will last for thirty minutes.'[19] This illustrates, then, how the Luftwaffe prisoners who still believed in an invasion were not just hopeless optimists yearning to escape: they had actually borne witness to the preparations for Sealion.

The Luftwaffe prisoners' purported timelines for Sealion did indeed correlate with some of the OKW's initial plans for the operation's commencement. A memorandum initialled by Keitel and Jodl on 3 September 1940 had stated that the earliest possible date for the transport fleets to initiate Sealion would be 20 September, with the actual landing taking place the next day.[20] Moreover, the use of smokescreens to open Sealion was confirmed by discussions between the Luftwaffe and the *Heer* in early September 1940, with the commander of Army Group A asking the German air force on 9 September 1940 for further joint exercises.[21]

A follow-up report stated that 'A smokescreen whose depth and density is sufficient to support a landing group and grant protection against observation from the coast can only be achieved by aircraft spraying or dropping smoke bombs [...] the smoke is sprayed by reconnaissance aircraft,' while 'The smoke bombs are dropped by fighter aircraft.'[22] The growing interconnectivity of the Wehrmacht could also be seen in the report's conclusion that 'The use of all smoke products of the army, Luftwaffe and navy intended for fogging a coastal section requires uniform regulations. It must be brought into line with the requirements of the *Kriegsmarine*.'[23]

Such in-depth planning for the operation by the *Heer* and *Kriegsmarine* shows just how thoroughly the Germans continued to look into Sealion during this time. By mid-September 1940, the German navy had completed its necessary preparations for the landing after converting nearly 2,000 barges into landing craft.[24] In total, 168 freighters, 1,975 barges, 100 coasters, 420 tugs and 1,600 motorboats had been assembled by 19 September 1940.[25] The plans also demonstrated the relentless pressure that was being placed upon the Luftwaffe to finally dominate Britain's skies and enable the rest of the Wehrmacht to successfully play their parts in the operation.

One of the inter-Wehrmacht simulations for Sealion which took place in early September 1940 threw up a few questions about the most effective integration of the German army, navy and air force during the crossing of the English Channel. These questions were documented in an AOK 16 report: 'Continuous close cooperation with the Luftwaffe regarding fighter and anti-aircraft protection during embarkation, crossing and landing on the coast is required. Requests in this regard must be reported to the army.'[26] The level of meticulous planning included the

German High Command launching an exhaustive survey into the capabilities of British guerrilla warfare on 7 September 1940.

Shared with the Luftwaffe, this army report noted that 'For guerrilla warfare, there are two special manuals in England that were captured during the fighting in Flanders ("*The Art of Guerrilla Warfare*" and "*Partisan Leader's Handbook*"), in which this combat activity is described in great detail. It is not clear whether these are official regulations.'[27] The report explained that a British '... organisation is planned that works in the form of gangs of eight to twenty men (possibly also in larger units) under a leader, or has acts of sabotage carried out by individuals'.[28]

Another example of the need for different branches of the Wehrmacht to work together and suppress any British resistance can be seen in an AOK 9 report outlining that, 'After a further thorough examination of the conditions on the enemy coast, the *Heer* requests that paratrooper units be made available to it – both to support the capture of Brighton and of the cliffs at Beachy Head.'[29] An overview of the preparations was provided in a supplementary army General Command XIII situation report:

> According to the information available, it is likely that the enemy will resist stubbornly and use all available means. Assuming absolute air superiority is achieved, the enemy is unlikely to be able to prevent a landing given the weakness of his land forces [...] it is also likely that the prepared enemy counter-measures may arrive late or be late due to the influence of our own Luftwaffe.[30]

That the *Heer*'s success in Sealion would hinge directly on the military performance of both the Luftwaffe's flying and paratrooper arms, then, was uttered in no uncertain terms by their army colleagues. By mid-September 1940, the *Heer* was coordinating with Air Fleet 2 and the Luftwaffe liaison officer of AOK 16 on providing aerial reconnaissance of British civilian streets and railways lines not just in London, but also in the southern English towns and cities of Maidstone, Ashford, Chatham, Canterbury, Dover, Sevenoaks, Tonbridge, Croydon, Uxbridge, Maidenhead and Windsor.[31]

The 45th Division of Infantry Regiment 134, meanwhile, decreed that 'From 9 September 1940, training courses will take place in Rotterdam as part of the training for *Unternehmen "Seelöwe"*: 1st course, 13–15 September; 2nd course, 17–19 September; 3rd course, 21–23 September.'[32] Concurrently, the *Kriegsmarine* had been tasked with shiploading, mine-laying and removing 'the previous visual signs of the war vessels (red painting of the turrets and protective shields, swastika on the fore and aft of the ship)' prior to the invasion.[33]

Thus, it was not just the Luftwaffe's war that Göring was potentially affecting by switching to the Blitz: the entire German armed forces would also be messed around if this newest phase of the *Luftschlacht um England* stalled. Unfortunately for the *Reichsmarschall*, though, the damaging ramifications of his tactical change quickly started to materialize. On 9 September 1940, the Junkers 88 bomber pilot Peter W. Stahl walked into the *Staffel* lounge for a briefing on his first ever sortie over Britain. His eyes were immediately drawn to a large map of London hanging on the wall, upon which the *Staffelkapitän* Erich Stoffregen began pointing out the areas to be attacked. But, upon noticing all of the puzzled faces peering back at him:

> Stoffregen did something that the superior of a task force never does, at least in the German Wehrmacht: he justified this new tactic with a long explanation. There were three arguments for this: 'The news says that war morale on the island has reached zero, so that all it takes is a few hard knockout blows to end the war. The British, for their part, carry out ongoing and planned terrorist attacks against the civilian population of German cities. In a modern war, the boundary between military objectives and non-military objectives can hardly be defined. The workforce

of an industrial plant is also essentially waging war against us through their work.'[34]

This briefing demonstrates, then, that Göring was instructing his commanders to present the Blitz as the final decisive blow that would vanquish the British. As the Luftwaffe airmen took off for the raid and crossed over the treacherous English Channel, Stahl was among the German bombers flying in formation: 'The fighter escort flies in a zigzag course next to, above and below us. Flak greets us quite well on the English coast. For the first time during the war, I fly over English soil. The visibility is good. We observe moving trains on the ground. This must be London, because you can see columns of black smoke reaching up to 5,000 metres above sea level.'

The next part of Stahl's testimony illustrates, however, that you can be a technically excellent and experienced pilot; being an effective combat operations pilot, on the other hand, is a whole other ball game. 'We quickly reach the outer flak belt around London,' he wrote tensely, adding 'The British shoot uncomfortably well. The whole formation becomes uneasy. It is hardly possible to hold your position anymore. First of all, I actually have to concentrate all my attention on flying to avoid colliding with another plane.' Stahl became increasingly unnerved by the disorientating experience of an early Blitz raid:

> The whole thing is completely new to me, so I have no idea what the best tactics should be. So, I just stay in the middle of the big pile. The flak here is really unpleasant. There is terrible lightning all around and above and below. I'm surprised that all the planes apparently continue to fly unmolested. At least the anti-aircraft fire seems to me to be an indication that there are no enemy fighters there.[35]

Trying to keep his mind firmly on the target, Stahl observed how 'The large formation now moves across the city with uncanny uniformity. The first bombs fall in front. I also press the red button. The aircraft makes its juddering release. We look down. The bends of the Thames, the docks, the entire huge city lies beneath us like a map.' He appraised the raging fires below with mixed emotions: 'Then comes the impact of our bombs, which we observe as we turn in a wide arc from the east to the south. It must be cruel down there.'

Stahl then spotted the '… fires in many places from previous attacks. The effect of our attack is a cloud of smoke and dust that shoots up from the target area in a wide stripe.' The Luftwaffe bombers had opened up portals to hell all across London. A brief flicker of hope leapt into his chest because 'It is inconceivable that a city or a people can withstand this gruelling constant strain for long!' Soon enough, however, the Ju 88 pilot got his ultimate baptism of fire over the heart of London, when he noticed fighter aircraft flickering in the corner of his eye.

'At first,' he recounted, 'I think they are our escort and wonder about their tactic of moving so unreasonably between the fighter planes until I realize that they are English. Damn.' Then, he heard the terrifying call – 'Hey, just open your eyes, they're Tommies!' – and suddenly, 'The peace and the feeling of security are over. I see tracer fire everywhere, whizzing through the air. A wild fight broke out between our 109s and the Spitfires and Hurricanes.'

Surrounded by the deafening staccato sound of gunfire, Stahl recalled that 'Everything is happening right among us. I hide in the thickest crowd and urge my crew to keep me constantly and precisely informed about everything that is happening around us. I myself am very absorbed in the flying. Hein [the radio operator] shoots with the machine-gun, and at first it frightens me.' Despite his operational naivety, Stahl managed to keep it together and get the Ju 88 crew safely back towards the French coast. He took his 'calming piece of bread', which he kept stored in the pocket by his knee, and briefly savoured the comforting spring between his teeth – but the danger was not over yet:

> Theo [the air gunner] taps me on the shoulder from behind and draws my attention to a wide streak of oil that runs across the wing behind the right engine. I reach for the switch to check the oil level and see that we have already lost 80 litres of oil. We still have 10 litres in the tank. This means that I have to stop the engine immediately if I want to avoid it breaking down. I let the engine run for a few more minutes and realized that there was really nothing left but to continue flying with a single engine.[36]

Eventually, the bombardier Hans nudged Stahl in the ribs and pointed to an airfield below. They made an emergency landing in Beauvais, France, and not at their base in Belgium – but, for now, they were safe.

These early Blitz raids were soon trumpeted in Nazi propaganda, but Baumbach noted that 'The German propaganda success reports, which were intended to cover up the failure of the Battle of Britain, had caused great bitterness among the Luftwaffe units deployed against England.'[37] In fact, members of his bomber wing, KG 30, were so furious about the airbrushing of the campaign that they complained directly to Milch about it when he came to visit the Luftwaffe bomber units in Holland during the Blitz. They also criticized this new phase of the *Luftschlacht um England* straight to his face:

> The *Staffelkapitäne* who had flown the attacks with their crews gave the *Feldmarschall* their unvarnished opinion: that it was not possible to carry out the ordered tasks with the existing aircraft, targeting devices and attack means, and that the English fighters were becoming superior to German bombers in the exact same way as the German fighters were to the British bombers; that a night bombing war against England cannot have a decisive effect on the war, since there are no night targeting devices at all, and that the combat strength of the bomber units is not sufficient.[38]

Rather than systemically attacking the RAF's airfields, radar installations, and related infrastructure across the full swathe of south-east England, as had been carried out during the main Eagle Attack, the *Kampfflieger* had now been forced to change tack with barely a moment's notice. The *Jagdflieger*, too, were highly frustrated by the *Reichsmarschall*'s unforeseen switch to attacking Britain's cities and civilians. Although they managed the occasional fighter sweep over Surrey, Kent and Sussex amid the early Blitz, they were now largely tied to escorting the bombers during the concentrated daylight raids on London.[39]

Flying over the British capital squeezed every last drop out of the Bf 109's fuel tanks, giving the escorting fighters a reserve of just ten minutes' flying time over the capital.[40]

At the same time, Göring had declared that 'We have no chance of destroying the English fighters on the ground. We must force their last reserves... into combat in the air.'[41] The *Jagdflieger*, then, were not only tasked with protecting the slower bombers whilst operating at the very end of their range; they were also expected to soon be on the offensive in order to eliminate the last remnants of Fighter Command.

This was rendered even more difficult by Luftwaffe intelligence still being riddled with discrepancies over the true remaining number of British fighters. Kesselring, for instance, asserted that Fighter Command had been all but eradicated, while Sperrle argued that it still had around 1,000 aircraft left.[42] Sperrle's assessment was much closer to the truth: on 6 September, Fighter Command wielded over 750 serviceable fighters and 1,381 pilots, 950 of whom flew Spitfires or Hurricanes.[43]

Even more crucially, the unanticipated switch to attacking British cities gave the RAF the breathing space it needed to recover from the relentless pounding of its airfields in late August. As Air Vice-Marshal Keith Park – the commanding officer of 11 Group – once said of the Blitz, 'It was burning all down the river. It was a horrid sight. But I looked down and said: "Thank God for that", because I knew the Nazis had switched their attack from the fighter stations, thinking they were knocked out. They weren't, but they were pretty groggy.'[44]

Pilot Officer Michael Constable Maxwell noted similarly that 'I never lost my total confidence in victory, but it was a relief when London was bombed. Awful as this was, it was much better than losing fighters on the ground or driving us further back to safer bases.'[45] After Fighter Command patched up its infrastructure, aircraft and pilots, it then began to shoot down two German warplanes for every British fighter lost between 7 and 15 September 1940.[46] It dropped from losing an average of nineteen-and-a-half aircraft a day during the heaviest phase of fighting to an average daily loss of fourteen aircraft.[47]

Despite this more favourable loss ratio, however, the Germans still came day and night to assail the island. The Luftwaffe ground crews worked around the clock to replenish their *Staffel*'s aircraft, with the *Soldat* from the bomber wing KG 51 writing on 13 September 1940 that 'At the moment, however, work is a little more strenuous than usual because we have to load machines until late every evening.'[48] He added determinedly, though, that 'We like to do this in the hope that everything will end soon. The Englishman has already hit quite a lot in Berlin in the last few days and is making trips in every day. The light of life must finally be blown out of him...'

It is not surprising that such men believed that the invasion could still take place. Despite the fact that Sealion was looking more and more infeasible, Keitel and Jodl directed on 14 September 1940 that 'all preparations are to be continued' for Sealion and that 'New orders will be issued on 17 September.'[49] What the RAF desperately needed, then, was to trounce the Luftwaffe so thoroughly that the Nazi leadership would be deterred entirely from invading Britain. Eventually, on 15 September 1940, fate chose to back the British.

Invigorated by their brief respite, Fighter Command reached an unprecedentedly high contact rate with the Luftwaffe that day: 77 per cent of the RAF fighter pilots who were scrambled engaged with the enemy, compared to just over 30 per cent on 15 August and 26 per cent on 30 August 1940.[50] It was later immortalized as 'Battle of Britain Day' by the euphoric British, and Student later wrote of it that:

> This premature operation against the docks district of London on 15 September led to the worst day of the entire air offensive. This day cost us fifty-six planes. The British fighters had only had a short but, especially at that moment, extremely desired respite and were suddenly back.[51]

Perhaps the most evocative recollection of this climactic day came from Bechtle on his own men's losses: 'Six planes went and two came back. That evening, the whole *Gruppe* was nearly demoralized. It was a drunken party, weeping for comrades. The survivors sang: "the steel wings are flashing; the Tommy has given us a hiding."'[52]

On 17 September 1940, the *Briesetal-Bote* newspaper desperately attempted to use the emerging Blitz to distract its readers from the Luftwaffe's recent unfavourable clashes with Fighter Command. In an article entitled 'British Fighter Defence Fails', it expounded upon the fact that not even the British Royal Family was immune from the thunderous reach of the Luftwaffe:

> The Germans sent 350 to 400 aircraft in waves against the capital and the entire south-east of the country. Fierce dogfights had also taken place over Maidstone and Canterbury and over the River Medway. Reuters [a British news agency] then had to admit the failure of British fighter defences and the penetration of German aircraft to London. A fierce battle then developed just above the heart of the English capital. A certain number of bombs fell in an area 'which is described as the poshest district'. According to the Air Ministry, Buckingham Palace was again

hit during the attack by the German formations. The Queen's chambers were damaged.[53]

To the German fighter and bomber crews venturing over Britain in this newest phase of the *Luftschlacht um England*, however, the searing flak guns and roaming night-fighters demonstrated that the island's light of life was far from extinguished. The very same day that the *Briesetal-Bote* article came out, Hitler finally postponed Operation Sealion indefinitely.[54]

After a gruelling two and a half months of the *Luftschlacht um England*, the intensive daylight tussles between Luftwaffe and RAF fighter pilots had now reached their climax.[55] Crestfallen by the disappointing results, the Germans quickly attempted to hiding the flagging Luftwaffe's failures from their international allies. The minutes of a discussion in Rome between the Reich foreign minister Joachim von Ribbentrop and Benito Mussolini on 19 September 1940 highlight the Nazi war leadership's desperate attempts to salvage the campaign's narrative:

> Germany had won air superiority, and was bombing England, and particularly London, by day and by night, while the British were at most sending a few planes over Germany at night, in order to drop bombs there at random. During the day no British plane dared to fly over German territory. Germany, however, was carrying out strong reprisal attacks by day as well as by night. On one occasion the German Luftwaffe had been ready for a large-scale attack as early as August. But this attack had to be called off because of bad weather.[56]

The meeting notes continued that 'A really large-scale attack had not taken place since, because the *Führer* wanted to accept the responsibility for this only when it was certain that such an attack would mean the beginning of England's destruction.' It was added that 'Although Germany had merely carried out reprisal bombings from the air, their results for England had already been extraordinarily serious. With a continuation of these attacks London would be in ruins within a short time.'

The minutes concluded that 'The British armament factories had been seriously affected, and important ports, such as Portland, had been entirely crippled. Moreover, all aerodromes from the South Coast to London had been made unusable.'[57] The Germans, then, were determined to save face and deny the fact that the *Luftschlacht um England* was beginning to spiral out of control. To claim that 'Germany had won air superiority' when the RAF was beginning to shoot down Luftwaffe aircraft at a ratio of 2:1 was a bare-faced lie, while describing the Eagle Attack as being solely 'reprisal bombings' gave the impression that the Luftwaffe had far more left in the tank, despite the fact that it was already running on empty.

Even before the first month of the Blitz was over, Göring was beginning to realize that his mammoth gamble had failed to pay off. Student recalled being invited to visit the *Reichsmarschall* at the Rominten hunting lodge in late September 1940. There, he found that 'Göring himself was feeling disappointed. Not so much about the outcome of his air offensive; on the contrary, he was very proud of his pilots. He was amazed at the great resistance of the English, but then said to me one time: "We have forgotten that the English fight best when their backs are pressed against the wall."'[58]

PART IV

DIVERSION

14

Great Dark Bloodstains

Because Operation Sealion failed to materialize, some historians have given the impression that the idea of invasion was dead in the water by 17 September 1940, never to be revisited. Granted, the German discussions surrounding an invasion of Britain would never reach the same fever-pitch level as they had in the summer. Yet, the idea of Sealion, though increasingly diluted, never fully disappeared for the Germans during the later stages of the *Luftschlacht um England*. Although he had postponed it indefinitely, Hitler was loathe to cancel the invasion outright, which would signal to the British that the aerial campaign had failed and also weaken his standing with General Franco, whose Nationalist Spain could prove useful to the Axis war machine.[1]

At the same time, however, he could not deny the fact that the chances of launching a successful invasion of Britain slimmed with every dimming hour of autumn. Thus, behind the scenes, Sealion bifurcated into two forms during the Blitz. The first was continued, albeit reduced, preparations in the vein of the original invasion, in the eventuality that the operation would be revisited in the spring of 1941. The second was to turn the embers of the original invasion plan into an in-depth deception operation, largely staged in Norway and in sight of the British coastline, with the intention of appearing to the British as if the German invasion could still take place at any time.

Before moving on to considering how the Luftwaffe began to unravel in the Blitz, it is important to fully appreciate the increasing administrative pressure that it was put under. This was caused by the fact that the Nazi war leadership continued to keep a multitude of options on the table. Whether or not Sealion could have been executed successfully, especially by the autumn of 1940, the Wehrmacht continued to plan both a real invasion and a feigned mobilization. Looking at how thoroughly they laid their plans gives a better understanding of how the Nazi war leadership began to overstrain itself heading into 1941. Despite what senior Luftwaffe commanders may have claimed after the war, some of them still entertained the possibility of an invasion well after the official postponement order.

Two weeks after Sealion had been formally adjourned on 17 September 1940, Kesselring and *Oberstleutnant* Walter Loebel, the *Kommodore* of the bomber wing KG 30 'Adler', signed an Air Fleet Command document entitled 'Special Arrangements for the "Sealion" Procedure up to and Including S-Day.'* A detailed breakdown was given for the week leading up to S-Day (the launch of Sealion). For eight days beforehand, there were to be attacks by VIII. Air Corps on British coastal batteries, and an intensification of previous *Kriegsmarine* minelaying by the 9. Air Division. Reflecting an international element, there would also be 'destructive attacks against ports occupied by warships on the east and south coasts, and attacks on moving

---

* Bestand 500, Findbuch 12488. 'Besondere Anordnungen für Ablauf "Seelöwe" bis zum S-Tag einschließlich', Luftflottenkommando 2, Nr. 7245/40, 29 September 1940. 'Ablauf' in German can also mean 'expiration' or 'termination' in English, but given the context of the document outlining the envisioned progress of Sealion, it would appear to refer to its other translation of 'procedure', 'course' or 'process' for carrying out the invasion – especially as the title speaks of carrying this out 'up to and including *S-Tag* ['Sealion Day']'.

warships by 9. Air Division, II. Air Corps, VIII. Air Corps and the Italian Division'.

Next, huge aerial reconnaissance sweeps would clear the operational area for the Air Fleets, until a day before *S-Tag*. Then, there would be the 'beginning of the interruption of the railways and access roads that lead from the area of the English operational reserve north-west and west of London into the combat area, in particular by blocking town entrances and exits and narrow passageways'. Finally, on S-Day itself, there would be a 'landing of the advance parts of VIII. Air Corps and stronger parts of the II. *Flakkorps,* reconnaissance troops for ground organization and a Luftwaffe construction battalion'. Reinforcing the international dimension to Sealion, an Italian division would hold down 'the flanking effect of Dover in cooperation with Artillery Regiment *'Hohmann'* by assaulting British ground organizations in the area'.

Part of the reason that such preparations continued appears to be due to a deliberate miscommunication from Hitler to some of his senior commanders. The confusion surrounding the true status of the invasion could be seen in a post-war testimony from *Generalleutnant** Heinrich Rauch, who was interrogated by the Americans about Sealion at Berchtesgaden on 30 July 1945:

> After the conversations and investigations of the three components of the armed forces, in which the leaders of naval warfare predominated, it was shown to be impossible for all

---

* Rauch had been made a *Generalleutnant* in March 1945, but the American version refers to him as a Major in his report within the Von Rohden Collection at the National Archives (College Park, Maryland). In the German Bundesarchiv version, however, Rauch has his rank of *Generalleutnant* attached to his testimony. It is likely that *'Major'* was put in from American translators to equate Rauch's rank with 'Major General' according to American ranks. This is indicated from the pencil markings on the Bundesarchiv version which state 'Major (Gen. Lt.) Rauch at the top of the report. See RL 2-VI/211, "Bericht über das im Jahre 1940 geplant gewesene Englandunternehmen (Deckname "Seelöwe")". – Bericht von Generalleutnant Rauch, 30 July 1945. Bundesarchiv, Freiburg-im-Breisgau.

conditions to be fulfilled at the same time at that time of the year. The undertaking was therefore given up by Hitler. This fact, however, was made known only to the commanders and chiefs of staff of Army Group 'A' and Air Fleets 2 and 3. All others, even the general staff officers of the operational sections, were not to know. The undertaking was worked upon further under conditions of special secrecy even after the 'Sealion' was long dead.[2]

Given that Kesselring commanded Air Fleet 2 during the Battle of Britain, it seems a mystery as to why he would sign off on Sealion orders after being informed of its postponement. It is possible that the meeting he mentioned had not yet taken place by late September 1940. Alternatively, perhaps Kesselring wanted to maintain the façade – either to more junior officers of the Luftwaffe or *Heer* – that Sealion was still being planned.

As late as 27 November 1940, the Luftwaffe continued to correspond with the *Heer* about different aspects of Sealion, albeit more vaguely. On that day a memorandum was sent by the Air District Staff z.b.V.300 to AOK 16 on Air Fleet 2's role in transport movement and supplies in Operation Sealion. The memorandum claimed that 'The Supreme Commander of the Wehrmacht has ordered preparations for a forced landing in England. Code name: "Sealion". The operation's timing is unknown.'[3] It added that 'All measures must be taken in such a way that the operation can be stopped twenty-four hours before the S-Time.'

The report further noted that the non-flying elements of Air Fleet 2 – such as anti-aircraft crews, air signalmen and airfield construction units – would be transported along with the 16th Army by the *Kriegsmarine*. Most disturbingly, however, the memorandum adds that 'The following vehicles are available: steamers, ferries, SS ferries. The vehicles are used to transport ready-to-fire anti-aircraft guns. The protection of transport movement is carried out by naval forces, air forces and mine defence.'

This demonstrates how closely the Luftwaffe, *Heer* and *Kriegsmarine* were expected to work with the SS units that would be transferred to Britain. Indeed, the memorandum further notes that 'Steamers (as well as *Prähme* ['ferry barges'] and SS ferries) will be loaded with Luftwaffe units and Luftwaffe supplies. Mixed loading together with army troops and army supplies is also to be expected.'

By the summer of 1940, the SS's *Einsatzgruppen* ('Task Forces', often known as mobile killing squads) had long been executing those who the Nazis deemed as 'asocials' within the occupied territories: Jews, members of the Roma, Sinti and Traveller communities, homosexual and transgender people, the so-called 'work shy', Black people, disabled people and political opponents. In addition, between September 1939 and January 1940, 20,000 individuals who were linked to the Polish intelligentsia, religion, arts and culture, as well as former military servicemen, were murdered by the SS in Operation Tannenberg.[4]

Had the SS stormed the beaches of Britain in 1940, between 370,000 and 390,000 Jewish people would have been put at immediate risk of persecution and execution, along with thousands more from the aforementioned communities.[5] Those groups were at the highest risk of SS violence, but as shown by the Wormhoudt massacre in northern France on 28 May 1940 – in which eighty-one British and French PoWs were murdered by the 1st SS *Panzerdivision Leibstandarte SS Adolf Hitler*[6] – any life was considered expendable by the SS.

It is important, then, to not lose sight of what a Luftwaffe victory in the *Luftschlacht um England* would have later meant had Sealion gone ahead.

Even if the invasion had become exceedingly unlikely by the later autumn of 1940, SS and *Heer* forces continued to make plans and conduct exercises for the endeavour. The XIII. Army Corps, for instance, looked further into the possibility of deploying their own special forces commando units in Sealion: the *Lehrregiment* 'Brandenburg', known as the 'Brandenburgers'.[7]

On 23 October 1940, *Hauptmann* (Captain) Menzel of the *Generalkommando XIII* passed on a report that the *Lehr-Rgt. 'Brandenburg' z.b.V. 800* special units would be assigned to the 'England mission' following an impressive showing in recent Sealion exercises:

> The operational order of 30 September 1940 reflects the versatile use of this force. The inspection carried out and the recognition expressed by the 17th and 35th Divisions show that this unit, intended for special tasks, has a high level of military training and stands out for its discipline and soldierly attitude. It was established that the difficult special military tasks set as part of the overall operation were handled correctly and thought through in an exemplary manner during the preparations (exercises and simulation games) [...] the army therefore believes that this special unit will serve the troops well in operations.[8]

This directive illustrates that despite Hitler's postponement orders well over a month beforehand, certain branches of the *Heer* believed that the *Luftschlacht um England* was still going well enough for the Luftwaffe that it was worth earmarking some of its elite units ahead of *Operation Sealion*. Another report on the Brandenburgers' potential deployment said:

> The fortifications on the edge of the hill near Sandgate are to be taken by parachuting in cooperation with gliders. Once these facilities have been removed, the small fighting facilities between Sandgate and Hythe must be put out of action. Connection with the right wing of the 17th Division (west of Hythe) should be sought. Dropping off must also take place following bombing raids. Contact with the VIII. Air Corps should be established for this purpose. Deployment: five groups of parachutists.[9]

This report detailed how 'The Lw. [Luftwaffe] units will generally be deployed until the very end,' and that 'The command authorities to which the individual troop units are subordinate will be asked to have accommodation prepared now in or near the relevant ports, which can be occupied for the last days before the invasion.'

In addition to accommodation provisions for the crossing, medical kit contents for the Luftwaffe units during the crossing were also outlined: 'Every officer and soldier must be provided with the prescribed dressing packs (one large and one small dressing pack each). Missing first aid packs must be requested immediately from the responsible unit. Likewise, every car and truck must carry the prescribed first aid kit.' These medical kits included 'Vasano tablets (two–three per man) to prevent seasickness are to be used before the crossing,' while for any servicemen who were particularly at risk, it was recommended that:

> The entire body should be rubbed with grease (Vaseline) to protect against the cold. Seasickness tablets are to be given out to soldiers on the day of the operation or at the earliest the evening before and are to be worn on the strap attached to the paper bag around the neck and under no circumstances in a coat pocket.[10]

The tablets that were to be administered to crossing Luftwaffe and *Heer* troops included Pervitin – a low-dose form of methamphetamine. By the spring of 1940, 35 million pills of this had been ordered by the *Heer* and Luftwaffe.[11] Although it soon became less popular among the Luftwaffe's airmen as it made them rather jittery when flying, the tablets were still prescribed to the non-flying arms as 'They strengthen cardiac activity and circulation, but at the same time calm you down and prevent premature fatigue, so that their administration increases the body's resistance to the dangers of distress at sea.'

Further provisions included that a *Luftgaustab* ('Air District Staff') England would be formed and that Luftwaffe mail for all units involved in Sealion would be transferred to St Omer.[12]

Clearly, then, the idea of Sealion pervaded the German air force well after the official postponement order. A Keitel-signed order from 12 October 1940, however, demonstrated that there was only a vague notion among the Wehrmacht that the operation could potentially be revisited when the weather cleared up next spring:

> The *Führer* has decided that the preparations for the landing in England are to be maintained from now until the spring only as a political and military means of pressure on England. Should a landing in England be planned again in the spring or early summer of 1941, the required level of readiness will be ordered in a timely manner. The military basis for a later landing must be further improved by then.[13]

The order shows that Hitler held to the vision of maintaining the appearance of an invasion as a means of military and political leverage while he waited out the winter. Given that the British had just been subjected to a months-long invasion threat, maintaining that psychological advantage – especially when the declining weather precluded any significant military activity – could possibly hold the island in suspense until Sealion could be reconsidered in 1941. On the same day as Keitel's order, an Allied Combined Intelligence Committee report mused that:

> The frequent reports from diplomatic sources of the postponing of the invasion are counter-balanced by excellent evidence that preparations for invasion continue. These reports of postponement may well, therefore, have been planted in an effort to reduce our scale of readiness. In any case it is unlikely that the final decision has been taken.[14]

Deception operations were not a new departure for the Wehrmacht. As early as 7 August 1940, the OKW's 'Guidelines for the Deception of the Enemy' had declared that 'Regardless of whether and when we land in England, this pressure on England's people and military forces must remain constant. It is important to create the impression that the main German operation is directed against the English east coast and that a landing in Ireland is also being prepared.'[15]

At the time, this deception was necessary because, in the week before Eagle Day (13 August 1940), the Germans were not prepared for the cross-Channel invasion. The report added that 'The [British] defence on the Channel coast is so strong that only large-scale mock operations are being prepared here, but a serious attempt at crossing over is out of the question.'[16] By the autumn of 1940, as Sealion appeared to be highly improbable, the Nazi war leadership doubled down on the deceptive aspect of the operation. As an OKW report from 12 October 1940 detailed, fake preparations were to be made at the following locations to create a climate of fear among the British: 'Norway and Denmark (landing on the east coast of England); the Netherlands (landing north of the mouth of the Thames River); Brest and Biscay (landing in Ireland).'[17]

The report added that 'After preparations have started, certain areas may be closed to civilian passenger traffic in order to increase the credibility of the preparations. The people in charge of the processing, apart from a circle to be determined by the High Command, must not know that it is a case of deception.' It appears, then, that the *Heer* were left in more ignorance than the Luftwaffe with regard to the invasion's postponement, given that they were not even to be told that the invasion they were preparing for was not real. In an OKW memorandum from 22 October 1940, it was further declared that 'The pressure of a landing on England should continue to be maintained. This should give the impression that the focus of the preparations is increasingly shifting to Norway.'[18]

Fatefully, the OKW believed the bluffing had been entirely successful, with a report dated 22 October 1940 noting that 'The preparations made at great expense on the Channel coast for a secondary (or fictitious) operation have, in the German opinion, fully achieved their purpose, i.e. the attention of the English has been diverted exclusively to the south coast.'[19] This claim notwithstanding, a British Home Intelligence report during the third week of October actually stated that 'There is neither fear nor expectation of invasion.'[20] This correlated with a considerable downsizing of the amassed Sealion forces, with the OKH document of 12 October 1940 ordering that 'The associations designated for the "Sealion" operation are available for planned reorganizations or other uses.'[21]

It is relevant to consider how the Wehrmacht maintained this deception operation because it demonstrates that, contrary to the way it is often portrayed, the Blitz was not conducted by the Germans as a purely standalone campaign. Rather, the Nazis hoped that pairing its deceptive operations with the Blitz would hasten the disintegration of British morale by continuing to present the German invasion as a potent and alarming possibility. Both of these elements, then, intended to put political and military pressure on the British and force them to concede defeat. Given the Nazi war leadership's oscillating approach to Sealion as the year edged towards winter, however, the Blitz appeared better placed to give immediate and demonstrable results to the German population.

From the RAF's perspective, the final phase of its Battle of Britain concluded on 31 October 1940 with the last Luftwaffe daylight raid over Britain. The Blitz, which had already started on 7 September 1940 with the bombing of London, then progressed until May 1941. For the Luftwaffe, however, the dates of the *Luftschlacht um England* are slightly more complicated. Galland roughly outlines the campaign phases as follows –

- Phase 1: The Channel Battle (early July–24 July 1940)
- Phase 2: Heightened German and British fighter clashes (24 July – 8 August)

- Phase 3: the lead up to and launch of the Eagle Attack, conducted by widespread fighter and bomber formations against the RAF (8 August – 6 September)
- Phase 4: the opening of the Blitz, which saw a mix of daylight and night-time Luftwaffe raids mainly concentrated on London (7 September – 20 October)
- Phase 5: the 'Night Blitz' offensive, conducted independently by German bombers from the end of October 1940 until May 1941.[22]

It should be noted, of course, that Luftwaffe veterans sometimes manipulated the timings and intentions of the campaign to present it in their favour. Baumbach, for instance, distinguished between 'The first phase of the Battle of Britain – the achievement of air supremacy in the summer of 1940 – and the second phase, the throttling and destruction of the British military's economic power'.[23]

For the bomber pilot Baumbach to describe this second phase as an attack on 'the British military's economic power' hugely sanitised the fact that attacking the nation's civilian population was one of the most integral elements of the Blitz. However, Galland's suggested timeline is helpful to follow because he allows for the neater insertion of a smaller phase sandwiched between phases 4 and 5: 'The first attacks by fighter-bombers, a novelty in the history of aerial warfare.'[24]

Technically, the first significant attack by *Jagdbomber* (or '*Jabos*') in the Battle of Britain had already come as early as 12 August 1940, when fighter-bombers of the test wing *Erprobungsgruppe* 210 were instructed to take out five Chain Home radar sites mostly dotted along the Kent and Sussex coastlines.[25] The bombs had been placed relatively accurately by the eight *Jabo*s, blowing up several huts, but the radar masts they had aimed for were only slightly damaged. Fortunately for the British, the key parts of the radar system – the transmitting and receiving blocks, as well as the 'watch office' – remained untouched.[26]

Nevertheless, the early *Jabo* Bf 109s caused quite the stir among the British, with the German fighter ace Hans-Ekkehard Bob recalling an amusing story he later heard about the British discovery of these new planes: 'The ground station shouted: "Attack the German bomber formations!" The British formation leader shouted: "There are no bomber formations here!" The ground station replied: "But they have dropped bombs!" The formation leader: "I see only Me 109s, and they cannot throw bombs!"'[27] When a formation of Bf 109 *Jabo*s attacked a railway line on 15 September 1940, the concept of fighter-bombers was branded as 'unfair' in the British press.[28]

In addition, the *Jabo*s flew at such a high altitude that it was difficult for both radar and the Observer Corps to keep track of them, while the two-stage supercharger that powered the *Jabo* aircraft gave them a considerable boost compared to their British fighter foes.[29] Bob noted how the *Jabo* pilots had come to the conclusion during the earlier *Luftschlacht um England* that 'The best results were achieved when we dropped our bombs from a shallow dive at a certain angle.'[30] Galland also noted of the *Jabo*s that 'A fighter-bomber in itself can be a quite effective weapon. This had been demonstrated by us in Spain and Poland, and through the repeated missions by *Erprobungsgruppe* 210 during the Battle of Britain.'[31]

Thus, by the commencement of what can be deemed the '*Jaboangriff*' from mid-September until late October 1940,* a third of the Luftwaffe's fighter aircraft had been converted into fighter-bombers to strike precise daylight blows against London.[32] The Bf 109 was deemed most suitable for the task, given its greater capacity to defend itself. Initially, success was observed on 20 September 1940, when the first *Jabo* raid to be conducted at 12,000 feet largely went unchallenged; for

---

* Luftwaffe *Jabo* attacks over Britain extended into 1941 and 1942, finally evolving into the infamous 'Tip and Run' campaign (1942–3) – but, during the *Luftschlacht um England*, they had largely ceased by October 1940.

the Luftwaffe's *Jagdfliegerführer* 2 (or *Jafü*, 'Fighter Leader'), only two Bf 109s were lost in exchange for eight British fighters shot down and four damaged.[33] It should be noted, too, that the *Jabos*' fighter escorts remained formidable by the end of the month. Just over a week after the above exchange, the Germans lost only three fighters after downing seventeen RAF fighters and badly damaging five more.[34]

Yet, as *Kommodore* Hanns Trübenbach of the fighter wing JG 52 explained of the *Jabo* operations, 'The ideal condition for an escort task for bomber wings was to have one *Gruppe* as close escort, one as extended escort and one as a 'free chase' *Gruppe*.'[35] Without sufficient protection, the *Jabos* became far more vulnerable to roaming Spitfires and Hurricanes during the autumn daylight raids against London. As with the Blitz, the *Jabo* cannibalization of the Luftwaffe's fighters was another incidence of Göring running away with an improbable scheme. Trübenbach added:

> *Reichsmarschall* Göring really believed that the rapid conversion of the individual fighter *Gruppen* to fighter-bombers would finish off the war over England in a short time (four days!). When Field Marshal Kesselring expressed this philosophy of Göring to the assembled *Kommodoren*, we naturally laughed heartily at this idiocy. But that was how Göring was.[36]

Galland noted, too, that 'We fighter pilots looked upon this violation of our aircraft with great bitterness. We had done everything possible to increase our performance in order to keep up with a progressive enemy. We had discarded everything dispensable in an attempt to squeeze another ounce of speed out of them.'[37] Having finetuned every detail of dogfighting against the British by the late Eagle Attack, the German fighter pilots were now forced to change their tactics to accommodate the *Jabos* and escorting the main bomber formations to London.

Most damagingly, however, the *Jabo* pilots themselves had

to adapt to the fighter-bomber role with very little warning. Just before the Blitz, the fighter pilots had received only basic training in using their gunsights as a bombsight.[38] Yet, as Galland further explained, 'The fighter-bombers were put into action in a great hurry. There was hardly time to give the pilots bombing training. Most pilots dropped their first live bomb in a raid over London or on other targets in England.'[39] The effectiveness of the *Jabos* was also reduced by the fact that, if they were attacked by Fighter Command, the German fighter pilots often gave in to their instincts to engage, jettisoning their bombs as if to shake off their shackles.[40]

Galland also noted that 'The Me 109 carried a 500-pound H.E. [High Explosive] bomb; the Me 110 carried two of them and four 100-pounders, all together 1,400 pounds. No great effect could be achieved with that.'[41] In addition, as Trübenbach admitted, 'On many days we sent only two *Jabo*s, or fighter-bombers, across just to frustrate the crowded traffic of the Londoners when we delivered our couple of small bombs.'[42] The *Jabo*s also constituted a blow to Luftwaffe morale because, as Trübenbach elaborated, 'At such a late stage of the battle, in October 1940, it could not meet the supposed military aims. It is quite possible that the "fighter-bomber offensive" that now began was primarily politically motivated, namely, to show that we could still drop bombs over London.'[43] In addition to their rudimentary aiming systems, another factor that hampered the accuracy of the *Jabo*s was the deteriorating weather over Britain in September and October 1940.

In fact, even just training new fighter pilots was proving to be increasingly perilous during the darkening autumn. According to a medical report compiled by the German Aviation Research Institute in 1941, some of the Luftwaffe's accidents and injuries up to the end of 1940 were due to relatively mundane issues such as bad weather, overshooting runways, malfunctioning engines, and baling out of damaged aircraft, but there were also freak accidents like being sliced open by running propellers.[44]

Some of the grisly injuries sustained included shattered legs, arms and skulls; protruding ribcages; testicular and lung ruptures; fractured jaws and kneecaps; gaping forehead wounds; full-body burns; corneal injuries due to splinters; knee injuries and internal bleeding in those who had parachuted; concussion and broken necks; bruised spines and kidneys; torn-off legs and resultant shock.[45] They sound unpleasant enough when listed in a clinical document; when they are described by the Luftwaffe personnel who witnessed them, it makes for especially nauseating reading.

Heinz Knoke, who was in the midst of fighter pilot training on 12 October 1940, recalled the moment that his trainee cohort were let loose on a real Bf 109 for the first time. One of the men on his training course, *Unteroffizier* Schmidt, '... took off without difficulty, which was something, as the aircraft will only too readily crash on take-off if one is not careful'. Knoke noted that 'A premature attempt to climb will cause it to whip over into a spin, swiftly and surely. I have seen that happen hundreds of times, and it frequently means the death of the pilot.'[46]

So far, it seemed that Schmidt had avoided this rookie error. Coming in to land, however, was a whole other matter. Knoke watched as Schmidt misjudged the speed and subsequently overshot the runway. No problem – it took some time to get used to this powerful new steed. They waited patiently as he came round for the second time, but he aborted the landing again.

'We began to worry,' Knoke recalled. '*Unteroffizier* Schmidt had clearly lost his nerve.' Third time lucky, perhaps? Knoke held his breath as Schmidt came in to land, quite literally, for the final time. As he made his descent, the engine stalled due to his low speed; the Bf 109 spun out of control, crashed into the ground and burst into a plume of flames. The horror-struck Knoke sprinted over to the crash site in the slim hope that Schmidt was still alive. He was – but it might have been far kinder if he were dead:

Schmidt had been thrown clear, and was lying several feet away from the flaming wreckage. He was screaming like an animal, covered in blood. I stooped down over the body of my comrade, and saw that both legs were missing. I held his head. The screams were driving me insane. Blood poured over my hands. I have never felt so helpless in my life. The screaming finally stopped, and became an even more terrible silence. Then Kuhl and the others arrived, but by that time Schmidt was dead.[47]

As Knoke left the scene, he 'noticed the stains on my flying-suit. They were great dark bloodstains, and I was frightened. It was a horrible, paralysing fear. I could only be thankful that there was no one else present to see how terrified I was.' Incidents like this were compounded by the fact that, as the senior Luftwaffe psychologist Dr Siegfried Gerathewohl noted, 'Due to an order of the GAF [German Air Force] High Command, in 1940, only candidates with the most limited aptitude were disqualified from training and finally, due to a shortage of flyers, even the totally unsuitable were accepted for training.'[48]

This illustrates how the calibre of Luftwaffe aircrews began to decline as the *Luftschlacht um England* wore on. The cream of the crop was being gradually skimmed off each month until they were reaching the bottom of the barrel during the Blitz. A *Hauptmann* pilot captured over the Thames, for instance, confessed that 'in my civil job I make petrol installations. That is my speciality'.[49] He also noted that he:

> ... was an observer in the air force in the last war. In this war I was with a ground organization to start with. Then I got a letter calling me to Berlin. I went there and they asked me whether I wanted to become a pilot. I said: "Yes", and at the end of a week I was transferred to a fighter "*Gruppe*".[50]

Thus, it was not as if all Luftwaffe pilots were like Galland or Mölders, who came with an esteemed pedigree from serving in the Condor Legion; others, like this Great War veteran who had to be pushing forty, followed more unusual pathways and had little time to consolidate their combat experience before they got shot down over Britain. Yet, no matter how well-trained its fighter pilots were, it quickly became apparent to the Luftwaffe that the daylight element of the Blitz had become increasingly untenable.

On 7 October 1940, one of the last major daylight raids that included a significant fighter escort incurred great losses when attacking the Westland aircraft factory in Yeovil, Somerset: two Ju 88s bombers and seven accompanying Bf 110 escorts were lost in the raid.[51] By the middle of the month, the mass-escorted daylight bombing attacks had all but given way to the daylight *Jaboangriff* and the widespread bomber raids at night. In his Daily Order of 18 October 1940, Göring attempted to stir his lethargic Luftwaffe by booming that 'You have inflicted devastating wounds on the British world enemy, especially in the last few days and nights in continuous, devastating blows. With your tireless, death-defying attack on the heart of the English Empire – the eight-and-a-half-million city of London – you have terrified the British plutocracy.'

Glossing right over the Luftwaffe's accumulating losses during the previous month, the *Reichsmarschall* declared that 'The losses you inflicted on the vaunted Royal Air Force in daring fighter battles are irreplaceable. Right now, the entire burden of the gigantic battle that the German people must wage against those threatening their freedom and independence is once again on your shoulders.'[52] To further pile the pressure onto his strained airmen, he added that:

> Not only the *Führer* and I, but the entire German people, look to you with deep admiration – but also with the highest expectation – as you tirelessly and courageously throw yourself against the enemy in these day and night attacks, which are made even more difficult by adverse weather. Do not forget for a moment

> that the air war that you are currently fighting at the risk of your lives is a fight for our Reich's freedom and independence. It is the highest honour for every airman to be on the front lines fighting in this gigantic conflict. During these days I have been able to convince myself again and again of your boldness and commitment. Comrades, thank you for this. I am proud to be at the forefront of such brave fighters.[53]

Having once berated his airmen for failing to win him the Eagle Attack, Göring now simpered and asserted that 'I know that you will do your duty and obligation as you did before. It will be our concern that the *Führer*'s word comes true: "This war will end with the proudest victory in German history!"' Thus, even though this newest phase of the *Luftschlacht um England* was not progressing as he had hoped, Göring nevertheless attempted to bolster the flagging morale of his fighters and bombers through his usual theatrical rhetoric.

That this statement was given as part of the Daily Order, which was 'banned from publication in the press, radio magazines, etc.', demonstrated that it was not entirely for show: the *Reichsmarschall* knew that if he failed to deliver on this next phase of the *Luftschlacht um England*, the *Führer* would have his guts for garters. Yet, by late October, the autumnal storminess which clung to the British Isles and the increasing failures of the daytime *Jabo* attacks meant that, as Trübenbach put it, 'The superiority of the English now grew from day to day. Our units had heavy losses, while the English were creating new squadrons.'[54]

On 29 October 1940, the last fighter-escorted daylight bombing raid took place on Portsmouth, alongside a smattering of *Jabo* attacks.[55] Then, on 31 October, the final German daylight bombing raid of the campaign commenced. However, in Galland's words, 'The [Luftwaffe] raids on England had become a question of prestige. The day bombing could not be continued, and night raids were only in preparation.'[56] As Trübenbach spoke of the *Jabos* by the beginning of November, 'We had already lost half of our pilots and the replacements from the OTUs and flying schools could not stabilize the fighting value of the fighter units anymore.'[57]

By this time, the Germans had lost 610 Bf 109s, 235 Bf 110s and 937 bombers in return for 403 Spitfires, 631 Hurricanes, and 115 Blenheims in the *Luftschlacht um England*.[58] In terms of fighters, then, the Luftwaffe had taken slightly lighter losses during the main dogfighting phase. However, from the beginning of the campaign in July until the end of October 1940, roughly 46 per cent of the Luftwaffe aircraft lost had been fighters.[59] The RAF's losses had not been insignificant, having shed 1,547 aircraft – over 66 per cent of which were from Fighter Command.[60]

Yet, with the British having successfully ruled out both German daylight raids and Operation Sealion for the foreseeable, the Luftwaffe's fighter wings had largely been clipped over Britain by late October 1940. From the Channel Battle to the Eagle Attack, and from the *Jaboangriff* to the opening of the Blitz, the *Luftschlacht um England* had not played out in the way that the Luftwaffe and Nazi Germany at large expected it to. Now, the Third Reich's hopes for the final victory over Britain would soon be pinned nightly onto its bomber crews alone. The means now looked very different to the earlier phases of the campaign, but the end goal remained the same: ruthlessly subdue the British until they could no longer militarily oppose Nazi Germany.

15

Dante's *Inferno*

In the ink-black hours of 15 October 1940, a Do 17 bomber crew from *Kampfgruppe* 606 was flying over the British Isles. They squinted through the darkness to look for their target below in Northern Ireland – but, as they neared the Irish coast, they were unexpectedly pounced upon by a British night-fighter. *Leutnant zur See* Havemann, a *Kommandant* and the observer on the Do 17, detailed the fraught exchange in a battle report which demonstrated the heightened dangers of having reduced visibility over enemy territory.[1] The pilot, *Feldwebel* Selig, was alarmed to discover that 'The oil pressure indicator on the port engine was completely at zero'. Their battle report continued: 'At first, we continued to push with both engines running until long red flames suddenly came out of the exhaust of the port engine.' Struggling to hold the Do 17 level at an altitude of 2,500 metres, they eventually broke through the cloud cover at 400 metres.

They finally managed to recover at an altitude of 100 metres, with Havemann and *Hauptgefreiter* Dorfschmidt, the flight mechanic, taking it in turns to press the starboard rudder down with their feet to take some of the flying exertion off Selig. Persevering through multiple rain squalls, and with one engine continuing to splutter, Havemann made the daunting preparations that no Luftwaffe airman wants to do on a stormy autumnal night: 'I had the dinghy, emergency transmitters, star flare ammunition and everything else made ready for disembarkation.' Havemann further noted that 'Although there was moonlight, visibility was poor under the thick cloud cover,' but despite the danger, he urged the aircrew to press on.

They passed over two large British convoys, then a destroyer and a torpedo boat, without incident. Having ditched fuel to make the Do 17 lighter, it became a little easier to handle as they tried to steer over to Land's End and onwards to Brest in Brittany, northern France. When they finally crossed over the French coast at 0150 hours, the German bomber crew sucked in a huge sigh of relief as they found their landing ground. To be back in one piece after eight hours in the air, five of which were flown on a single engine, was a remarkable achievement by the aircrew. Before they were safely on the ground, however, Havemann noted that the Do 17 landed with just under 90 litres of fuel left, but that 'The starboard motor starts to vomit as soon as it taxiis to the hall.'

Leaping out of the aircraft the minute they could, they examined the damage to the port engine: 'A single cannon hit was recorded as having gone through the surface of the aft edge, further through the surface into the oil tank, against the measuring tube in the oil tank, out of the front of the oil tank, out of the surface into the engine cowling, out again a little further forward and finally through the exhaust pipe.' Given that just a single hit from the enemy could disrupt a German bomber crew's daylight sortie, the fact that this Do 17 aircrew had to navigate safely back to the French coast at night-time demonstrated the new challenges that came with flying at night. The bomber force had never been designed to carry out a protracted night-time bombing campaign over the course of nine months.

Admittedly, some German airmen preferred operating in the dark – with Fink claiming that 'During the day attacks, it was a horrible feeling when they came down on you, as you can see them; at night, not so.'[2] Others, however, were petrified of venturing through the darkness and above the tarry seas. *Hauptmann* Rudolf Lamberty of the bomber wing KG 76, for instance, commented that he 'had some experience of night attacks and didn't like it – hours of seeing nothing, fog, strain, etc.', and that he preferred 'more hectic day attacks'.[3]

Such was their heightened reactions in the darkness that Fink recalled seeing 'the air mechanic of a plane ahead fire wildly and shoot holes in his own tail'. Fink tutted that when they had all landed, the mechanic 'was excitedly pointing to all the holes in the tail of his aircraft and saying what a narrow escape he'd had: "Look at all the hits we've had!" I walked up to him and said: "That was *you*."'[4]

The most perilous aspect of flying Blitz operations, however, tended to be taking off and landing in the dark. In a British intelligence report from November 1940, it was learnt from a captured Luftwaffe officer that 'Quite a large number of aircraft are lost during take-off at night':

> He had himself examined crashes and found that, although flying from an excellent aerodrome, the pilots had tried to take off with full load before flying speed had been reached. It was estimated that bomber losses on night operations average ten aircraft a night, including aircraft lost over this country and those who crashed when taking off or landing. A pilot of *Gruppe* 606, returning from a recent raid on Birmingham, was ordered to land at Cherbourg, as Brest was covered with mist. Three of the aircraft, including that of the *Gruppenkommandeur*, overran the aerodrome and fell into the sea; the crews were all saved. A fourth crashed into the cliff and the crew were killed. At this time, two Ju 88s and one He 111 were already in the sea.[5]

Another potent danger by the autumn of 1940 was actually the growing sophistication of the German aircraft. Rolf Schroedter, an engineer who served as a technical officer in the Luftwaffe, was in charge of testing planes and investigating aircraft crashes.

He reported that 'During the campaign in France in 1940, the altitude requirements of the aircraft were still comparatively low. In the fall of 1940, however, they grew by leaps and bounds. The formations attempted to outclimb each other to obtain a better attacking position... new types of aircraft, with higher altitude ranges, were needed.'[6]

This could become especially hazardous, as the report notes, since 'At this time, most pilots had very little knowledge and experience concerning correct conduct in high altitudes, and accidents were caused by the slightest carelessness or defect.' To counter this danger, Schroedter continued, 'It was of considerable importance that, through a clarification of the problems involved in parachuting from heights of up to twelve kilometres, the men at the front were able to receive essential instructions about conduct while parachuting from aircraft.' But, even if the men were able to bale out, they were not guaranteed a safe landing – especially when the English Channel dropped to perishing temperatures by the close of 1940.

A Luftwaffe manual written during the Blitz gives an insight into the protocol when a German airman or aircrew found themselves drifting in the English Channel. Their procedures included using dye bags to form a luminescent ring around their location, as well as launching a red distress flare every half hour – or a white flare at certain times of the day, as sometimes this could be seen more clearly from above.[7] Fighter pilots and Stuka crews tended to choose which kind of life-vest they wanted to wear, as their narrower cockpits meant that they had limited mobility compared to the men baling out from bombers.

Their first choice was the slimmer and smoother Model 10-30 *Schwimmweste* ('life jacket'), which 'does not hinder movement as much as the kapok life vest', and which was normally favoured by fighter pilots as 'When ditching while wearing the kapok life vest, it is extremely hard and sometimes impossible to get out of the capsized plane because of the immediate buoyancy.'[8] Indeed, the Model 10-30 *Schwimmweste* was so popular among fighter pilots that even a number of RAF pilots ditched their own Air Ministry life jackets for it![9] Nevertheless, the 10-30 came with its own risks, as the Luftwaffe manual cautioned: 'When punctured or otherwise damaged it becomes useless, which is not the case with the kapok life vest. In contrast to the kapok life vest, [the 10–30] first develops its buoyancy when blown up; without action on the part of the wearer, it is not ready.'[10] So, it required the wearer to blow it up, and keep it topped up with additional puffs on the inflation tube.

This fact notwithstanding, the benefit of the lighter air vest '... apart from the fact that it is easier to move around in, is that it is advantageous in case the aircraft capsizes or when fuel is burning on the surface of the water. If it is not blown up, you can duck under the water and swim from the danger area.' The alternative – the Model 10-76A *Schwimmweste* kapok life vest – which had been issued at the beginning of the war, was characterized by its ribbed surface that looked like a string of butcher's sausages. It hindered free movement due to its bulky design, and could be harder to swim in, as it inflated on contact with the water.

However, as the Luftwaffe manual pointed out, 'When punctured or otherwise damaged it does not become useless. In contrast to the air life vest, the kapok life vest always maintains buoyancy and it is buoyant without any action having to be taken.'[11] In order to rescue more German airmen who had been forced to ditch into the sea, the Luftwaffe air-sea rescue teams tried to set themselves up in areas where they would have a good view of the English Channel and were directly under the flight path of their aircraft, as this was often where Luftwaffe planes that had been hit ended up having an accident at sea.[12]

Quite frequently, the rescue service could see a stricken German plane crashing, and would dutifully scoot over on a small motorboat to retrieve the casualties. Sometimes, though, Luftwaffe airmen who had baled out over the sea had to save themselves by making their way to one of the German rescue buoys – bobbing platforms that were anchored in the English

Channel with the intention of providing temporary protection to downed airmen. In September 1940, the Luftwaffe issued a handbook which noted that:

> The buoy is equipped to shelter four people for several days. If necessary, more aircraft crews can be accommodated simultaneously. On board are all necessary aids, namely: four beds, wardrobes with clothing, stimulating and strengthening substances, fresh water, emergency signalling devices, etc. The buoys numbered above thirty-five are furnished with emergency radio transmitters.[13]

Under the waterline, they were painted grey. Above it, they were painted a bright yellow while the boarding tower was adorned with four Red Crosses.[14] The buoys had a floating line attached not only to give the airman something to grasp onto and drag himself towards the buoy, but also to demonstrate the direction of the water currents so that the downed flyer could avoid getting swept away by the sea.[15] To retrieve a Luftwaffe crew or airmen, the rescue teams had to act fast. A Luftwaffe special duties instructional leaflet from November 1940 noted that there were two different procedures depending on whether the casualty was conscious or unconscious.

For the former, the procedure was 'dry rubbing by bending, rubbing and brushing the skin; wrapping in warm blankets and applying heat carefully', although it discouraged 'heating too quickly, e.g. avoid the immediate vicinity of an oven!'[16] The leaflet also suggested administering a 'careful intake of hot liquids (alcohol in the form of grog or cognac with sugar)' and 'mushy warm to hot food in small divided quantities'. The conscious patient would finally be bestowed with 'one tablet of caffeine and one tablet of Cardiazol [a cardio-pulmonary stimulant]; if necessary, repeat the administration of the tablets after two hours!'

For unconscious casualties, on the other hand, there was a swift attempt to resuscitate, with the patient immediately receiving injections containing Cardiazol and caffeine. This was alternated every half an hour with Cardiazol and Lobelin [a respiratory drug] until the patient awoke. Retrieval of downed Luftwaffe crews in the long autumn and winter nights of the Blitz, however, was rendered far more difficult than in the summer months.

Indeed, the night-time hazards presented by the latest *Luftschlacht um England* phase could be seen in a British intelligence report for the month of November 1940, which stated that 'A total of 158 GAF [German Air Force] personnel were reported, as either PoW [prisoner of war] or dead, by this section. Of these fifty-four were living, and 104 dead, which is a considerable reversal of the proportions previously encountered.'[17] It could be tricky to identify which Luftwaffe units the men had come from, so the British personnel who retrieved their bodies sometimes had to piece together their last movements as if it were a crime scene. On 7 November 1940, Flying Officer Pelham E. S. Toll, an interrogation officer attached to RAF Digby in Lincolnshire, provided a report on three dead German airmen who had been brought into Grimsby the previous day:

> They had not been in the water long, the bodies being quite fresh, and the mortuary staff in Grimsby stated that they had possibly been in the water since the previous day only. The watch worn by one had stopped at 5.53 British Summer Time. No evidence whatever could be found on any of the bodies to indicate the Squadron to which they belonged, or even the aerodrome from which they had started. Fairly large sums of French money were found on two of the bodies, and odd coins of Belgian, German, Austrian and Czech currency were found on these bodies. One body produced no money whatsoever. Letters found were of recent dates and were addressed to field post numbers at Brussels.[18]

It was these small clues that gave the RAF pathologists an idea of the Luftwaffe crew's lives. Some of the personal effects that were reported included 'On one body was a letter addressed to him at Gr. *Kampffliegerschule, Tutow b/Demmin/Pomm*, and another body produced a Post Office Savings bank book with this same place recorded for about the same time, May 1940.' A curious item was 'A ring, depicting a biplane over the sea, with "Firth of Forth" inscribed below it,' presumably to honour his previous raid or campaign there. Reflecting the young age of the bomber aircrew, the surviving identity card helped the RAF piece together that 'The ages of the crew were twenty-one, twenty-two and twenty-two years, and their ranks were *Unteroffizier, Gefreiter* and *Gefreiter* respectively.'

Flying Officer Toll noted, too, that 'Two of the bodies bore no traces whatever of wounds, but the third was shot up in the buttocks. All three bodies indicated that they had died from drowning.' This was the end that was dreaded by most, if not all, of the Luftwaffe's aircrews who ventured forth over the English Channel every single night. They particularly feared being badly wounded before falling into the sea, when they would be helpless to stop themselves from drowning.

Some of the Luftwaffe air crews that had been drowned amid the main action of the Battle of Britain continued to wash up on British shores during the Blitz, with the British intelligence report adding that 'In addition to the figures given above, thirty-six bodies were washed up from the sea, in such a condition as to indicate that they were killed before November began. This figure was made up of eleven pilots, eight each observers, W/T operators and aircraft mechanics, and one rear-gunner.'

Sometimes, their bodies even washed up as far as Ireland,* with the *Irish Independent* newspaper reporting on 28 October 1940 that 'The bodies of two German airmen, both wearing the Iron Cross, were found on the strand, one at Mosney, Laytown, Co. Meath, and the other about ten miles away at Clogher Head, Co. Louth. They were Horst Felber (twenty-five),

---

* The Irish name for Ireland was (and is) Éire; it became a republic on 18 April 1949.

Ponmern; and Walter Hoppmann (twenty), apparently of Tilberg. The bodies appeared to have been in the water some weeks.'[19] Flying over the British isles in the dark, menaced by night-fighters, harassed by flak there and back, the German bomber crews began to experience critical levels of burnout. Such was the psychological impact of switching to widespread flying at night that Luftwaffe welfare officers issued two-page well-being questionnaires to the airmen about their combat experiences.

Otto-Wolfgang Bechtle recalled how his colleague at the bomber wing KG 2, *Hauptmann* Walter Bradel, '... took the first page, drew two lines across it (like a crossed cheque) and wrote "Shit", turned it over, drew two more lines across the back page, wrote "All Shit." And signed it: Bradel, *Hauptmann*'.[20] In fact, the nerves of the German airmen were so stressed by the *Luftschlacht um England* that suicide rates within the Luftwaffe skyrocketed during the campaign. Between mid-September and October 1940 in particular, an unprecedented level of seventy-nine suicides and suicide-attempts occurred; the issue became so out of hand that the Luftwaffe surgeon-general had to issue a special directive in 1941 that dealt exclusively with the prevention of suicide.[21]

During the early years of the war, suicides in the Luftwaffe were broken down into different categories: twenty-five due to 'insanity', twenty-seven due to having a 'psychopathic personality', seven out of an 'exaggerated sense of honour', nineteen out of a 'fear of punishment', twelve out of 'alcohol intoxication', and ten others with no ascribed cause.[22] The presiding doctor, Professor Luxemburger, attributed those who had a 'psychopathic personality' and those with 'an exaggerated sense of honour' to the military culture of the Luftwaffe, claiming 'that individuals with a strong drive for personal dominance ['*Geltungsbedürfnis*'*, or a 'craving for recognition'] wilfully had

---

* The English translation of 'personal dominance' by the post-war translator is not quite precise enough here – '*Geltungsbedürfnis*' translates more accurately to 'a craving for admiration or recognition', which is more consistent with the psychological context of the original German report.

worked their way into the GAF'. This was offered by him as an explanation of the figures.²³

He also noted that 'The inflated sense of honour which rendered even the slightest failure in military conduct unbearable, and incompatible with life, was a noticeably frequent cause among young and immature officer candidates.'²⁴ He further recalled that 'In the Luftwaffe, 91 per cent of all suicides were committed by non-officer personnel. After eliminating cases of "insanity", the statistics showed that 60 per cent of the individuals involved had military service records with above-average rating.'²⁵ It is unclear which arms of the Luftwaffe were most affected, but such statistics demonstrate the immense pressure that its servicemen – especially among its oft-ignored lower ranks – placed themselves under during this phase of the *Luftschlacht um England*. Deibl's earlier observation regarding the 'instilled harshness in us, especially against ourselves' was starting to exact costs from the Luftwaffe in a new way.

Amid this bleak picture, by the late autumn of 1940 Luftwaffe intelligence scrounged around for any indication that the British were close to folding in the newest phase of the aerial campaign. German intelligence officers continued to interrogate British prisoners of war, whose numbers rose as Bomber Command ventured further into the Reich in retaliation for the Blitz. A Luftwaffe intelligence report of 26 October 1940 claimed that:

> Recently brought-in prisoners describe the difficulties of feeding the London population, which arose due to the local destruction and traffic difficulties despite supplies still being available. If their usual grocery stores were destroyed, customers would be forced to shop in other streets; supplies would not be sufficient to meet the increased demand and the population would then try to move from store to store to obtain the bare necessities.²⁶

The report added scornfully that 'the English people are such slow-thinkers that they cannot yet fully imagine the consequences of the current situation in London. However, the prisoners are of the opinion that something unforeseen must happen within the next two to three months because the current conditions are unsustainable in the long term.'[27] In other cases, Luftwaffe intelligence relied on information fed back to them through Axis sources assessing how bad the damage was on British soil.

One Luftwaffe intelligence report dated 17 October 1940 noted that 'The destruction caused in the English capital is very great. It is therefore impossible to enumerate the damage in detail.'[28] This might have been partly to obscure the fact that the Germans lacked proper quantitative evidence regarding the damage caused. Nevertheless, the report added that 'A businessman who returned fourteen days ago after an eighteen-year stay in England described the conditions in London as extremely difficult.' Delineating the businessman as a *Gewährsmann*, or an informant,* he reported that:

> [People were] standing for eight hours in air raid shelters and subway tunnels because of overcrowding, and no hot food was available for days because of gas shortages. He considers the evacuation of London to be impractical due to the disrupted railway situation and lack of organization.[29]

The report adds that 'The informant confirms that there is a shortage of aircraft crews and emphasizes the inevitable weakening of the British Air Force as a result of short-term aviation training.' The Luftwaffe intelligence report concluded that 'when he returned, he still believed in the British final victory; now that he could see the actual situation, he had no doubts about the German final success. His previous misjudgement and

---

* *Gewährsmann* can also be translated as a 'source of information', but it is normally used to refer to an informant, which is the more likely scenario when read within the full context of the report.

the attitude of the English people are based on England's isolation from news outside England and the incorrect presentation of the situation by the British government.'[30]

The source had, of course, overstated the despair of the Londoners in the Blitz, and it is not known what his nationality was or where he returned to – but this extraordinary report demonstrates how German intelligence was able to obtain detailed insights into the daily lives of the beleaguered islanders during the *Luftschlacht um England*. Indeed, German intelligence even tried to intercept British post during the Blitz as a means of assessing the country's morale during the bombing. A Luftwaffe intelligence report from 31 October 1940 noted that within a captured letter from London:

> [...] there is the following passage, which sheds a significant light on the social attitude of the letter's writer: 'It is unbearable – the air raids are repeated several times a day and bombs rain continuously during the night. There isn't a street in the West where several houses haven't been blown up. Add to this the hail of anti-aircraft shells, which, according to official estimates, have fired one shot per second over the last fifteen nights. There are no theatres or cinemas, and all the well-known entertainment venues have long been closed or empty. Last Friday, a bomb fell in a dance hall, but fortunately there were only "third-class" deaths because the explosion took place in the kitchen and the dancing couples were uninjured.'[31]

The dismissive final line of this intercepted letter highlights the genuine social tensions that simmered in Britain during the Blitz, with the entrenched class system often being derided by Nazi propaganda to present the nation as a stuffy and rank-conscious kakistocracy. In addition, Luftwaffe intelligence also scoured the international press for clues as to how the British capital was holding up during the Blitz, noting in a specialized report on Great Britain of 18 October 1940 that 'There is no point in trying to deny the fact that London has endured terrible attacks. A few weeks ago, you could walk through very large areas without seeing any witnesses to the bombings. Today that is no longer the case. The world-famous streets such as Oxford Street, Piccadilly Circus, those historic buildings such as St Paul's Cathedral and others, everything bears telling signs of damage, some significantly more than just smashed windows.'[32]

In a Luftwaffe situation report three days later, it was claimed that 'According to a reliable source, the mood is said to be very depressed. All hope is in the new leader of the RAF, Air Marshal* [Charles] Portal. An industrialist said literally as follows: "Now everything is falling to pieces, factories are disorganized, damage is significant and worker morale is declining, cursing terribly at the government."'[33] Yet, although the Luftwaffe was now two months into its sledgehammer blows against the British capital, London refused to yield. British citizens across the country kept picking themselves up and dusting themselves off from the increasing Luftwaffe air raids during the late autumn of 1940. With the overture having long faded out to awkward silence, and with a triumphant postlude still to be written, the bomber crews turned to composing the Luftwaffe's next movements – including a most devastating sonata.

Between August and October 1940, the city of Coventry in the industrial West Midlands of England had sustained seventeen relatively minor raids from the Luftwaffe – although 176 people were killed as roughly 200 tonnes of bombs rained death over the city.[34] After Bomber Command attacked Munich on 8 November 1940, however, the Luftwaffe determined that Coventry would be a suitable city for retribution. During the early years of war, there had been an increasing maturation of an experimental 42–48 MHz radio beam navigation and bombing system: *Knickebein*

---

* Portal was promoted to the temporary rank of Air Chief Marshal, and appointed as Chief of the Air Staff four days after this report was written.

('Crooked Leg').³⁵ This system used two radio beams – one as a navigational line and a second that intersected the first beam at the target – in order to set the course for incoming German bombers to fly in and drop their bombs. It had been developed from the earlier *Lorenz* system, which had become the standard navigational aid for commercial flying before the war.

If the flyers deviated too far to the right, long Morse signals warbled in the headphones of the pilot or observer; if they drifted too far to the left, small chirps of Morse signals alerted them to correct their course.³⁶ The key benefit of Crooked Leg was the fact that all German bombers were already fitted with the *Lorenz* system, which operated on the same frequencies of 30, 31.5 and 33.3 MHz, and they could, at their best, pick up Crooked Leg signals from 270 miles away.³⁷ However, the new German system had been rumbled by the British, with the Scientific Intelligence Officer Reginald Victor 'R. V.' Jones and the Secret Intelligence Service Branch at the Air Ministry having already examined and identified Crooked Leg equipment on a crashed German bomber in June 1940; he later confirmed it as a bombing aid after bugged conversations from German prisoners of war.³⁸

British jamming equipment was in use as early as autumn 1940. Their system emitted a specialized signal known as 'Aspirin', which mimicked Crooked Leg signals in an attempt to confuse the German bomb aiming.³⁹ By the beginning of November 1940, however, a far more sophisticated radio navigational system had been put into action by the Luftwaffe: the *X-Verfahren* ('X-System').* Using its *X-Gerät* ('X-Apparatus') required specialist knowledge to operate, but it was a notable

---

* Paul George Freer points out that historians often erroneously refer to the entire navigational system of *X-Verfahren*, which includes the ground transmitting stations, as *X-Gerät* – but this latter term only refers to the equipment carried in the aircraft. See P. G. Freer, 'Circumventing the law that humans cannot see in the dark: an assessment of the development of target marking techniques to the prosecution of the bombing offensive during the Second World War', University of Exeter [PhD Thesis], August 2017, 132. Available online: www.ore.exeter.ac.uk/repository/bitstream/ handle/10871/30059/FreerP. pdf?sequence=1 [Accessed 20 February 2024].

improvement on previous systems. Its bomber crews had initially struggled to distinguish between genuine and decoy targets during the Blitz, with Rieckhoff noting that:

> As subsequent aerial reconnaissance showed, the overall successes were low. This was partly due to the use of fake installations and fake fires, which repeatedly tempted many crews to drop their bombs on these points. In its first attack on Liverpool, Air Fleet 3 pelted a mock facility south of the city with its entire combat force of 400 bombers and reported great success due to the very strong fires. None of this was true! Liverpool had not been reached at all![40]

With the X-System, however, the increased navigational accuracy made it far less likely for German bombers to get distracted by the decoy fires below and veer off course from their intended target. Covering a range of around 300 kilometres and humming at a frequency between 66 and 75 MHz, the X-System system consisted of a single coarse approach radio beam ('Weser') which the German bomber located in order to be led roughly towards the target. Their course was further refined by a more accurate fine beam leading towards the target, with a 'spare' fine beam in reserve in case the first faltered due to technical malfunction.[41]

These beams were then intersected diagonally by three others – the 'Rhein', 'Oder' and 'Elbe', all named after German rivers – which gradually denoted how close the bomber was getting to the target.[42] After crossing the 'Oder' beam, the first hand on a special clock was set in order to determine the bomber's true groundspeed. When it passed through the third beam five kilometres from the target, the first hand on the clock was stopped and a second hand was then started. This was three times faster than the first clock hand, because of the smaller distance between the third beam and the target.[43] When the two hands came together, it closed a pair of electrical contacts which automatically dropped the aircraft's bombs over the target.[44] Carrying out forty raids with the new *X-Gerät* equipment over three months, the Luftwaffe's experience using the system expanded considerably in the run-up to the Coventry raid, which was scheduled for 14–15 November 1940.[45]

However, although *X-Gerät* was used to great effect in setting up the Coventry raid, *Generalmajor* Paul Weitkus of the bomber wing KG 2 noted that 'I didn't use it, [as it was] such a clear night' – demonstrating the terrible aptness of the raid's name, *Unternehmen Mondscheinsonate* ('Operation Moonlight Sonata').[46] 'What we experience at the destination', as the Ju 88 pilot Stahl recalled of the raid over Coventry, 'exceeds anything one can imagine. The whole city seems to be burning. And we are just the beginning – there are still many aircraft behind us that have no worries about finding their way in the light of the fire.'[47] With pre-dropped flares scintillating all over the city, too, Stahl easily located his target for the night.

'Below, we can already see details of burning streets and large wildfires,' he mapped out in his memoirs. 'Then the flak caught me and forced me to turn away again. I now wait until the anti-aircraft fire concentrates on another aircraft and use this moment to fly steeply downwards and at high speed towards my release point. My flight path takes me once again right over the burning city. When I set off my bombs at exactly the designated point, I was only 2,000 metres high.' Stahl was then painfully distracted by a thorny jab to the ribs: one of his colleagues, Hans, pointed worriedly at the drooping altimeter. Stahl urgently pulled up his Ju 88 – but the flickering city remained seared into his mind:

> We are in an inferno we have never experienced. Below us are red embers of fire, from which a cloud of smoke rises, which turns the airspace into a ball of fire because the firelight makes it glow, as it were. There are flashes of bomb impacts, anti-aircraft firing and detonations below and above and all around. We're flying through hell![48]

While Stahl and his men had opened the deadly air raids on

Coventry, the Luftwaffe bomber pilot Jan Klatte and his air crew were among the bombers to complete the city's destruction. 'We were the last planes to take off before dawn,' he informed his mother in a letter dated 16 November 1940.⁴⁹ 'The flak was firing like crazy. But detonations continued to light up in and around the fire site, starting new fires. The whole city was pure hell. It must have been a catastrophe like the Last Judgment for the people below.' The bomber pilot added, though, that 'One shouldn't think about the fate of these people in a total war, and hopefully this attack will help to give a better insight.'

He claimed rather disappointedly to his mother that 'We do not yet have any daily newspapers or reports about it,' but he assured her that 'The Wehrmacht report and the intelligence service spoke of Coventry.' By the end of Operation Moonlight Sonata, 554 people had been killed, a third of the city's houses had been destroyed, its waterworks and gas lines had been largely severed, and the historically revered Coventry Cathedral had been gutted.⁵⁰ Indeed, the extent of the damage was such that the German verb '*coventrieren*' (sometimes '*coventrisieren*') was coined to describe a city being utterly razed to the ground.⁵¹ Göring subsequently proclaimed in his Luftwaffe Daily Order from 21 November 1940:

> Comrades! You have been destroying the British Empire for months in non-stop, ruthless attacks. Tremendous successes were worth your bold and dogged efforts. No enemy resistance, no matter how severe and threatening the weather, has held you back from carrying out your attacks day and night. Over the last few nights, I have given you two special tasks: the destruction of the city of Coventry and the destruction of the war industry in the city of Birmingham. Both goals were of decisive importance in the war. Your success shows me that each and every one of you understood what was important here. Only an air force like ours, in which leadership and followers are so welded together, could carry out these attacks so completely and

perfectly. I thank you as your commander-in-chief for your commitment, your courage and your tenacity...⁵²

In this newest phase of the *Luftschlacht um England*, Luftwaffe aircrews looked to the destruction they wreaked over British cities as a means of boosting their morale. The fifty-seven consecutive nights of major bombing over London, in particular, seemed to constitute an unstoppable onslaught that, they thought, would surely force the British government to finally sue for peace. Otto-Wolfgang Bechtle, the Do 17 pilot with the bomber wing KG 2's 9. *Staffel*, spoke of witnessing the mounting devastation as he flew over the capital on 29 December 1940: he saw 'A concentration of fire in London for once, instead of single sparkling lights. A Dante's Inferno over the city – red glow on clouds, searchlights [and] flak.'⁵³

At the same time, however, the fraying nerves of the German bomber crews struggled to hold out as the Blitz edged into 1941. Stahl noted that his bomber unit were told 'The leadership has reliable information that it is only a matter of weeks before the island people are ready to give up.'⁵⁴ Yet the Luftwaffe airmen's immense frustration at the relentless campaign was seen in the fact that Stahl's bombardier, Hans Fecht, blurted out before their next raid '... so loudly that those around him could hear that "perhaps the Fat Man and his red-trousered pen-pushers* believe in something like that – but they should fly over themselves first, then they'll see for themselves who's going to collapse soon!"' Stahl recalled that 'I give him a poke in the ribs. "It's true!" he shouts at me. The room becomes dead quiet until Stoffregen says: "Well – safe travels!"'

---

* It is likely Fecht was referring to the ornamental carmine *Lampassen* ('trouser stripes') worn by certain ranks of the general staff, although the colour was different according to the service and branch.

16

Flying, Sleeping, Eating, Waiting

The Luftwaffe's Blitz has long been presented as an entirely one-sided endeavour – its bomber crews vastly outstripping RAF Bomber Command in terms of navigational aids, the range of cities targeted and the level of destruction caused. Given the huge discrepancy between the two sides in the numbers of civilians killed and the amount of industrial damage wreaked in this period, this characterization is not unfounded. Nevertheless, if we become too preoccupied with the number of payloads dropped and acres destroyed, we are in danger of ignoring the stories of the Luftwaffe personnel who defended the Reich amid the final phase of the *Luftschlacht um England*.

Initially, the Germans were most concerned by Bomber and Coastal Command striking at the possible embarkment points for Operation Sealion along the Continent's shorelines. On 2 October 1940, an OKW report noted that:

> The recent enemy air raids on the Channel coast have repeatedly resulted in disproportionately high losses of soldiers and Wehrmacht assets. The damage was significantly worsened several times by hits on ammunition storage facilities and trains, including looted ammunition. The civilian population

is also severely affected as a result. The *Führer* and Supreme Commander of the Wehrmacht has ordered that all measures be taken to immediately preclude such serious consequences from enemy air attacks.[1]

Soon, however, the rise in British air raids on Berlin and Hamburg led to German children being evacuated from major German cities on an unprecedented scale in October 1940. As the schoolboy who later became a Luftwaffe flak assistant wrote, 'The frequent air raid alarms, which disrupted the nights, ensured that lessons at Berlin schools became more and more irregular. It often stopped in the first few hours, sometimes completely. This prompted the authorities to look for a way out. And this was called "*Kinderlandverschickung*" ['Evacuation of Children'], abbreviated as KLV.'[2] He added that:

> From the beginning of October 1940, as part of the 'extended evacuation of children', widespread evacuations of entire school classes from the large cities and industrial regions that had been particularly affected by bombing raids to rural areas of Germany began. On 3 October, the first 3,000 children from Berlin and Hamburg were sent to KLV camps. The evacuation, which was voluntary at the time, was not to be described as an 'evacuation' in official announcements. The plan was to send as many students as possible from a school to less threatened areas in order to then continue teaching there under much better conditions. We students actually didn't see much of the preparations themselves.[3]

Since German children now ran the risk of being caught up in British air raids, the attitudes of some German citizens hardened dramatically towards their aerial enemy. How far the stakes had raised for the German public during the Blitz could be seen in a letter from a mother in Darmstadt, who wrote on 11 October 1940: 'I no longer dare to believe in a ceasefire by Christmas. Of course, I really wish and hope so. England will likely need to be destroyed because it is only open to violent means in the long term. It must be terrible in London. But I lack any feeling of compassion, even for the children, even though I have a child myself.'[4]

The mother brusquely explained that 'It's like this: you or me. And the debt that we have to settle with England has accumulated over the course of centuries. We should even have the toughness as "masters" to let the entire English people starve. You always have to imagine how the British would act in the opposite case. We have alarms more frequently now.'

During these more regular alarms, the British air raids exposed some key deficiencies within German air defences. On 28 November 1940, a recently captured German pilot in the 2. *Staffel* of the fighter wing JG 26 was overheard recalling the current difficulties of patrolling Germany's skies, summarising thus:

> The general opinion at that time among the men themselves appears to have been that any interception was purely a matter of luck. The men were nearly all fairly inexperienced, having come straight from fighter schools, and this period was rather a nightmare – at least two officers were killed; one for an unknown reason and the other split his skull when landing.[5]

For the Luftwaffe flak crews defending the Reich, too, the last quarter of 1940 proved a long and weary grind. A newly conscripted servicemen in a Hanover searchlight unit quickly realized how ceaselessly the Luftwaffe flak crews had been deployed throughout the year. 'The *Wachtmeister* asked us yesterday when we wanted to go on vacation,' he wrote on 4 December 1940. 'I said, for Christmas. He then explained that this was not to be expected at all. There were so many in the battery who had only been on vacation three or even two times in the last year, so we could easily calculate when it would be our turn.'[6]

The intensity of being in a Reich-based flak crew during late 1940 can be seen when one *Kanonier* ('gunner') spoke of how, 'Although we are initially being trained here, we are still involved in battery operation and have to do everything we can in the event of an emergency. This means we are on call every night and sleep next to our equipment at night. Once a week we are free from midday until eleven in the evening.'[7] In addition, by the New Year of 1941, the British air raids continued to be troublesome enough over Reich-occupied territory that Kurt Gerhard Raynor, who had just been transferred to Le Havre with a Luftwaffe anti-aircraft unit, noted that 'British air-attacks against targets in occupied France became decidedly more frequent and severe, and the High Command deemed it necessary, for its defence, to place 2-cm anti-aircraft guns near our Western Headquarters in Versailles.'[8]

To avoid inviting further retaliation over Reich-occupied territory, then, the aerial campaign over Britain had to succeed once and for all. Yet, by the opening of 1941, the Luftwaffe's activity in the winter Blitz started to become inconsistent. Having launched a total of 3,884 sorties in December 1940, the Luftwaffe dropped to just 2,465 sorties flown in January 1941 and then down to only 1,400 in February 1941.[9] The 'knock-out' blow which had been promised for so long in the final, gruelling Blitz phase of the *Luftschlacht um England* continued to evade the Germans as the war trudged into another year. As Baumbach put it:

> Bombing raids at night against industrial centres, military installations, airfields, shipping targets and the mining of the English east coast were intended to be directed against British military strength and supplies. However, the Luftwaffe units were badly hit. The Western weather conditions with fog and other navigation difficulties meant that the bombing campaign against England in the winter of 1940/41 was only effective against large area targets.[10]

Attached to the bomber wing KG 30 during the Blitz, the Ju 88 pilot Stahl also seemed to perceive no weakening in the British resolve. In early 1941, he logged in his diary that 'I'm getting more and more respect for our English opponents. The country has had to endure an air war by a superior enemy for over six months now and still no weakness can be felt. On the contrary, the defence is strengthening from week to week. During the day, we can really only be seen over there when the weather is really bad, otherwise we are inevitably the certain victims of the Spitfire.'[11]

For the German *Heer* units that were stationed on the French coastline, however, the Luftwaffe's ongoing air raids did not hint at its increasingly dire straits in the *Luftschlacht um England*. At the beginning of 1941, eighteen-year-old Werner Mork had been transferred to St Omer, France, as an army signalman. Mork's unit often went out to watch the endless stream of Luftwaffe aircraft taking off to pummel Britain in the Blitz, his heart swelling with pride as his airborne brothers delivered the blows that his own men could not yet manage:

> Standing on the beach in Calais, I saw several *Staffeln* of German planes flying to England to drop bombs there. But I just felt a certain satisfaction that the English cities were also being bombed. I didn't think much about the fact that this would also cost victims 'over there', including innocent and defenceless civilians, as well as women and children. What I thought was just that the English should stop the senseless war against us on their own. Then they would have peace and quiet again on their island, the current bombings on their cities were their fault, and they were also a response to their terrorist attacks on *our* cities! The way of thinking we had within us was that simple and seemed completely normal to us.[12]

So, rather than resenting the Luftwaffe for the delayed conclusion of the Blitz, some of the German soldiers blamed the British for drawing out the campaign. Mork further recalled the attitude that Britain was '... also responsible for the fact that the German air war against England was not as successful as everyone had once believed. There were even significant losses of aircraft and crews, which could not be hidden. But the victory fanfares that kept resounding in the [German] special reports did not raise such doubts; the great successes were more impressive.'[13]

Mork's recollection demonstrates how Nazi propaganda continued to assuage German concerns over the inconclusiveness of the *Luftschlacht um England* by early 1941. With both the *Heer* and *Kriegsmarine* unable to directly take the war to the defiant island, they increasingly looked towards their flying comrades to break the political deadlock and pave the way towards an amphibious landing. 'To us soldiers in France,' as Mork recalled, 'the invasion of England seemed urgently needed to finally get the English willing enough to conclude the much-needed armistice with us, even if Churchill did not appear to be prepared to do so.' He added that:

> The change of location of our headquarters to St Omer also seemed to us to be proof that things would soon 'start' and also proof that we would be involved in moving to our new location on the island. With this very firm assumption, we viewed our service in St Omer as a waiting period until the time when we would cross over to England.[14]

As an old man, Mork asked his readers, 'Pretty naive? You can probably see it that way, because we were really naive in our views and opinions, even if they weren't necessarily absurd because they corresponded to what came down to us from the very top.' Indeed, in the opening months of 1941, even the Luftwaffe had not entirely abandoned the hope of Operation Sealion. On 20 March 1941, Air Fleet Command 5 sent a memorandum to Army Group A in which Sperrle – as the chief of Air Fleet 3 and commander in the West – confirmed that 'the preparations for the Sealion operation are confirmed to have started again.'[15]

Although the memorandum stated that operational readiness of the other Wehrmacht forces would not be required at that time, as 'There is no need to initially orientate the other leaders and the troops,' it does note that the 'Commanding generals of the Air Corps are authorized to immediately brief the *Geschwaderkommodoren* on the intentions and orders for this scenario.' In a possible reflection of how the Luftwaffe's faith in the Ju 87 had not been entirely lost during the *Luftschlacht um England*, the memorandum noted that 'Another Stuka *Staffel* will be added to *Jafü* 3. Accommodation of this *Staffel* in the area south of Cherbourg and on the Channel coast south of Le Havre is to be explored. The result is to be reported to Air Fleet Command 3 by 1 April.'

The memorandum further details that Air District Command 5, Western France, 'is preparing to stock the places intended for Operation Sealion', while Sperrle ordered that the 'Air Corps, *Jafü* 3, Air District Command (Western France), and the leaders of the air-sea rescue service will report on the status of the preparatory work every Saturday, starting on 29 March 1941.' Sperrle even demanded that 'The preparations for Operation Sealion must be started more quickly.'[16] Thus, even seven months into the Blitz, the invasion had not been fully discarded by the Luftwaffe. However, this was more likely to be symptomatic of increasing desperation at the dissatisfactory course of the *Luftschlacht um England* than wholehearted belief in the feasibility of Sealion.

To make matters worse for the Germans, the RAF's Bomber Command appeared to be hitting back harder than it ever had. On 8 April 1941, Kiel had suffered its worst attacks to date: sixty British aircraft had set out to pummel the port city, resulting in 84 deaths and 184 injuries.[17] The next night, more than 200 people were killed in the city, with 650 buildings destroyed and 850 damaged, including 30 schools.[18] The acute psychological effect of this raid could be seen in bombing statistician Detlef Boelck using the term '*Panikstimmung*' ('a panicked mood') for the first time in his report on Kiel's morale.[19] The *NSV Frauenwerk*, a German equivalent of the Women's Voluntary Service in Britain, attempted to calm frayed nerves with a hot drink and a soothing hand.

Nevertheless, there was little they could do about the sudden exodus from Kiel, with 8,000 people evacuated and thousands more fleeing from the city.[20] The Reich's capital, too, saw a number of British air raids around this time. A member of the Luftwaffe's Flak Regiment 22 later wrote in a letter that 'I actually wanted to write about Easter, but the air raids, especially the one on the night of 9–10 April 1941, caused a lot of work for us on the night.' He described how 'There was a lot of banging, and when I heard the English eggs whistling and crashing 40–50 metres away, I was lying on my face in a way that I can't even describe. And afterwards the splinters rattled into the treetops and fell to the ground next to me, so that I collected some nice souvenirs. But we got quite a few of the men down.'[21]

Berlin, Hamburg, Kiel and Bremen remained key targets for Bomber Command during that month. On 10 May 1941, an *Unteroffizier* based with the 6. Battery of Luftwaffe Flak Regiment 49 in Bremen wrote of the increasing air raids that 'There was quite a bit of damage in Bremen itself. Tommy is here every night. Although they are often just passing through to Hamburg, Kiel or Berlin, people sit in the basement from 12 p.m. to 4 a.m. ... it is now shortly after 4 a.m. and the alarm for Bremen has ended.'[22] An *Unteroffizier* stationed with the Luftwaffe transport ship *Bukarest* wrote towards the end of May 1941 that 'We are currently back in Świnoujście [north-west Poland] and will be able to get away from there. All the ports are overcrowded since all the ships from Hamburg and Kiel have left because of the air raids...'[23]

This made it all the more frustrating that Britain continued to stand firm under German bombs. In a pre-sortie briefing ahead of a May 1941 raid on Hull in northern England, Stahl recalled Stoffregen saying that Göring 'had personally ordered the night operations against the island not only to continue with the previous intensity, but also to intensify them using all possible means'.[24] Part of the reason for this renewed ferocity was allegedly to mask the Wehrmacht's intentions for the upcoming invasion of the Soviet Union in Operation 'Barbarossa'. Yet, as evidenced by Stahl's diary, the growing strain on the Luftwaffe to deliver results was evident:

> Flying, sleeping, eating, waiting – that's how it went these last few days. Liverpool, Glasgow, Glasgow again, then Hull... night after night! And it was always a success. But over there, night-fighting is becoming more and more unpleasant, and the anti-aircraft fire control system has also been improved to such an extent that once they catch you, it becomes very difficult to escape. I have now completed around sixty attacks on the island and almost all of them at night. Sometimes I shudder to think that the average crew survives between three and four of these night missions. Today, it's Hull's turn again.[25]

The logistics and energy required to pull this off was astronomical: in total, Air Fleets 2 and 3 ended up launching just over 27,000 sorties during the Blitz.[26] Even as the Luftwaffe became increasingly committed to fighting in the Balkans and North Africa, the bomber crews of both air fleets were still flying an average of 4,442 sorties a month over the British Isles throughout the spring of 1941.[27] This final phase of the *Luftschlacht um England* gave a more sustainable loss rate for the Luftwaffe amid its other demands, but it was nevertheless losing 12 per cent of its overall bomber strength per month by May 1941.[28]

Then, as Lehweß-Litzmann noted, the Luftwaffe's attack on London during the city's 'Hardest Night' on 10–11 May 1941 marked 'The last large-scale attack and the official end of the *Luftschlacht um England*, although I still remember individual attacks until the third week of May.'[29] He added that his last ventures over Britain led him to '… the conclusion that a military invasion across the Channel would be a hopeless undertaking. We did not have air supremacy and our army units would not be adequately supported during a landing and the possible advance'.[30] His frustration was shared by a Luftwaffe *Unteroffizier* of the 6. Battery of Flak Regiment 49 in Bremen, who wrote on 1 June 1941:

> We are still sitting in Bremen and in a different position. It's a terrible thing with us. We lie here ready to march, newly equipped with everything, even protective shields on the guns, and we can't get away. No vacation, no exit, uncertain what will become of us; just awful. The entire battery could have been sent on vacation during this time. Our only wish is to put an end to all this crap as soon as possible and pay Tommy back for everything. It's better to have a difficult final battle now than to sit around like this for another year…[31]

Thus, while the 'Wonder of 1940' had been a roaring success, the *Luftschlacht um England* went out with a whimper by the summer of 1941. True, the Luftwaffe had wreaked unparalleled destruction on a whole host of British cities and towns, including Cardiff and Swansea in Wales, Clydebank and Glasgow in Scotland, and Bristol, Plymouth, Southampton and Portsmouth in the south of England. In addition, Liverpool and Hull were among the worst hit coastline targets in the Blitz, with the former losing 4,000 civilians by the beginning of 1942 – the most outside of London – and the latter seeing 95 per cent of its buildings damaged by the Luftwaffe's bombs.[32] England's industrial powerhouses in the Midlands and the North, such as Manchester, Sheffield, Leeds, Birmingham and, naturally, Coventry, had also been constantly attacked.

In total, around 43,000 civilians were killed and between 139,000 and 200,000 were injured in the main offensive of the Blitz from September 1940 until May 1941.[33] Some 24,500 industrial buildings were severely damaged, with more than 18,000 tonnes of bombs dropped on London alone.[34] With an average 1.5 per cent loss rate of the sorties flown during the Blitz phase of the *Luftschlacht um England*, the price that the Luftwaffe paid appeared to be relatively cheap for the destruction it wreaked.[35]

This is particularly evident when compared to Bomber Command's average loss rate of 1.75 per cent from all sorties flown between 13 October 1940 and 12 March 1941, for far less damage caused in the Reich and other targets in occupied Europe.* In addition, Fighter Command had lost many of its most seasoned commanders during the campaign in 1940, and became undermanned for its size.[36]

Nevertheless, Fighter Command still wielded more aircraft than it had entered the Battle of Britain with, increasing its operational strength of pilots by 40 per cent in October 1940.[37] From August to December 1940, meanwhile, German fighter strength had been whittled down by 30 per cent and bomber strength by 25 per cent.[38]

Overall, by the end of May 1941, British civilian morale was enervated but intact. The country's wartime industries stood relatively unaffected in the long run, and, most importantly, north-western Europe had not fallen entirely to the *Führer*. Great Britain's goal throughout the *Luftschlacht um England* was to stay in the war; Nazi Germany's objective was to knock her out of it. Thus, as hostilities continued between the two sides during the summer of 1941, it was evident which side had achieved their original aim.

Further evidence that the air campaign over Britain had been a significant defeat for the Luftwaffe can also be seen in how, after

---

* Everitt and Middlebrook, *The Bomber Command War Diaries*, 122; 130. This average comes from combining Bomber Command's overall 1.9 per cent loss rate for all sorties flown between 13 October 1940 and 10 February 1941 with its 1.6 per cent loss rate between 10 February 1941 to 12 March 1941. It should be noted though, that a handful of these losses were also due to minelaying operations.

having been unable to cover up the failure of the *Luftschlacht um England* by shifting its timeline, the Nazi propagandists now performed mental gymnastics that would have scored them tens across the board in the Olympic Games.

On 18 December 1940, Hitler had secretly outlined in his War Directive No. 21 that 'The German Armed Forces must make preparations to crush Soviet Russia in a lightning campaign (Operation Barbarossa).'[39] The *Führer*'s haste to attack his Eastern adversary was shown by his intention for it to take place 'even before the conclusion of the war against England'.[40]

Thus, on 22 June 1941, Hitler further unleashed his bloodlust for *Lebensraum im Osten* with the invasion of the Soviet Union. The launch of Barbarossa gave the Nazi press an excuse to hide the failure of the Battle of Britain and the Blitz at the end of May 1941 by giving it a strange and spurious link to the attack on the Soviet Union a month later. As Lehweß-Litzmann pointed out:

> When all hopes were dashed by the enemy's growing strength at the end of September 1940, the Wehrmacht leadership used the air battle, which was also hyped up by propaganda, to camouflage the preparations for the campaign against the Soviet Union. At the beginning of 1941 [the *Luftschlacht um England*] was described as 'the greatest deception in the history of war' in order to distract from the defeat and even turn it into a strategic victory.[41]

Throughout the *Luftschlacht um England*, the Nazi press had published antisemitic passages which presented British 'demoplutocrats' as being seedy bedfellows with Jewish bankers and 'stock jobbers'. Now, on 23 June 1941, the German *Sorauer Tageblatt* newspaper ran the headline 'The British–Bolshevist Conspiracy' to announce the opening of Barbarossa. The article lamented Russo–British collaboration during the First

World War, before citing Hitler's claim that his 'peace deal' in the autumn of 1939 'was rejected by international and Jewish warmongers'.

The apparent reason for this rejection was that 'England still had hopes of being able to mobilize a European coalition against Germany, including the Balkans and Soviet Russia'.[42] To cover up the disappointment of the *Luftschlacht um England*, then, the Nazi propagandists exaggerated the threat of its political enemies and represented the supposed collusion between Britain, the USSR and international Jewish financiers as much more dangerous than the German leaders had initially predicted. The impact of these messages should not be underestimated: they shaped the minds of some Luftwaffe personnel, even if a greater proportion may not have necessarily subscribed to those ideas.

For instance, a Luftwaffe *Unteroffizier* based at the airfield in Lyon, France, wrote on 23 June 1941 that 'Now Jewry has declared war on us across the board, from one extreme to the other, from the London and New York plutocrats to the Bolsheviks. Everything that belongs to the Jews stands in a front against us. The Marxists are fighting shoulder to shoulder with high finance – just like in Germany before 1933.'[43] Although the British, with their large proportion of Anglo-Saxon ancestry, were not subjected to the same dehumanizing rhetoric that was aimed at people of Slavic and Eastern origins in Nazi propaganda, the island nation was nevertheless used as yet another medium to stoke the 'war of ideologies' and further justify the Soviet invasion to the German public.

As the anti-aircraft gunner Raynor confessed, though, Barbarossa actually 'stunned the German population as much as it must have the outside world. I am sure that it was thought by many Germans to be a grave mistake'.[44] Lehweß-Litzmann, on the other hand, later argued that 'The Luftwaffe did not go east in any way weakened, as its superior initial effect proved. The losses over England were made up for by the combat experience gained by the units and the commanders, of which I was a member.'[45]

This statement, however, is demonstrably incorrect. Strategically, the failure to knock Britain out of the war ensured that when the Wehrmacht launched Barbarossa off the back of the unsuccessful Blitz in June 1941, it risked fighting a war on two fronts.

In addition, the Germans knew that a failure to subdue the RAF would have negative repercussions for the war in the Mediterranean, Balkans and North Africa, with the Luftwaffe having been assigned to protect General Erwin Rommel's *Afrikakorps* in February 1941. The increasing Luftwaffe commitment to the Balkans as the Blitz came to a close was seen on 6 April 1941 with the launch of Operation Marita in Greece and Operation 25 in Yugoslavia, and then Operation Mercury over Crete on 20 May 1941. The protracted *Luftschlacht um England* had forced the Luftwaffe to undertake these operations with barely a moment to catch its breath.

From the beginning of July until 31 October 1940, the Luftwaffe lost around 1,887 aircraft of all types in combat – 1,014 of which were bombers and 873 fighters.*[46] Then, during the Blitz, a further 600 German bombers had been lost by May 1941, thus constituting a total of 2,487 Luftwaffe aircraft lost across the entire *Luftschlacht um England*.[47] Along with these losses, nearly 2,000 members of flying crews were killed and a further 2,600 went missing during the campaign.[48]

In 1941, German aircraft production now struggled to balance out these losses.[49] Churning out just 950 more aircraft than it had managed the year before – a meagre increase of 8.8 per

---

* There is naturally some discrepancy in numbers depending on the source, with the Deutsches Historisches Museum claiming that 2,265 German aircraft were lost in the *Luftschlacht um England* (see A. Scriba, 'Die "Luftschlacht um England"', *Deutsches Historisches Museum*, 19 May 2015. Available online: https://www.dhm.de/lemo/kapitel/der-zweite-weltkrieg/kriegsverlauf/luftschlacht-um-england-194041 [Accessed 27/12/2024].) However, some of these figures can depend on whether losses that were not due to combat, such as during training or crashing whilst landing, are included. In addition, some of the figures given in older works for Luftwaffe losses can sometimes be based on either RAF overclaims or Luftwaffe underclaims, making them somewhat unreliable.

cent – this would prove inadequate for the heavy demands being increasingly placed upon the Luftwaffe.[50] Admittedly, Lehweß-Litzmann was right about the German air force's 'superior initial effect' during the opening months of Barbarossa, with 1,800 Soviet aircraft crumpled in their unprotected hangars under the first German assault.[51]

However, the Luftwaffe's bomber strength was hugely depleted by the end of the opening phase as a result of not fully replacing its losses from the *Luftschlacht um England*. In addition, its losses had been further exacerbated by the Balkans campaign. At the end of 1941, it could only call upon 47.1 per cent of its authorized bomber strength; of this, just over half were actually in commission.[52]

Thus, by 6 December 1941 – two days before the United States of America entered the war – the Luftwaffe had a mere 468 serviceable operational bombers, constituting only 24 per cent of its authorized bomber strength.[53] The true significance of the *Luftschlacht um England*, then, is that it created a butterfly effect which did not guarantee Nazi Germany's defeat, but which made a swift and easy Axis victory far more difficult.

The German press knew that far too much had happened over Britain to pretend that those months had not been important. The newspapers often harkened back to the *Luftschlacht um England* whenever new raids were conducted against Britain, such as the Baedeker Raids on historic British cities in the spring of 1942 and Operation 'Steinbock' – the 'Baby Blitz' bombing campaign against London, Hull, Bristol and Cardiff – from January until May 1944. During the latter, for instance, the *Luftschlacht um England* was evoked in the Nazi press with a rose-tinted perspective, as shown by an edition of the *Sorauer Tageblatt* newspaper that reported on a Luftwaffe sortie from 24 January 1944:

> They roar off into the night while we follow the start and course of the other comrades. It is an image that gets the heart

racing, reminding us of past missions and at the same time pointing promisingly towards the future. It runs through our minds: pilots, bombs and airfields – these are weapons that hit the enemy here in the West. Our thoughts wander backwards. We created this place over three years ago. From here we flew to England. Then came the battle in the East. Here we are again, as armed as ever. Time is running out. A glance at the clock and the map reveal that the first aircraft must now be above the target.[54]

The Nazi propagandists linked this 'Baby Blitz' to the earlier aerial campaigns over Britain in a manipulative attempt to present Steinbock as a natural successor to the supposedly triumphant *Luftschlacht um England*. That same year, however, Otto-Wolfgang Bechtle set about creating a more honest appraisal of the original air war over the island nation. Entitled 'The Use of the Luftwaffe against England, its Tactics and the Lessons Learnt 1940–3', his lecture on 2 February 1944 constituted one of the earliest German reassessments which acknowledged the chronic squandering of Luftwaffe resources and advantages.

Bechtle's lecture provided further indication that the Germans fell notably short of their aims in the *Luftschlacht um England*. He declared that 'After winning the Western campaign, the goal of German warfare was to quickly pursue a definitive conclusion to war with England.'[55] Reflecting the immense thought which had been paid to the prospect of Operation Sealion, he confirmed that 'It hoped to achieve this through an invasion of England, the prerequisite of which was to be German air supremacy over the British island.' Bechtle noted that 'By occupying all areas of Europe opposite England with their airfields and their facilities, every possibility of a comprehensive deployment was given.'[56] 'Despite such a favourable starting position,' as he further cogitated, 'achieving air supremacy presented a very difficult task for the German Luftwaffe, which was superior in numbers and often also in terms of materiel.'

He conceded that 'England's insular location also makes it more difficult to launch an attack. In order to reach the attack targets, long distances had to be flown over sea, which was completely unfamiliar to the majority of the German flying crews.' This applied '... especially for the fighter pilot, who was dependent on a single-engined aircraft; flying over sea required him to overcome a lot of internal resistance'.[57] He acknowledged, too, the quality of aerial opponent that the Luftwaffe had faced in Fighter Command:

> Of the air forces of all the enemy powers that were at war until 1940, the English one had the greatest combat value. The battles between German units and English fighter formations of the Spitfire and Hurricane type during the Western campaign, especially above the Channel at the time of the English retreat to Dunkirk, were the toughest they had to endure so far.[58]

Bechtle also made a similar claim to Deichmann that 'For the first time in the history of war, an air force was to carry out an offensive against enemy territory alone and independently of the operations of other parts of the Wehrmacht, with the aim of destroying the enemy air force and thus creating the starting point for deciding on a war.'[59] Some of the less controllable factors which had exacerbated the Luftwaffe's monumental task, then, had been identified. Yet, Bechtle also pinpointed some key deficits within the protection afforded to the Luftwaffe bomber formations by their fighter colleagues:

> At the beginning of the day's attacks, the principle was to provide the bomber units with immediate fighter protection limited only to what was absolutely necessary, in order not to deprive the bulk of the fighter arm from its actual task of destroying the enemy in free hunts. Already, in the first days of this air war, it became clear that the numerous, doggedly fighting

British fighter pilots, supplemented by volunteer formations from the nations defeated by Germany, made the deployment of the combat bomber and dive-bombing units so difficult that it became necessary and usual for the strength of their immediate fighter escort to be double or even triple the strength of the formation they were protecting.[60]

Bechtle's lecture is particularly significant because it shows how the Luftwaffe perceived the failed aerial campaign over Britain while the war was still ongoing. Unlike so many other accounts, it was compiled while the events were fresh in the author's memory. In addition, it was not prepared as Nazi propaganda, nor was it written while the author was in captivity, being influenced by overt or subtle pressures. As such, it perhaps constitutes one of the more nuanced and thoughtful accounts of how the Luftwaffe digested the campaign. It is inaccurate to state, then, that the Luftwaffe did not identify any lessons from *Luftschlacht um England*; but, as rumours of an Allied invasion swirled by 1944, they had learnt them far too late.

17

The Other Faculty

Just over a year after Bechtle's lecture was delivered, the Second World War finally ground to a halt. As the Allied post-war interrogations sought to make sense of the cataclysmic disaster that had just befallen humanity, the Luftwaffe's first opportunity to recount their experiences of the *Luftschlacht um England* to the Allies emerged. *Generalleutnant* Heinrich Rauch's interrogation report from 30 July 1945 was perhaps the earliest account of Operation Sealion written in peacetime. He was chosen as he had served with Air Fleet 2 under Kesselring, before being transferred as a Luftwaffe liaison officer between Air Fleet 2 and Army Group A in the preparation and execution of Sealion. Some of the details in Rauch's report confirmed exactly why the Germans never felt emboldened enough to attempt the operation.

The optimal conditions that would have been needed for Sealion to be carried out included 'consistent good weather, a calm sea, a clear night and moonlight, an incoming tide at early dawn, and a maintenance of the surprise element'.[1] He added that 'It was assumed, however, that the British Navy would employ its numerous destroyer formations recklessly, and would let the first units sail over the mines in order afterwards to break through with the other destroyer units and then rip to pieces and bottle up our defenceless barge train, if good weather did not prevail and our flying formations were not in the sky.' The calm sea, meanwhile, was vital due to 'the special nature and makeshift character of our transport craft', and if the night were not sufficiently illuminated, 'the barge trains would not arrive at the prescribed points'.

In addition, Rauch pointed out that the Germans 'did not have quick moving craft, all with their own motive power, but many were attached to tow cables, the assembling of which therefore required time'. That such temperamental and unsuitable barges were so hastily put together shows that, although in some areas the Germans plotted Sealion in minute detail, in other ways their plans were completely haphazard. It was also noted by Rauch of the incoming tide at early dawn: 'This condition also did not exist every day, and at the same time had to coincide' with the good weather and clear night. Rauch's account clearly details just how perilous and difficult an undertaking it would have been for the Germans – and that it required a nerve-wracking number of variables to go absolutely right just for the beginning of the invasion to take place.

Along with the post-war testimonies given by captured Luftwaffe personnel, the Allies were able to retrieve some of the most incriminating passages from Wehrmacht and German air force documentation, as noted in a report of the British and American Combined Intelligence Committee from 4 September 1945: 'As a result of the capitulation, certain German official and secret documents relating to the projected invasion of the United Kingdom in 1940 have fallen into our hands. There is some interest in comparing official German intentions with the deductions and appreciations made at the time on the evidence then available.'[2]

It should be noted, however, that the initial post-war interrogations on Sealion sometimes produced a slanted view of the campaign owing to lingering German disappointment and anger over the outcome of the war. Nowhere was this more keenly observed than in the Allied interrogations of Göring, in which none of the *Reichsmarschall*'s former associates were safe from criticism. Characteristically, he could not stand to take the blame and besmirched the military capabilities of his once-beloved *Führer*: 'In the early years when I had supreme command of the Luftwaffe, I had definite plans, but in 1940 Hitler began to interfere, taking air fleets away from our planned operations. That was the beginning of the breakdown of Luftwaffe efficiency.'

When asked on 10 May 1945 by his interrogator, General Carl A. Spaatz of the US Strategic Air Forces (USSTAF*) in Europe, why he maintained such rigid formations of fighters and bombers in the Battle of Britain, Göring replied that 'It was necessary to cover the bombers because their firepower was low (not like your bombers). It was also necessary for our fighters to closely cover each other. You see, it was a question of equipment.' Again and again, Göring deflected the blame onto anything else – equipment, other leaders, aircraft, pilots – to explain the shortcomings of the Luftwaffe's air war. When asked by Spaatz if the Ju 88 had been designed for the Battle of Britain, Göring's answer was that 'The Ju 88 was primarily a commercial aeroplane which had to be adapted for the Battle of Britain along with the He 111 because we had nothing else.'

His answer about the Ju 88 displays gross ignorance, as that aeroplane was in fact developed from a specification by the Reich Air Ministry in 1934 for a multi-role military aircraft and later a dedicated fast bomber – thus demonstrating that, even when his mind had been freed from the ravages of morphine addiction, the *Reichsmarschall*'s aeronautical knowledge still left much to be desired.[3] Spaatz went on to pose an intriguing question to Göring regarding how susceptible the Nazis felt Britain would have been

---

* Spaatz commanded the USSTAF in the Allied bombing campaign over Germany from January 1944 until the end of the war.

to Operation Sealion. In his typically supercilious manner, the former commander-in-chief replied that:

> To me, this is a difficult question. Germany was prepared for war and England wasn't. I was forced by Hitler to divert air forces to the East (which I always opposed). Only the diversion of the Luftwaffe to the Russian front saved England. She was unable to save herself and unable to bomb Germany.[4]

Once more, the *Reichsmarschall* determined that it had been Hitler's blunders, and not his own, that consigned his Luftwaffe to an irrevocable defeat. Here, we can see that the Luftwaffe's post-war repackaging of the *Luftschlacht um England* and its repercussions was already beginning to appear. Nevertheless, even Göring was unable to dispute the superiority of the British when it came to the aircraft detection and interception abilities of the Dowding system. He acknowledged these as key reasons that the Germans had been unable to obtain air superiority in the Battle of Britain:

> [The RAF] had committed themselves even more strongly to the defensive character and built up accordingly. Above all, it was an exemplary air interception system that they had built up. As a result, they had made progress in the field of radio measuring devices, and had developed it considerably. The interception of the German aircraft took place in perfect form. The guidance of their fighters could be carried out easily and precisely, guided and controlled from the ground.[5]

Göring also claimed, though, that 'I was not in favour of engaging in the Battle of Britain at that time. It was too early.'[6] Not all Luftwaffe personnel shared this apparent reticence. Kesselring, for instance, asserted in a post-war interrogation on 21 April 1945 that he had 'urged strongly the invasion of England. It was generally realized in Germany that England was in a critical position'.⁷ In any case, Student testified as to Göring's viewpoint on Sealion that:

> Göring was visibly disappointed by Hitler's hesitation and wavering. Göring always exerted a restraining influence on Hitler as long as he could. How often he said to me in the first years of the war: 'after his constant successes, the *Führer* tends to be careless. He risks too much. Don't encourage him in your airborne operations.' But in the fight against England, Göring was always the driving force.⁸

Before long, however, Göring would find himself defending far more than his conduct during the Battle of Britain. At Nuremberg, the International Military Tribunal opened war trial proceedings on 20 November 1945 until 1 October 1946. It sought to prosecute former Third Reich individuals who had planned, enabled or committed everything from military atrocities to crimes against humanity, which came to encompass victims of the Holocaust.

Having once been Hitler's 'most loyal paladin', Göring was the highest-profile Luftwaffe individual to stand trial at Nuremberg. Accused of four charges – committing crimes against peace, war crimes, crimes against humanity, and conspiracy to commit them – he was subsequently convicted and sentenced to death. Instead, the *Reichsmarschall* committed suicide in October 1946.⁹ One of the defence exhibits presented on Göring's behalf at Nuremberg was the German Foreign Office publication entitled 'Documents Pertaining to England's Sole Guilt'.¹⁰ Published in 1943, this document tried to exonerate the Luftwaffe of aggressive conduct during the *Luftschlacht um England*, stating that:

> The *Führer* used the conclusion of the operations in the West to once again make an 'appeal to reason in England too' in a major speech to the German Reichstag on 19 July 1940. The appeal was answered by Churchill with the first British air raids

on German cultural sites, the Goethe House in Weimar and the Bismarck mausoleum in the forest of Friedrichsruh. A few days later the first night attacks on Berlin took place. After the British attacked Berlin from the air for the first time on 26 August 1940 and continued their terrorist attacks on residential areas of the Reich capital as planned in the following nights, the German Luftwaffe moved on to attack London with strong forces on 7 September 1940.[11]

Thus, in one of his final public comments on the *Luftschlacht um England*, Göring attempted to portray the campaign as a necessary evil that came in response to apparent British aggression. The publication maintained that Hitler called multiple times for peace during the *Luftschlacht um England*, claiming that he 'constantly warned the English and how he only struck when his warnings against the use of the night bombing campaign against the civilian population propagated by Churchill were interpreted as a sign of German impotence'. Naturally, it did not make a shred of difference to Göring's defence. But it *did* constitute one of the earliest cases, and certainly not the last, of former Luftwaffe personnel twisting the *Luftschlacht um England* in order to reflect more favourably upon themselves.

For instance, in his memoirs *Soldat bis zum letzten Tag* ('*A Soldier to the Last Day*'), Kesselring retrospectively identified a number of leadership failings in the Luftwaffe that he had pointed out in a 1946 study he compiled on Operation Sealion. He claimed that 'The main mistake, however, lay in the lack of a well-thought-out war plan at the beginning of the war. The fact that the successes or failures ultimately determined the manner in which the war was continued must be viewed as a mistake in the conduct of the war on the German side. Adolf Hitler is to blame for this.'[12] A second mistake that he pinpointed, which perhaps reflected his hybrid leadership as Wehrmacht Commander-in-Chief South from November 1941, was that insufficient thought had been given to the paratroopers and gliders: 'The most striking thing was that when considering Sealion, the experience of the German airborne operation in Holland was completely ignored and the support of the parachutists was not wanted.'

He added that 'With appropriate planning, enough paratroopers and gliders would have been available [...] to dig up a defence and radar base on the relevant coastal front and to gain airfields that would have made it possible to land one or two airborne divisions.' So, after the war, Luftwaffe commanders began to agonize over what had gone wrong for the Germans in the Battle of Britain – but they partially did so to save face. Kesselring, for instance, claimed that:

> The *Luftschlacht um England* also suffered from this unclear programme for 'Sealion'. All sensible circles, including Hitler, were clear that England could not be brought to its knees by the Luftwaffe alone. One cannot therefore speak of a failure of the German air force if the unattainable goal was not achieved. It was also clear to us Luftwaffe commanders that although we would temporarily achieve air supremacy, permanent rule without taking possession of the country was not possible.[13]

Even if it were true to some degree, claiming that the Battle of Britain was unlikely to succeed in the first place allowed Kesselring to write off the Luftwaffe's failings as inevitable from the outset. Indeed, he hammered the point home once more in his memoirs, claiming that 'It would be absurd to speak of a failure of the German Luftwaffe in the Battle of Britain. Taking into account the level of development at the time, the number of aircraft usable in war and the duration of their use in the most adverse weather conditions, all criticism must be silenced.'

Even more vehemently, he added, 'It is a historically unprovable interpretation that the invasion ("Sealion") had to be abandoned because the German air force failed and was not up to the task.' If this had really been the case, they could not have carried out nine months of continuous, uninterrupted bombing raids against Great Britain immediately following the calling off of Sealion. Kesselring's assertion smacks of deflection, given that the demands on the Luftwaffe were very different in the Blitz compared to the Channel Battle and Eagle Attack.

Student, meanwhile, sought to downplay the importance of the Battle of Britain in relation to the cancellation of Sealion, arguing that:

> The English have every reason to be very satisfied with the result and to be particularly proud of their fighter pilots. But it is a myth to believe that the Battle of Britain decided the fate of the planned invasion and thereby saved England. This air offensive against England was only loosely related to Operation Sealion. This is clear from the entire content of the 'Instructions for the Conduct of the Air and Sea War against England', which was issued from the *Führer*'s headquarters on 1 August 1940 and triggered the intensified continuation of the air offensive. From this, Göring made an attempt and undertook a test of strength to defeat England or make it willing to make peace through a pure air war alone.[14]

Baumbach, on the other hand, who published his German account of the *Luftschlacht um England* in 1950, emphasized the favourable situation that the Luftwaffe wielded at the beginning of the campaign and offered a more critical perspective on how the Luftwaffe had squandered its opportunity to force a decisive outcome. Indeed, his notes as a bomber pilot during the Battle of Britain appeared to reflect that at least some of the Luftwaffe personnel recognized the gilded scenario they found themselves in at the time:

> Both flying arms of the Luftwaffe have bases at their disposal for their attritional attacks against the citadel of the British Empire and the vital access roads to it, which are not too far away from the island. These are the Dutch–Belgian–French ports on the

Channel and on the Atlantic as well as the entire hinterland. On the Channel coast opposite England and in the flank position in Norway, on the fourth bank of the North Sea, Germany has the most valuable defensive areas for its armed forces and military power core, and on the other hand an excellent starting point offensively.[15]

Already, then, we can see how some Luftwaffe memoirs sought to paint very different pictures of the *Luftschlacht um England* in order to deflect blame and to change the parameters of the campaign in favour of their air force. Other writers, like Baumbach, elected to give a more balanced, albeit too optimistic account of the campaign. Wherever they fell upon the honesty spectrum, however, what afforded former Luftwaffe personnel the platform to discuss their Battle of Britain experience beyond the necessary legal processes at Nuremberg was the Allies' genuine and understandable human curiosity to review their war through the eyes of their former enemy.

The RAF veterans were aware that the Germans, too, knew how it felt to have survived the air war, and their insider knowledge promised to offer clarity and catharsis on the most defining event of their lives. As a result, this relative tolerance created a favourable political climate for the first wave of German recollections from the Second World War. Memoirs like Knoke's *I Flew for the Führer* (1953) and Galland's *The First and the Last* (1954) were voraciously consumed by English-speaking audiences – especially the sections which covered the momentous Battle of Britain. Such memoirs gained traction and credibility within English-speaking circles, which helped other Luftwaffe veterans gain the necessary profile to also share their wartime experiences in English after the war.

That accounts which covered the Battle of Britain were even published in English by Luftwaffe war criminals like Kesselring – who was convicted after the war for carrying out the Ardeatine massacre, in which 335 Italian civilians, Jews and political prisoners were murdered in Italy on 24 March 1944[16] – demonstrates the full extent of this fascination. Kesselring ended up contributing a great deal of his knowledge and experiences to early literature on the Battle of Britain, quickly publishing his memoirs in German and English.

Thus, what had started off as military interrogation regarding the air campaign over Britain now mellowed into earnest interest from both sides to understand the other's perspective. The dialogue between Luftwaffe and Allied personnel also came from a genuine desire for post-war reconciliation and, in some cases, even warm friendship. Aviation bonded the airmen in a way that powerfully transcended any language barriers.

Steinhoff noted how the *Jagdflieger*, the fighter pilots, 'in contravention of all military dress regulations, had given free rein to our individual tastes, wearing floppy fur-lined boots and marvellous leather jackets taken from the enemy so as to mark ourselves off from the earthbound military'.[17] Fighter Command veterans identified with much of the unapologetic panache that these Germans displayed – particularly as their similarities were now more evident after the tension of war had dissipated.

In some ways, then, the Luftwaffe and the RAF had a number of common points over which they were able to bond after the Second World War. Such was this reconciliatory dialogue by the 1960s that the ultimate RAF–Luftwaffe collaboration would occur at the end of that decade: Guy Hamilton's classic 1969 film *Battle of Britain*. Keen to provide a true-to-life account of the iconic campaign, Adolf Galland, *Oberst* Hans Brustellin and *Major* Franz Frodl were brought in to serve as Luftwaffe technical advisers on the film.

We should not be fooled by the title of 'technical adviser', however: in addition to his eagle-eyed spots on insignia, aircraft and uniforms, Galland in particular exerted a significant level of influence on how the Luftwaffe was portrayed in the film. The collaborative process initially hit some turbulence in August 1967, when he protested strongly against a proposed scene in which an RAF airman dangling from a parachute was strafed by the Germans.[18] He was also irked by a number of technical inaccuracies and resentful of the failed promise to present the story of the campaign without prejudice against the Luftwaffe, as were four other ex-Luftwaffe pilots to whom he had shown the script.[19]

When Hamilton agreed that the script ought to be revised after Galland threatened to quit as the technical adviser, the new script opted for a more multidimensional depiction of both sides during the campaign.[20] Although it is unsurprising that Galland wanted to move away from demonizing the Germans and towards a more favourable perspective of the Luftwaffe, it is easy to see how a sanitized presentation of the aerial campaign was reinforced by the film. Galland long maintained that the Battle of Britain '... was tough but it never violated the unwritten laws of chivalry'.[21]

He claimed, too, that 'To shoot a pilot parachuting would have seemed to us an act of unspeakable barbarism.' When Göring once asked him if the Luftwaffe men had considered it, he claimed to have replied, 'I should regard it as murder, *Herr Reichsmarschall*.'[22] It is certainly true that most pilots on both sides largely regarded the shooting of each other's parachuting airmen to be morally reprehensible, but it did occasionally happen. As Squadron Leader Peter Townsend noted, 'Sergeant [John Holt] Dickinson had been shot dead by an Me 109 while coming down by parachute. By the rules of war, it was justifiable to kill a pilot who could fight again. But few of us could bring ourselves to shoot a helpless man in cold blood.'[23]

Equally, on 4 October 1940, a captured German bomber gunner was overheard claiming that 'Near Calais, a Spitfire shot down a German bomber and when one of the crew baled out with his parachute, the Englishman shot him with his M.G. That is clear proof that they are swine. A German would never do a thing like that.'[24] Dowding, on the other hand, felt that the Germans 'were perfectly entitled' to fire on descending British airmen, although he was 'glad to say that in many cases they refrained'.[25] Indeed, the choice very much came down to the individual, as Flight Lieutenant Charles M. Lawson explained:

> I recall one of the pilots on my squadron, a very nice guy, who claimed that if he saw a German airman floating in a rubber dinghy after being shot down, he would attack him from the air. His rationale was that total war pulled no punches. Personally, I would not have done that, but I guess there are two sides to every argument.[26]

It is important to remember, then, that in the Battle of Britain one man's chivalry was another man's idiocy. It should be pointed out, though, that it was often the ground crews who shot at parachuting airmen, not the pilots. Flight Lieutenant James Nicholson and Pilot Officer M. A. King of No. 249 Squadron, for instance, were fired upon by an officer of a Royal Artillery detachment who believed they were enemy parachutists; with the rifle fire shredding the cords on King's parachute, he plummeted to the ground and tragically died in the incident.[27]

Nevertheless, when Galland excluded the scene with the strafed RAF airmen, the dogfight scenes were presented as being more restrained and less morally dubious. In return, a scene proposed by the Luftwaffe veterans, in which a German flying boat painted with Red Cross markings was shot down by the RAF, was also removed.[28] The British exclusion of this event is perhaps one of the most sanitizing aspects of the film, given that the policy of shooting down German air-sea rescue aircraft was a routine practice during the Battle of Britain. To point this out is not to villainize the British: instead, it is to dispel the sanitizing myth that, in the words of Steinhoff, it was solely 'a battle in which very young men fought like medieval knights' and where 'both sides not only recognized, but also adhered to the rules of fair play'.[29]

A couple of days before *Adlertag*, Steinhilper recalled helping the leader of his swarm to finish off a Blenheim after it had 'been positioning to attack our Heinkel 59 sea rescue aircraft, which was unarmed'.[30] He noticed that the German rescue plane was 'clearly painted white with the bold red crosses on the wings and fuselage, there could be no possible confusion with any other aircraft'. As the Blenheim closed in on the hapless air ambulance, Steinhilper noted that 'We really began to wonder what kind of people we were fighting who would attack such helpless targets which, ironically, were often engaged in rescuing downed British airmen as well as our own.' He recalled his blood-boiling rage at seeing a gaggle of Spitfires now closing in to finish the job on the He 59:

> We shot [the Blenheim] down and, unbeknown to us, the crew escaped. A Heinkel tried to pick up these men (it may have been the one which we were originally protecting) when it was attacked by Spitfires. These Spitfires were, in turn, attacked by more of our fighters, with two of them being shot down. It was all very confusing and, to us, utterly sickening. To me and the other German pilots, these attacks were viewed as nothing short of murder. Eight-gun fighters and fighter-bombers tore into these rescue aircraft which were armed with nothing more lethal than an air pistol. Even more sickening was when we saw these attacks driven home by multiple passes over a downed or damaged aircraft, ensuring that there were no survivors.[31]

Steinhilper wrote after the war that 'Apparently the orders for this criminal behaviour had come direct from Churchill, claiming that the sea rescue plane was flying reconnaissance on convoys.' German anger at Churchill's justification for bringing down Red Cross aircraft is quite understandable, and it ought to be acknowledged that there were certainly a few instances where Churchill distorted the narrative surrounding the shooting down of German rescue aircraft. Indeed, he received sufficient criticism of the practice that he felt the need to address the subject after the war:

> German transport planes, marked with the Red Cross, began to appear in some numbers over the Channel in July and August whenever there was an air fight. We did not recognize this means of rescuing enemy pilots who had been shot down in action, in order that they might come and bomb our civil population again. We rescued them ourselves whenever it was possible, and made them prisoners of war. But all German air ambulances were forced or shot down by our fighters on definite orders approved by the War Cabinet. The German crews and doctors on those machines professed astonishment at being treated in this way, and protested that it was contrary to the Geneva Convention.[32]

Here, we can see that Churchill had departed from the initial excuse of shooting down the aircraft because they were potentially carrying out reconnaissance work. Instead, he had changed the story to the air-sea rescue planes presenting a clearer and more imminently lethal threat that could have bombed the British population. One of Churchill's most misleading claims, though, was his insinuation that the sea rescue service was largely inactive during the rest of the Battle of Britain: 'They soon abandoned the experiment, and the work of sea rescue for both sides was carried out by our small craft, on which of course the Germans fired on every occasion.'

This claim did not correlate with the fact that the German sea rescue service remained very active throughout the Battle of Britain. Indeed, as a Luftwaffe intelligence document outlines, there were seventy-one rescues made using German aircraft (Heinkel He 59, Dornier Do 24, Dornier Do 18, and Arado Ar 196) in August 1940. A further fifty-nine rescues were made that month using motorboats and other vessels from all three branches of the Wehrmacht.

September 1940, meanwhile, saw a further forty Luftwaffe

airman rescued by aircraft and fifty-six by boats, meaning that a total of 226 airmen were successfully picked up by the German air-sea rescue service or smaller Wehrmacht transportation during the peak months of the Battle of Britain.[33] Indeed, such was the problem of fending off British attacks on these flying ambulances that throughout the *Luftschlacht um England*, German fighter pilots were sometimes tasked with serving in an air-sea rescue patrol role.

As a captured pilot from the 3. *Staffel* of the fighter wing JG 51 confirmed to his British interrogators in November 1940, he had '... made a number of sea rescue patrols over the Channel. His function was to spot survivors and call up an He 59 of the Sea Rescue Service. He would then protect the He 59 while it picked up the survivor.'[34] Thus, from the German perspective, the British shooting of Luftwaffe air-sea rescue aircraft would have been seen as a deplorable practice, especially given the strong historical precedent of using Red Crosses to denote the medical nature of a vehicle.

So, it would appear that, in order to justify the contentious RAF attacks on the rescue aircraft, Churchill diminished the legitimate rescue work being carried out by the German sea rescue service and presented the rescue craft as a cover for more sinister motives. Nevertheless, the British practice of shooting down German rescue aircraft – and their decision to exclude such scenes from the 1969 film – must also be considered within the full context of the time.

Steinhilper's argument that 'We didn't need the Heinkel 59s to fly reconnaissance for us' as 'There were ample 109s for that, some already fitted with cameras for that purpose'[35] is presumptive and unreasonable, given that Britain was very much fighting for her life in the face of a severe and unrelenting invasion threat from the Luftwaffe.[36]

In a campaign where German feints and deception were commonplace, the British naturally remained suspicious of Luftwaffe aircraft marked with Red Crosses. There was a very real fear that the rescue planes might capitalize on any hesitation by the RAF, and airdrop paratroopers over the island. Moreover, Churchill was undoubtedly correct that 'There was no mention of such a contingency in the Geneva Convention, which had not contemplated this form of warfare.'[37]

Also, given that the Hague Rules of Air Warfare from 1923 – which would have protected flying ambulances marked by a Red Cross[38] – were never formally ratified by both countries, there *was* no genuine international law which truly encompassed widespread air-sea rescue like that witnessed in the Battle of Britain.[39] Churchill added, perhaps a little more dismissively, that 'The Germans were not in a strong position to complain, in view of all the treaties, laws of war, and solemn agreements which they had violated without compunction whenever it suited them.'

This demonstrated, then, how ill-feeling regarding this subject was able to bubble up on both sides – therefore illustrating how morally nebulous the Battle of Britain could actually be. Nevertheless, it is important to recognize that the 1969 film was meant to be about reconciliation, not incrimination. Both sides ultimately agreed to drop their most stinging criticisms of the other, which allowed the film to move forward, and led to the Luftwaffe's image being somewhat rehabilitated in British post-war memory.

Thus, the British public were exposed to a Luftwaffe that was proud, confident and deadly, but also sensitive, humorous and jovial. With the unmistakable tinkling and thrumming of the 'Aces High' march, audiences began to comprehend the life of the Luftwaffe in the Battle of Britain on an unprecedentedly detailed level. Some of the scenes were particularly memorable in encapsulating human emotions on the German side, such as the scene of the Luftwaffe airmen being served their meal in silence and looking forlornly at all the empty spaces at the table left by their dead.

Other moments, however, are quite a bit more subtle. Up until the *Battle of Britain*, it was rare for the suffering of the 'other side' to be directly shown during the 'pleasure culture' era of British war films.[40] In the 1955 film *The Dam Busters*, for instance, the 1,293 German civilians and foreign prisoners of war who drowned in No. 617 Squadron's breaching of the Möhne and Eder dams are not depicted. In *Battle of Britain*, however, we begin to witness the strains of war on the Luftwaffe and even some of the panic caused by a British air raid in Berlin, even if no deaths are shown in the same heart-stopping detail as the Blitz on London.

Instead of faceless German aircraft picked off cleanly by Hurricanes and Spitfires, we see a Stuka being shot down by a Spitfire and breaking up with a wounded howl of shock from the pilot. Another Stuka pilot and his rear gunner freeze as their damaged Stuka crashes into a shed. German bombers fan out like a startled flock of birds, with the immortal line from the British fighters: 'Help yourself, everybody – there's no fighter escort!'[41] A rather grisly end awaits one Luftwaffe nose gunner when bullets rake his eyes. There is a German bomber crew trying to fight through to the escape hatch as their Heinkel lurches sickeningly towards the sea; an explosion rips through one airman, who slumps over as one of his crewmates desperately tries to bale out.

They are portrayed as being just as vulnerable as members of Fighter Command, with beads of sweat running into their widened eyes – one pilot being momentarily distracted as his gunner is shot, before receiving deadly bullets to the chest himself. A German bomber pilot trying to care for his badly wounded gunner and balance out the stricken aircraft over the sea puts out a protective hand across the latter's chest, with the gunner clinging onto his arm for comfort. The film also shows injured Luftwaffe airmen being loaded into ambulances while the ground personnel are breaking into a shattered cockpit to bring others out.

Occasionally, there are points where the Germans have the upper hand – there is a wave of relief and satisfaction from them as a Bf 109 pilot downs a Spitfire. At other points, such as the famous 'Battle of the Air' sequence, we see the sheer panic of the Bf 109 pilot whose windscreen ends up getting splattered with oil, thus obscuring his visibility. The film also highlights the concern that Bomber Command caused in Berlin in late August 1940.

It shows parts of Berlin, including the U-Bahn station Ruhleben near the Reichstag, past which two airmen are being driven. They are about to be reprimanded for having dropped their bombs over London during that stage of the war, but they cannot help but fondly smile at being home in their own country, with young women swishing in their skirts and Wehrmacht personnel milling about in the city.

Suddenly, however, their faces drop as the air raid sirens began to caterwaul and white searchlights sweep through the night sky. Whistles blare and panicked civilians run for cover as the British close in to attack Berlin, while the two German airmen look on in trepidation. This was not far from reality – Rökker, who was still undergoing pilot training in Berlin during August 1940, recalled how:

> Alerted by the noise of the anti-aircraft guns, we stood in front of our accommodation and observed the magic lights from a distance. The English planes were flying at high altitudes and were forced to turn away beforehand by the strong anti-aircraft defence on the outskirts of the city. Some bombers came into the anti-aircraft searchlights. If one searchlight caught the plane, the other searchlight beams swung onto the target and the anti-aircraft gun tried to shoot down the bomber.[42]

It does not go so far as to show any German casualties, but it nevertheless displays the climate of fear caused among the German civilian population by the RAF during the *Luftschlacht um England*. Thus, the 1969 film really gave a face to the Luftwaffe in the campaign – whether it is the smiley young joker, the swaggering fighter pilot or the glazed concentration of a rear gunner.

Battle of Britain, then, remains one of the most enduring depictions of the Luftwaffe. It did not shy away from the death and disfiguration they had inflicted upon one another, but it also presented both sides as having endured the same anxieties and hardships during the iconic campaign.

As a result of the public interest garnered by the film, there was a renewed wave of curiosity in Britain about Luftwaffe veterans. Finally matching faces to the illustrious names, veterans on both sides now exchanged stories and hearty slaps on the back instead of bombs and bullets. To dismiss the film's more human portrayal of the Luftwaffe entirely, then, is to ignore stark evidence of the shared kinship that blossomed between the two sides after the war. On the other hand, to take the airbrushed cinematic portrayal entirely at face value is to be oblivious to a darker side of the Luftwaffe that fought in the *Luftschlacht um England*.

18

Better Liars than Flyers

Flight Lieutenant John Simpson – who served in No. 43 Squadron during the Battle of Britain – knew first-hand how chivalrous some Luftwaffe airmen had been in the campaign. On 19 July 1940, he had taken off in his Hurricane to intercept a gaggle of Bf 109s roving over the south coast of England. With beads of sweat glinting on his furrowed brow, the disciplined burr of his Browning machine-guns plunged two of the aerial invaders into the sea. Simpson's heart soared with a sense of accomplishment – but, before he could revel in the victory, two more Bf 109s harried his Hurricane from the rear. 'I twisted and turned, but they were too accurate,' he later recalled. 'I could hear the deafening thud of their bullets. Pieces of my aircraft seemed to be flying off in all directions. My engine was damaged and I couldn't climb back to the cloud where I might have lost them.'[1]

Suddenly, he felt a searing pain radiate through his left foot as bullets from the German fighters hit their mark. With the Hurricane lurching forward towards the sea, Simpson was horrified to discover that the control column was lifeless in his frantic hands. As black smoke trailed behind his aircraft like ink from a threatened squid, he knew he was done for. An immediate wave of acceptance melted his hunched shoulders away from his ears, and his hands acted through instinct rather than terror. 'I was at about 10,000 feet, some miles out to sea. I lifted my seat, undid my straps and opened the hood. The wind became my ally and the slipstream caught under my helmet. It seemed to lift me out of the cockpit which was a pleasant sensation.'

As he drifted to Earth, Simpson recalled 'Floating down so peacefully in the cool breeze. I had to remind myself to pull the ripcord and open my parachute. When the first jerk was over, I swung like a pendulum. This was not so pleasant.' His descent was abruptly interrupted by one of the Bf 109s suddenly reappearing, causing Simpson to temporarily freeze. 'The pilot circled round me, and I was just a little alarmed. Would he shoot?' To his immense relief, however, the Luftwaffe fighter pilot did nothing of the sort. 'He behaved quite well,' Simpson concluded. 'He opened his hood, waved to me, and then dived towards the sea and made off towards France.'

The Luftwaffe fighter pilot Erich Rudorffer spoke of a similar incident on 31 August 1940, when he 'saw another Hurricane coming from Calais, trailing white smoke, obviously in a bad way. I flew alongside him and escorted him all the way to England then waved goodbye. A few weeks later the same thing happened to me. That would never have happened in Russia – never.'[2] Rudolf Lamberty of the bomber wing KG 76 recalled that 'There was a certain amount of chivalry at that time between the air forces. I thought the average of British fighter pilots were better than the Germans, although some Germans had the highest scores.'[3] Such knightly language was often reinforced by both sides after the war, with Flight Lieutenant Tom Morgan – also of 43 Sqn – noting how 'As far as the fighters were concerned, we'd go in against each other and it was a joust to see who came out the best.'[4]

The Luftwaffe fighter pilot Heinz Lange seemed to confirm this view of the *Luftschlacht um England* being conducted in a gentlemanly fashion, recalling that 'On the whole, the battle was conducted fairly, neither side shooting aircrew hanging from parachutes, but the strain on our nerves was immense.'[5] Given the taboo subject of shooting parachuting airmen it is possible that the fighter aces on both sides had simply not heard of such a practice within their immediate unit or wing. Nevertheless, as the previous chapter determined, there were in fact isolated moments where fighter pilots were strafed by the other side – but the fact that many Luftwaffe and RAF veterans vehemently denied that this could have ever happened illustrated their emotional attachment to presenting the Battle of Britain as an honourable duel.

What also won respect on both sides were the undeniable acts of courage they observed among the Luftwaffe aircrews. On 25 September 1940, fifty-eight Heinkel He 111 bombers set out to attack the Bristol Aeroplane Company, located towards the south of RAF Filton in South Gloucestershire. Hurricanes and Spitfires were desperately scrambled to intercept them, but it was too late: plumes of acrid smoke were already swirling above the Bristol works, inside of which seventy-two people had been killed and a further 166 injured.[6] This death toll rose when a further nineteen people died of their injuries in the subsequent days; outside of the factory, meanwhile, fifty-eight people died and 154 were seriously injured.[7] Finally, the RAF fighters managed to sneak up on the returning Germans as they headed over Bath.

Pilot Officer Dudley Williams, based with No. 152 Squadron, closed in on the leader of the Heinkel formation and pitted it with machine-gun fire.[8] Suddenly, he noticed both of its engines spluttering with smoke. Williams arched away from the afflicted Heinkel and watched it plunge towards the ground. As it slumped towards a nearby village, the pilot – Helmutt Brandt, a thirty-three-year-old *Hauptmann* – managed to keep the fallen Heinkel up enough to miss a church and then crash into a nearby field. Brandt, his twenty-three-year-old observer Gunter Wittkamp, and their twenty-two-year-old flight engineer Hans

Mertz, all launched themselves desperately into the sky with their parachutes. Brandt drifted safely down to earth, though he suffered grisly injuries in the process.

Wittkamp and Mertz, on the other hand, jumped with insufficient time for their parachutes to fully deploy. Dropping like stones through the open air, they hit the ground with a sickening crack; their fractured bodies were found 400 yards away from the Heinkel. But what of the rear gunner and radio operator? The former, twenty-year-old *Gefreiter* Rudolf Beck heard the panicked calls of his crew mates as they baled out; he felt the slain Heinkel take its last roaring breaths – but he did not care. His watchful eyes remained on the Spitfires witnessing their fateful descent. Jaw tightly clenched and fingers trembling, Beck pulled his trigger at the Spitfire formation.

Even as the Heinkel was about to slam into the ground, the young rear gunner was seen to continue peppering his aerial foe with bullets. One of his final bursts sent a lone Spitfire, piloted by Sergeant Pilot Kenneth Holland-Ripley of No. 152 Squadron, crashing elsewhere in the field. For Beck, too, his time had run out, but he was not alone at the end. Rudolf Kirchoff, the crew's twenty-nine-year-old radio operator, had also switched to a machine-gun to assist Beck. Staying firmly at their posts, both men shattered alongside their Heinkel. When confronted with such evidence of aerial heroics on both sides, then, it is hardly surprising that the two sides began to view each other with genuine respect – both during and after the *Luftschlacht um England*.

Nevertheless, it must not be overlooked that it also benefited each side, both during and after the war, to present the other as an honourable opponent in the air. On 15 September 1942, the *Bradford Observer* covered a gathering of RAF fighter pilots to commemorate the second anniversary of the Battle of Britain. The newspaper quoted Dowding as saying that the Germans were 'worthy foes' and how 'That fact adds to the glory of our fellows beating them. The Germans had to take a terrific gruelling [pounding], and they stuck it for a long time, and I know of several acts of chivalry on their part.'[9] Though this sentiment was largely heartfelt, presenting the Luftwaffe fighter pilots as noble foes instead of ideological brutes also helped Fighter Command to further define its 'Finest Hour' against the very best that the Germans had to offer.[10]

For the Luftwaffe too, being presented as noble adversaries by the RAF was desirable, as it helped to gloss over the organization's long and incriminating affiliation with the Nazi regime. Some of the Luftwaffe's most senior commanders, for instance, had been aware of early Nazi concentration camps well before the outbreak of war. Kesselring was asked by a prosecutor at the Nuremberg Trials, 'Have you any recollection when concentration camps were first established in Germany?' 'Yes,' he confirmed. 'It was in 1933.'[11] Kesselring added that:

> I remember three concentration camps, but I do not know exactly when they were established: Oranienburg, which I often passed by and flew over; Dachau, which had been discussed vehemently in the newspapers; and Weimar-Nora, a concentration camp which I flew over quite frequently on my official trips. I have no recollection of any other concentration camps; but perhaps I may add that, as a matter of principle, I kept aloof from rumours, which were particularly rife during those periods of crisis, in order to devote myself to my own duties which were particularly heavy.[12]

As a soon-to-be-convicted war criminal, 'Smiling Albert', as Kesselring was nicknamed by the Allies, was naturally an unusual case. But that same defence was commonly used by Luftwaffe personnel after the war – that their military service proved too all-consuming for them to be able to give thought to any peculiar sights they came across. Even the Luftwaffe fighter aces who emphasized the chivalry of their air war used this excuse, with Steinhoff once noting that 'On one occasion during a stopover in

Poland prior to the Warsaw Ghetto uprising, I did hear that the SS had devised a "clever" method of eliminating Jews. I couldn't assess that information. I had spent five and a half years serving as a fighter pilot, concerned primarily with daily operations.'[13]

Such testimonies illustrate how the Luftwaffe's commanders were not entirely isolated from the Nazi leadership's persecution of its racial, social and political enemies. With Luftwaffe personnel emerging from the Second World War knowing far much more than they should of Nazi atrocities, latching onto the chivalric aura of the Battle of Britain after the war proved a most convenient means of distancing themselves from the crimes of the Third Reich. Some of the Allied fighter pilots who had come from Nazi-occupied countries, however, were less willing than their British counterparts to accept a more rose-tinted perspective of their opponent in the Battle of Britain. Flying Officer Witold A. Urbanowicz of the Polish Air Force, for instance, refused to believe in a key pillar of the 'Knights of the Sky' narrative: the high number of kills claimed by the Luftwaffe's fighter pilots.

'In the Battle of Britain 1940,' he recalled, 'I noticed that often singular German fighters detached themselves from their formation and returned without fighting to France. Also, the escort of German bombers by their fighters left a lot to be desired.' He added that 'Sometimes the fighters were too far or too high from their bombers, or got lost simply in space. Then they gave their horrendous "victories", in this way misleading the top Luftwaffe authorities.' He concluded that 'After the war, the German bomber crew's answer to my question, "Do you really believe in such horrendous victories of the German fighter pilots in the air?"' was that '"These German fighter pilots with such high victories were better liars than flyers."'[14] Where Urbanowicz particularly refused to pull his punches regarding the Luftwaffe in the Battle of Britain, however, was regarding its political ties to Nazism.

Having witnessed the violent subjugation of his homeland by the Wehrmacht, he was in a better position than British RAF pilots to acknowledge the murderous repercussions that came once the Luftwaffe had laid the foundations for a successful invasion. He once noted on the morality and motivation of each side that 'RAF Fighter Command was incomparably better in this regard than the German Fighter Command. German (Nazi) fighters were pupils of a totalitarian regime, a tool of terror and murder.' Urbanowicz added that, in his view, 'Allied flyers fought for the freedom of man. German (Nazi) fighter pilots lost the Battle of Britain in 1940, and thus decided the defeat of Germany.'[15]

Urbanowicz never relented in calling the Luftwaffe airmen 'Nazis', showing that he was not willing to differentiate between the two in the same way that some of his British colleagues did. Certainly, both flying and non-flying Luftwaffe personnel did sometimes relish the destruction they carried out during the Polish campaign of 1939. A captured Luftwaffe observer named *Leutnant* Pohl was overheard by British intelligence claiming that 'On the second day of the Polish war I had to drop bombs on a station at Posen. Eight of the sixteen bombs fell on the town, among the houses, I did not like that. On the third day I did not care a hoot, and on the fourth day I was enjoying it.'[16] Pohl added that:

> It was our before-breakfast amusement to chase single soldiers over the fields with M.G. [machine-gun] fire and to leave them lying there with a few bullets in the back. The fighter pilot asks him, 'But always against soldiers?', to which Pohl replied, 'People (civilians) too. We attacked the columns in the streets. I was in the "*Kette*" [formation of three aircraft]. The leader bombed the street, the two supporting machines the ditches, because there are always ditches there. The machines rock, one behind the other, and now we swerved to the left with all machine-guns firing like mad. You should have seen the horses stampede!

In a blatant disregard for human life, *Leutnant* Meyer – a fighter pilot listening to him talk – chastised him not for attacking Polish civilians, but that it was 'disgusting, that with the horses', to which Pohl replied, 'I was sorry for the horses, but not at all for the people. But I was sorry for the horses up to the last day.' Admittedly, relishing the killing of the enemy was not limited to any one side; the blatant murder of civilians, on the other hand, does not correlate with the honourable 'knights of the sky' narrative that particularly clings to the Luftwaffe, especially with the fighter pilot showing disgust only for the murder of horses and not of Polish civilians.

In addition, there were also a number of Luftwaffe personnel who had enjoyed participating in antisemitic violence even prior to the Second World War, especially if they had earlier links to the SS or the SA. One captured Luftwaffe pilot, for instance, was overheard boasting that 'I took part in all that business with the Jews in 1936 – these poor Jews! (Laughter.) We smashed the windowpanes and hauled the people out. They quickly put on some clothes and (we drove them) away. We made short work of them. I hit them on the head with an iron truncheon. It was great fun.' The pilot further revealed that:

> I was in the SA at that time. We used to go along the streets at night and haul the Jews out. No time was lost, we packed them off to the station and away they went. They were out of the village and gone in a flash. They had to work in quarries, but they would rather be shot than work. There was plenty of shooting, I assure you. As early as 1932, we used to stand outside the windows and shout: 'Germany awake!'[17]

Other Luftwaffe airmen were initially able to keep their sinister pasts more quiet before they were found out. A *Leutnant* pilot of the bomber wing KG 53, whose Heinkel He 111 had been brought down near Bridgnorth in Shropshire on 16 November 1940, disclosed that he had previously served in the Austrian army. However, it was uncovered that during that time, he was also secretly a member of the SS. According to his interrogation report, 'While Schuschnigg [the Austrian chancellor] was negotiating with Hitler at Berchtesgaden, this man was posted at the frontier on special duty, but spent his time drinking with the Nazis on the German side of the frontier.'

With the *Leutnant* then volunteering for the Luftwaffe five months after the *Anschluss*, some German air force personnel harboured a fervent Nazi past which they brought unapologetically into their wartime service. Then, increasingly drunk on power in occupied territories, there was a small number of Luftwaffe personnel who went on to perpetrate the 'Holocaust by bullets'[18] through participating in executions by firing squad of local Jewish populations in the autumn of 1939.*[19]

Oberleutnant Fried, a captured Luftwaffe transport pilot, was overheard telling a fellow prisoner that 'When I came into contact with the war myself, during the Polish campaign, and I was making transport flights there, I was at Radom once and had my midday meal with the *Waffen-SS* battalion who were stationed there.'

He recalled how 'An SS captain or whatever he was said: "Would you like to come along for half-an-hour? Get a tommy-gun and let's go."' 'So,' he admitted, 'we went to a kind of barracks and slaughtered 1,500 Jews. It only took a second, and nobody thought anything of it.'[20] As an air signalman with a Luftwaffe aircraft warning station in Poland testified, the 'volunteers' were sometimes even rewarded for this heinous violence in September 1939:

> An NCO came and said: 'Any volunteers? We've got Polish prisoners rioting. To get them under control.' I said I wasn't going, but some... one man volunteered, an Austrian, and another man. The NCO was a real hooligan, he'd already got a tough

---

* This was the name of Father Patrick Desbois's study on the one and half million Ukrainian Jews who were murdered in the Soviet Union between 1941 and 1944 by German *Einsatzgruppen*. For further information, see P. Desbois, *The Holocaust by Bullets: A Priest's Journey to Uncover the Truth Behind the Murder of 1.5 Million Jews* (London: Palgrave Macmillan, 2008).

reputation during training. A Bavarian, Kern was the name, I'll never forget it. They'd shot perhaps thirty Poles, and they fetched Jews to bury them. They had to dig the graves. There was a young Jewish lad among them, I heard later, about twenty years old. He punched Kern in the face. So, he killed him, hit him over the head with his gun, so that his brains came out. The Jews had to dig the graves, then they threw the Poles in and finally the Jews were shot as well and thrown in with them. [Kern] came back and was promoted to the rank of sergeant. We got back to Vienna at the end of October. [Kern] got the Iron Cross, 2nd Class, as did all of them who took part.[21]

The depravity of some Luftwaffe personnel in the Polish campaign could also be seen when a *Hauptmann* bomber observer was overheard telling a *Leutnant* bomber pilot in October 1940: 'Of course, abortion is flourishing in Poland, so done by German doctors that it is supposed to help the poor girls. The foetus is removed. But it is taken out so thoroughly that they cannot bear children again! Moreover, if a Pole is found drunk a few times, the first time he gets a warning,' but 'If a man is found out on several occasions, he is sterilized.'[22]

It should not be forgotten, then, that the chivalrous fighter pilots who bade their foe goodbye with a friendly wave did not cancel out the small pool of ruthless killers who already lurked in all branches of the Luftwaffe by the summer of 1940. In addition, both the Luftwaffe fighter pilots and bomber crews who the RAF faced in the Battle of Britain were already benefiting from slave labour. The more powerful the Luftwaffe became through its various conquests, the more the organization could fund its research, aircraft and equipment – to the detriment, of course, of those in occupied territories. This is not to directly 'blame' the German aircrews that genuinely had no knowledge of their war leadership's most surreptitious activities, but to demonstrate the sinister level of power that its organization had accumulated through their military successes.

From as early as September 1939, Milch had begun to carry out the 'Deportation, enslavement and mistreatment of millions of persons', eventually culminating in 'criminal medical experiments upon human beings, and murders, brutalities, cruelties, tortures, atrocities and other inhumane acts.'[23] He had been present at Hitler's conference on 23 May 1939, which had declared four months before the invasion of Poland that 'The population of non-German areas will perform no military service, and will be available as a source of labour.'[24] Then, in a letter between Göring and Hans Frank, the Nazi's governor general of Poland, it was agreed on 25 January 1940 that:

> Supply and transportation of at least one million male and female agricultural and industrial workers to the Reich, among them at least 750,000 agricultural workers of which at least 50 per cent must be women, in order to guarantee agricultural production in the Reich and as replacement for industrial workers lacking in the Reich. The shipping will take place early enough to be completed early in the course of April [1940]. Jewish dealers* who can be freed for this purpose from forced service and so forth, may also be engaged.[25]

This persecution of the local Jewish population in Poland continued during the *Luftschlacht um England*: on 4 January 1941, for instance, an *Obergefreiter* in a Luftwaffe Signals Company in Radom spoke of how 'You feel rich against these backward, filthy people. The Jews here are still running around freely and only have armbands as symbols, [but] ghettos have been set up in Litzmannstadt [Łódź], which is better.'[26] Such a

---

* This comes from the above Nuremberg Trials document that has been translated into English, so it is not possible to find the original German word used for 'Jewish dealers'. It can be assumed from the context that this is referring to Jews whose businesses or trades of some kind had been closed down by the Nazi occupiers, though a derogatory meaning cannot be ruled out.

statement illustrates that some Luftwaffe personnel had their eyes open to the active Nazi subjugation of Jews – and were entirely in favour of it – as the tail end of the *Luftschlacht um England* began to play out.

In France, meanwhile, Jews were being violently treated when carrying out construction work for the Luftwaffe during the last months of the Blitz. 'It's the Jews who work for us,' as a Luftwaffe transport *Stabsfeldwebel* wrote on 23 April 1941. 'But you always have to urge them on. They are characters who could be pitied if they weren't Jews. Hardly anywhere else in the world do you have the opportunity to see such starved and filthy figures as you do here. But this includes all people, not just the Jews.' He concluded that 'You can't even make a comparison with our people in Germany because the difference is far too big. It is therefore no wonder that this *Volk* has so often been ruled by another foreign power throughout history. When you observe everything here, you involuntarily ask yourself whether you want or can make something out of these people.'[27]

That he could not spare any pity for their plight illustrated just how dehumanized Jews had become in this *Stabsfeldwebel*'s mind.

However, it must also be pointed out that during the *Luftschlacht um England*, maltreatment of the occupied Nordic populations was also carried out in territories which were seen as being racially superior by the Nazis, was also observed – something which flying Luftwaffe personnel were certainly not shielded from. Lehweß-Litzmann, who remained stationed in Norway during the spring of 1941, recalled his mixed feelings at seeing violent treatment of the local Norwegian population in Svolvær by the SS. Following Operation 'Claymore', when British cruisers and several destroyers had unexpectedly appeared in the Westfjord and sunk a number of German fishing ships in early March 1941, British and Norwegian commandos had stormed the small fishing town and forced a German surrender.[28]

The Norwegian population had cheered the British sailors who came ashore in the belief that they were being liberated; when it became apparent that the war was ongoing, forty young Norwegian men enthusiastically volunteered to join the Norwegian forces in Britain. In retaliation for the Norwegian support of Claymore, the Nazi politician Josef Terboven and his police chief in Norway, *SS-Obergruppenführer* Wilhelm Rediess flew out to Svolvær with a detachment of German police at their side. Lehweß-Litzmann was ordered by the Luftwaffe *Generaloberst* Hans-Jürgen Stumpff to pick up the men, with Terboven in the co-pilot's seat. 'There was talk of a "tremendous mess" and that something like this had to be stopped at all costs,' Lehweß-Litzmann recalled. 'Terboven had set out from the start to set an example.'

He added that 'Immediately after his arrival, he had residents randomly arrested at the harbour and on the main street of the small town. Everyone who had allegedly shown themselves helpful to the English was arrested.' Lehweß-Litzmann recalled that around a hundred men and women had been thrown aboard a small Norwegian steamer as a makeshift prison, trapped below deck overnight and 'The destination of the transport will certainly have been a German concentration camp. The properties of those families from which someone had gone with the English were burnt down.' In a clear demonstration of how Luftwaffe personnel could hold conflicting feelings about such incidents, however, Lehweß-Litzmann noted that:

> I was depressed: this kind of terrorization of the civilian population disgusted me. This had nothing to do with my ideas about fair fighting. But outrage or even protest didn't cross my mind. I convinced myself: if Germany wanted to be victorious in its war, tough action would have to be taken, which ultimately included certain deterrent measures. At least this experience helped me to no longer view warfare so naively.[29]

This example from just one small Norwegian fishing town as the Blitz continued over Britain, just because the residents of Svolvær had dared to back the British, clearly illustrated that the Luftwaffe which pounded Britain from above could be willing to turn a blind eye to violent SS reprisals in any occupied country. Given, too, that the Norwegians were regarded as being the most racially 'pure' of all Nazi opponents due to their Nordic heritage, the idea that such reprisals would not have also happened in Britain due to Hitler's apparent inclination towards the Anglo-Saxon stock of the nation is wholly naïve and infeasible. The existence of the SS' 'Black Book', which listed all British persons of interest for arrest by the *Einsatzgruppen* upon the launch of Sealion, is further proof of this.*

Even more tellingly, though, it shows how Luftwaffe personnel – as with other Wehrmacht servicemen – could easily justify the SS's heavy-handed treatment of local populations even if it conflicted with their better judgement and personal values. Not long after, Lehweß-Litzmann began his armed reconnaissance operations over Britain late into the *Luftschlacht um England*. Here, then, we get a terrifying glimpse into what a Luftwaffe victory during the ongoing aerial campaign would have meant for the British population – and why it was so fundamentally important that the RAF managed to hold out during the campaign. This became even more disturbingly apparent as the Blitz entered its final weeks, when certain nefarious figures in the Luftwaffe medical corps began to turn towards the idea of using human subjects to conduct horrific aeromedical experiments.

---

* The infamous *Sonderfahndungsliste G.B.* ("Special Search List Great Britain") listed all Britons of interest who could either prove useful to the Germans or who could make occupation of the island very difficult. These included key politicians, political agitators and ideological opponents; luminaries of British arts, literature and culture; journalists, critics and historians; scientists and mathematicians; military and youth commanders; and any other individuals who posed a threat to prolonged occupation. See LBY 89 / 1936, *Die Sonderfahndungsliste G.B* [the Black Book] (Berlin: Reichssicherheitshauptamt, May 1940). The Imperial War Museum, London.

The main Luftwaffe doctors who perpetrated these barbaric endurance, low-pressure, seawater and freezing water submersion experiments on concentration camp inmates – including Soviet and Polish prisoners-of-war, members of the Roma and Sinti communities, political prisoners, Jews and convicted criminals – were later charged with war crimes and crimes against humanity in the Nazi Doctors' Trial at Nuremberg. These men, many of whom held formal positions in the Luftwaffe's medical corps, included Staff Surgeon Hermann Becker-Freyseng, Professor Wilhelm Beiglböck, Surgeon General Gerhard Rose, Senior Staff Surgeon General Oskar Schröder and Senior Field Doctor (Dr) Georg August Weltz. In addition, Dr Hans-Wolfgang Romberg, Dr Siegfried Ruff and the particularly infamous Dr Sigmund Rascher also conducted some of these aeromedical experiments.

Although many of their crimes were committed during the later years of the war, it must be recognized that Luftwaffe doctors were already proposing to carry out horrific aviation experiments on human victims during the tail-end of the Blitz. Their plans were directly inspired by the lessons learnt from the *Luftschlacht um England*, and aimed solely to improve the survival rate of Luftwaffe fighter pilots and bomber crews. Rudolf Emil Hermann Brandt, an SS officer and civil servant, testified after the war that:

> I first heard of the plan to experiment on human beings in May 1941. The idea originated with Dr Sigmund Rascher, Luftwaffe staff surgeon and later *Hauptsturmführer* SS. At that time, Rascher was attending a course in aviation medicine at the Air District Command VII in Munich. He wrote to Himmler suggesting that concentration camp inmates be placed at his disposal for experimentation to determine the effects of extreme altitudes on the human body. Volunteers could not be expected, as the experimental subjects might die.[30]

Indeed, in his letter to Himmler dated 15 May 1941, Rascher

mentioned his current Luftwaffe medical course being undertaken at Munich: 'During this course, where researches on high-altitude flights play a prominent part (determined by the somewhat higher ceiling of the English fighter planes*), considerable regret was expressed at the fact that no tests with human material had yet been possible for us, as such experiments are very dangerous and nobody volunteers for them.'

Rascher openly wrote that 'I put forth, therefore, the serious question: can you make available two or three professional criminals for these experiments?' Rascher continues that such experiments would take place at the Luftwaffe Testing Station for Altitude Research in Munich, and that 'The experiments, from which the subjects can, of course, die, would take place with my cooperation.' Even more disturbingly, Rascher argued that human subjects would be needed as:

> They are essential for research on high-altitude flight and cannot be carried out, as has been tried, with monkeys, who offer entirely different test-conditions. I have had a very confidential talk with a representative of the Luftwaffe surgeon who makes these experiments. He is also of the opinion that the question could only be solved by experiments on human persons (the feeble-minded could also be used as test material).

Brandt reported that 'Himmler had me answer this letter from Rascher, informing him that prisoners would be made available for the research.'[31]

Thus, just days after London had suffered her 'Hardest Night' in the Blitz, the most deranged corners of the Luftwaffe's medical corps were already conspiring to subject prisoners and people with mental disabilities and learning difficulties to

---

* Both the Hurricane and Spitfire originally had lower ceilings than the Bf 109 during the main Battle of Britain, but this changed after the introduction of Mk II Spitfires in late 1940.

horrific aviation medical experiments for the benefit of German airmen against their British counterparts. Rascher had directly identified the technological and physiological shortcomings of the Luftwaffe during the 1940-1 campaign over Britain; now, he implored Himmler to test his grotesque experiments to improve aeronautical performance and aircrew endurance among the German air force.[32]

In addition, Becker-Freyseng testified that 'From February 1941 until August 1941 I was assigned to the 1st platoon for low pressure chamber tests of the Luftwaffe, which was temporarily stationed in Romania. My task chiefly consisted in assisting the demonstration of instructional experiments, which included a test of the reaction of fighter crews at altitudes of 12,000 metres.'[33] This demonstrates, again, how Luftwaffe medical tests were not just carried out by an obscure branch of the organization, but were in fact directly relevant to the fighting experience of its airmen. Naturally, this was planned within the darkest bowels of the Luftwaffe; the overwhelming majority of German airman would not be aware of such top-secret experiments.

Nevertheless, that this kind of research was quickly supported and funded by the Luftwaffe demonstrated the horrific lengths it was willing to go to rectify the mistakes it had made over Britain during the 1940-1 aerial campaign. It shows, once again, that we should not fall into the trap of perpetuating the 'just like us' myth regarding the Luftwaffe airmen, when they would go on to have their combat experience shaped by the knowledge obtained from the horrors of medical experimentation – the seeds of which were already being sown during the *Luftschlacht um England*. It is time, then, to depart from the 'clean Luftwaffe' myth that has long been upheld by the urbane German fighter aces who flew in the Battle of Britain.

Certainly, we should not be so doggedly revisionist as to disregard the admirable bravery and chivalry that was frequently observed among the German airmen during the campaign, which played such a vital role in the healing of relations between the

RAF and the Luftwaffe after the Second World War. Nevertheless, we should also not perpetuate the misleading post-war notion that the German 'knights of the air' were wholly detached from the crimes that their nefarious political regime committed, when such crimes were occurring within the flying and non-flying arms of their very organization. It is important, then, to not dismiss the countless tragedies and crimes that likely would have occurred against the British population had Sealion taken place. All of these would have been directly enabled by a Luftwaffe victory in the Battle of Britain – whether its airmen felt they were, in Steinhoff's words, 'fighting sportsmanlike' or not.

Conclusion

At a time of unmitigated disaster and paralysing uncertainty, it is human nature for the mind to desperately clutch at straws as to what could have been done to avoid the unpalatable situation at hand. This applies whether one is a regular person on the street, or one of the most evil dictators to ever pollute the earth. As the 'thousand-year Reich' disintegrated before Adolf Hitler's very eyes, it turned out that even the devil has his own demons. *Generalfeldmarschall* Ferdinand Schörner, an ardent Nazi and ruthless *Heer* general, witnessed the fall of the Reich before fleeing via his personal aircraft to Austria on 8 May 1945. Infamous for having executed thousands of his own men, he was later convicted of war crimes. An Allied interrogation session with Schörner recorded that:

> Before his death, Hitler, as the field marshal [Schörner] reported, regretted that he had not dared to carry out the England operation. He overestimated the strength of the English troops – or, at the very least, their strength on paper was passed onto him as if it were the reality. There was no doubt that the planned air superiority had not been achieved in the air area of the Channel. The British air force had become increasingly stronger. But the

actual strength ratio between the German and English troops was much more positive than assumed.[1]

Schörner's testimony demonstrates how quickly the German air force's campaign over Britain became one of the biggest 'what-ifs?' in military history. This, in turn, reinforces the importance of examining what the *Luftschlacht um England* had meant for the Luftwaffe and Nazi Germany: despite all of the pivotal moments that had occurred since 1940, it was allegedly the inconclusive Battle of Britain which haunted the *Führer* until his dying breath. The conclusion that the air campaign constituted Hitler's first real sticking point during the Second World War is not unusual; but the extent to which the aborted Sealion operation tormented Hitler certainly was.

Some historians have argued, however, that the *Luftschlacht um England* did not ostensibly change the course of the war – that Hitler largely retained the same military options after the aerial campaign that he had wielded before it,[2] or that the Luftwaffe perceived their failure to gain air superiority over the island as a mere bump in the road.[3] Others go so far as to claim that to call the Battle of Britain a decisive moment in military history is to fall prey to Anglocentric mythology, consigning major campaigns like the submarine war in the Atlantic or the Battle of Stalingrad to the background.[4] Lehweß-Litzmann also disagreed with placing great importance on the aerial offensive:

> Many historians today glorify the Battle of Britain as one of the decisive battles in the Second World War. But that is probably fundamentally wrong. Although the battle was lost for the German Luftwaffe and had materiel and, above all, personnel consequences, it did not mean a decisive change or weakening of Germany's strategic position. The march against the Soviet Union was never in danger for a moment; the postponement of the attack from May to June 1941 was due to the situation in the Balkans, especially in Yugoslavia.[5]

Indubitably, it would be inaccurate to argue that the RAF's defensive victory in the Battle of Britain sent the Luftwaffe into an irrevocable five-year tailspin towards defeat. If a 'decisive' victory is defined as one that is conclusive, then it is true that the RAF did not achieve this. But Fighter Command's aim during the campaign was to *endure* the Luftwaffe, not eradicate it. That it achieved this and more with flying colours was seen in how – contrary to Lehweß-Litzmann's claims – the *Luftschlacht um England* not only greatly complicated the Luftwaffe's war on a strategic level by risking battle on multiple fronts, but it also proved highly detrimental to its overall endurance and operational strength by 1941. Even Lehweß-Litzmann had to admit that:

> The fight against Great Britain ended in a fiasco. The self-proclaimed 'strongest air force in the world' proved to be too weak against an equal opponent. It could not bomb the British fleet from its ports or even persuade England to continue to stand still. I, too, was one of the masses of people in Germany who did not recognize this defeat at the time.[6]

Thus, although the campaign may have not been a decisive RAF victory that rendered the Luftwaffe's eventual defeat to be inevitable, it was certainly a momentous turning point. Both during and after the Second World War, a number of Luftwaffe personnel identified the Battle of Britain as a spanner in the works that the Germans were unable to fully bounce back from. Rieckhoff noted that 'The first blow from which [the Luftwaffe] never recovered came in the late summer of 1940, during the fight against England. This was followed by the winter war against England in 1940/41, in which it slowly continued to bleed out.'[7] As Bechtle noted in his lecture from 1944, 'Until now, every opponent had been defeated quickly and devastatingly in so-called "*Blitzkriege*".'[8] Steinhoff also left no doubt to the *Luftschlacht um England*'s significance:

> I personally believed that the outcome of this air battle was a major turning point in the war. From that moment on, I knew we had lost control of the third dimension: the air space over Europe. I had already taken part in the campaign in France, the destruction of Rotterdam, and the occupation of Scandinavia. In terms of air warfare, all of this had been child's play. Now the situation was deadly serious.[9]

Being so closely associated with the campaign, however, it is not surprising that Steinhoff wished to present it as a major and defining moment of military history. Yet the Luftwaffe anti-aircraft gunner Raynor added perceptively that 'There was no way to engage in a major operation across the Channel and/or the North Sea. Seen from the tactical point of view, another mistake by the German High Command and/or Hitler. We had left the British Isles like an untreated skin cancer, to develop and spread its deadly load. England became the gigantic aircraft carrier and garrison for its own, and, eventually, the American forces.'[10]

Indeed, as this book has extensively documented, the blow to Luftwaffe morale in being unable to subdue the 'pirate island' in the Battle of Britain was monumental – causing the irreconcilable chasm that opened up between the *Reichsmarschall* and his frustrated pilots, as well as the rocketing suicide rates that plagued the Luftwaffe as the campaign progressed. To cover up their disappointment, some of its personnel attempted to diminish the stakes of the aerial campaign in the first place, with Deichmann claiming in a Bundeswehr Staff College study from 1959 that:

> From the beginning, the idea of the air battle against London was to make England sue for peace... therefore the aim of the Battle of Britain was POLITICAL, not MILITARY. Fighting for air superiority was, in the first instance, designed to clear the path towards obtaining a political decision.[11]

This rendition of the campaign, however, does not compute with the fact that both the Luftwaffe and her sister Wehrmacht branches devoted a significant amount of brainpower to planning Operation Sealion. Such was the Luftwaffe's thoroughness in the endeavour that they had done everything from conducting flight tests with smokescreens and participating in joint war exercises for the landing, to planning the relocation of Luftwaffe mail closer to the coast and the establishment of a dedicated *Luftgaustab* ['Air District Staff'] to control England.

Thus, even if the Germans ultimately did not elect to roll the dice on Sealion, this did not stop them from giving it a few good shakes – although, of course, consideration is not the same thing as commitment. Irrespective of how likely the invasion was to succeed, it cannot be denied that in the *Luftschlacht um England*, the Luftwaffe was worn out by the divided aims of Sealion and, in Deichmann's words, forcing a 'political decision'. This ongoing exhaustion later exacerbated the detrimental effects of its overcommitment the following year. But, even if the reader is still not convinced that the Battle of Britain had a critical impact on the way the Luftwaffe's fortunes unfurled during the Second World War, it cannot be denied that the battle continues to command more public attention than any other aerial campaign in British military history.

Such a densely populated subject must be approached innovatively in order to prevent stagnation and complacency settling in – particularly when the men on both sides who fought in the campaign themselves have largely passed on. *Eagle Days*, then, has striven to provide a truly exhaustive account of life and death for the Luftwaffe in the Battle of Britain – from its flying arms to its ground personnel, and from the lowly *Gefreiter* to the blustering *Reichsmarschall*. More broadly, this book has also invited its readers to consider the campaign from a fresh and unique perspective: not just documenting the wartime German experience with unprecedented depth and immersion, but also pointing out the transnational dimensions of the combat zone itself.

This has greater implications for how we approach the study of military history, demonstrating the exciting new information which can be uncovered on well-trodden topics when historians strike a careful balance between documenting operational minutiae and not losing sight of the bigger picture. The same can be said of the Luftwaffe personnel themselves, whose similarities with the British airmen are often fixated upon at the expense of examining their greater connection to the overarching criminal regime they served. Without understanding this broader socio-political influence, it would be hard to fathom why the German airmen – who so often expressed their respect and admiration for their British rivals before, during, and after the war – would continue to risk life and limb to fight them.

This can still be acknowledged without erasing the fact that the Luftwaffe's experiences did often mirror those of the RAF in the campaign. These experiences were particularly well captured in a 1973 novel based on the Battle of Britain by Valentin Mikula, who had served as a Stuka pilot during the Second World War. Entitled *Stuka: Tatsachenbericht* ('Battle Report'), it tells the story of the *Luftschlacht um England* through the eyes of two young friends serving in the Luftwaffe. Certain sections on the campaign, however, are more serious, and appear to reveal the tired pilot behind the penmanship:

> The airmen had it really good while they lived, but most of the time they didn't live long. The others didn't know this exactly; they kept seeing them flying. But it was only a small fraction of the Luftwaffe that had to go to the enemy, and this fraction, time and time again, dissolved into dust and appeared as sand before the eyes of the angels of heaven. The ground troops were doing much worse, but they lived a little longer. Flying personnel were consumables, and the few days that separated them from death were made pleasant for them; they had to have it, otherwise they wouldn't have been able to fly; you can't fly on an empty stomach.[12]

CONCLUSION

It is hardly surprising, then, that the RAF and the Luftwaffe saw so much of themselves within each other. Aviators had a unique understanding which had the power to sometimes transcend the squabbling of politicians. For the sake of European harmony after the Second World War, too, it was just as well that they recognized each other's similarities instead of their differences. Yet, as the war slips increasingly from living memory, it is imperative that we reflect upon how such catastrophic events came to be in the first place.

Only by unflinchingly laying bare the ways in which air forces can be persuaded, cajoled, encouraged or coerced into doing the bidding of politicians can we be brutally honest about how such conflicts are able to emerge, despite a general sense of commonality and respect among their aviators. This reminds us, then, of the eternal lesson that political megalomania can usher even seemingly brave, decent and intelligent citizens down the most unsalvageable and unthinkable of roads.

What *Eagle Days* has done is to write the rest of the Luftwaffe back into the story of the Battle of Britain – not just the illustrious fighter pilots and their well-known generals, but the entirety of the infamous organization.

Writing individuals back into history is not always about the vital practice of giving voices to victims, or recognizing the valour of unsung heroes: sometimes, it reiterates to us that history is not always black and white. So, *Eagle Days* fully documents how the Luftwaffe was able to sustain its morale during its first protracted campaign, but also highlights the full cost – psychological, military and political – that it paid in the process. The book also counterbalances the lingering public notion in Britain that the campaign was an honourable joust, with some observers preferring to regard the Battle of Britain as a cautionary tale instead of a jingoistic ideal.

Though peppered with admirable moments of chivalry and the occasional softening of wartime attitudes, there is no doubt that the air battle was a rancorous fight to the death. For the defending island nation, it was easy to encapsulate what the Battle of Britain meant – it was their 'Finest Hour', which, as the Second World War progressed and the Allies slogged through to victory, became even more significant as a pivotal moment. It was 'The Few' – Bomber Command, Coastal Command and, most enduringly, Fighter Command – who rallied to defend the island and succeeded in securing a defensive victory that changed the tides of the Second World War.

In post-war Germany, of course, memories over Wehrmacht draws, victories and defeats are very different. But this does not mean that the Battle of Britain did not mean something to the Germans at the time. This was meant to be the Luftwaffe's own 'Finest Hour' for another reason: laying the foundations for one of the most ambitious and unprecedented amphibious crossings. The 'Wonder of 1940' had been militarily impressive enough; to prove once and for all that Britain was no longer an island, though, would have been – in Göring's words – 'the proudest victory in German history'.

To fully elucidate the Luftwaffe's role in the RAF's 'Finest Hour', it has been necessary to faithfully document the human experiences of the Luftwaffe. This, however, has not intended to be the same thing as 'humanizing' it. The former explores what it means to be human and all of its messy implications; the latter implies that there is a subtle ulterior motive to present such individuals as relatable, understandable, sympathetic and perhaps even reasonable.

Clearly, with a politically incriminated air force that left much horrific death and destruction in its wake, that is not an outcome which this book has sought to achieve. What it *has* endeavoured to convey, however, is that it is important not to completely strip these infamous young men of their humanity, because that is what drove them in their *Luftschlacht um England* – not to mention, it is also what allowed them and their British counterparts to heal together, and even to greet old foes as new friends. Human nature is what kept these men ticking over – fighting for their comrades, their families and their very survival, though some did not extend that sense of humanity to all racial members of the Reich and the foreign territories they pitilessly occupied.

Human nature also underpinned the Luftwaffe's rampant impatience at the Nazi leadership's failure to guide them towards a final victory over Britain. They had addled the Luftwaffe until its men were divided on whether the proposed invasion was a means to an end, or if bombarding the British Isles was the end in itself. Despite the first serious test of the German public's loyalty to the regime, the Nazi propagandists were able to salvage the situation. Yet, like a neurotic overachiever, the German air force had been heavily shaken the minute that it came up against its first great failure – the Battle of Britain. Thus, where the RAF nursed its 'Finest Hour', the Luftwaffe cursed its 'Frittered Hour'.

Acknowledgements

Writing may feel like a solitary pursuit at times, but it is more akin to a relay race: it is the hard legwork and heartfelt encouragement of others that really gets you over the finishing line. I would like to extend my warmest gratitude to the illustrious Dan Jones and Charles Spencer, who took a chance on a young and upcoming historian to represent their new military history series. The same is to be said of my fabulous agents, Georgina Capel Associates – thank you so much to George, Polly, Rachel and Irena for giving me the space and courage to grow as an historian.

I am hugely indebted to Head of Zeus for publishing this book – especially to Richard Milbank for passionately believing in it from the very beginning, and to Iain MacGregor for being such a dedicated and insightful editor. Thank you also to Ellie Jardine and to the marketing team at HoZ for working hard to promote *Eagle Days*. *Ein großes Dankeschön* goes to Peter Eisenbach for sharing parts of Kurt Scheffel's diary with me and also to Andy Saunders – one of my earliest champions in the field – who kindly volunteered other parts of Scheffel's diary and a sneak preview of his excellent new book on the Bf 109.

This book has also benefited enormously from the sagacious critiques of Professor Daniel Todman and Dr Stephen Moore

– they are both at the very pinnacle of the profession, and are just as wonderful as people as they are historians. I would also like to thank my dear friend Dr Jan Tattenberg, who reviewed an early draft and whose opinion, as a German historian, was invaluable in ensuring that a balanced assessment of the subject matter had been achieved by the book.

I wish to extend my thanks to a number of British and German archives that provided some of the materials used – especially the tireless staff at the National Archives (Kew), the Imperial War Museum (London), and the Royal Air Force (RAF) Museum at Hendon in London – especially Dr Harry Raffal, Lucia Constance, Gary Haines, Andrew Dennis, and Ewan Burnet – after patiently trawling through the museum's extensive collection for me!

Over in the United States of America, I am so grateful to the staff at the National Archives (College Park) in Maryland for locating the relevant microfilms from the German Captured Documents Collection. I am also very thankful for the staff at the Smithsonian National Air and Space Museum, especially for the guidance provided by Dr Mike Hankins. I would also like to thank staff at the United States Holocaust Memorial Museum for their assistance with aviation psychology and experimentation content.

I would like to honour the attentive German archival staff who found the best material at the Deutsches Historisches Museum, Berlin (especially Thomas Jander); the Bundesarchiv-Militärarchiv, Freiburg im Breisgau; the Militärhistorisches Museum Flugplatz Berlin-Gatow; and the Museum für Kommunikation, Berlin (particularly Gunnar Goehle). Particularly warm thanks go to Tobias Thelen at the Bibliothek für Zeitgeschichte, Stuttgart, who was incredibly helpful in locating the relevant Luftwaffe *Feldpost* (Fieldpost) for me and also was thoughtful enough to put out some additional collections that I was not aware of.

Sincere thanks to two of the key History mentors in my life: Mr Crocker, the 'gateway' GCSE History teacher who helped spark my lifelong obsession with the subject, and Dr Peter

Grieder, my MRes and PhD supervisor at the University of Hull, who spotted my potential before anyone else did – especially myself! A warm thank you also to the History departments at the University of Hull and King's College London – especially Dr David Jordan – for aiding enormously in my professional development and being most understanding of my book deadlines.

I must now turn to my dearest souls who somehow manage to love me through every (frazzled) writing stage! My love is infinite for my best friends – Dr Laura Burkinshaw, forever the steady aircraft carrier to my scrambling fighter jet; Dr Megan Kelleher, my almighty wingwoman whom the RAF Museum is blessed to have as their Historian and Academic Access Manager; Dr Tasnim Hassan – loyal, brave and an unrelenting warrior for the social change we desperately need in this world; and Dr Alice Whiteoak, whose kind words and hearty laugh reverberate wherever she goes.

There is no achievement, however, that I have ever made without my beautiful, devoted and nurturing family by my side. Their hard work – often against the odds – proved to me early on that you can always make something of yourself, no matter your background. To my grandfather 'Guv', who survived the Luftwaffe's Blitz himself as a young boy, and who sadly passed away shortly before this book was published: your colourful life, forged by 'Taylor tenacity' and an unwavering *carpe diem* attitude, serves as an eternal inspiration. Your 'Fab 1' loves and misses you deeply.

Finally, to my parents, Peter and Elizabeth Taylor – thank you, Dad, for being a fighter and a provider, no matter what the world threw at you. Thank you, Mum, for being a golden ray of loving sunshine in everyone's life – you always succeed in dissipating my storm clouds! Thank you, both, for exemplifying strength, resilience, compassion and kindness. Never, in the field of military history, was so much owed, by a grateful daughter, to her loving family.

Further Notes & References

Introduction

1. B464 – 'Collection of reports (in German) by German aircrews involved in the attacks on RAF stations Kenley and Biggin Hill', August 1940. Lw. Kriegsberichterkomp. (mot.) 4, Goybencourt, 26.8.1940, Nr. 502. 'Grossangriff gegen England. Einige Besatzungsmitglieder erzählen ihre Erlebnisse'. RAF Museum, London.
2. H-E. Bob, 'Lecture as dialogue at Air Command and Staff College in Maxwell/Alabama USA. Subject: Battle of Britain', Private Papers of Major H-E Bob. Imperial War Museum, London.
3. J. Müller-Bauseneik, 'Mythos "Blitz"', *Militär & Geschichte Extra*, Nr. 13, (March 2020), 3.
4. J. Weigelt, 'Falsche Vorbilder – Hängepartie um Marseille-Kaserne', NDR, 13 May 2020. Available online: www.ndr.de/nachrichten/info/sendungen/streitkraefte_und_strategien/Falsche-Vorbilder–Haengepartie-um–Marseille–Kaserne, streitkraefte602.html [Accessed 24 May 2021].
5. See J. Althaus, 'Die Sturzkampfbomber starben wie beim Moorhuhnschießen', in 'Luftschlacht um England: Unternehmen "Adlertag"', *Die Welt* (13 August 2020); G. Kramper, 'Hitlers Flieger hätten den Luftkrieg über England gewinnen können – sagt die Mathematik', *Stern* (3 February 2020); and *Spiegel-Geschichte*, 'Luftschlacht um England – Drei Tage der Entscheidung' (2023) – the German re-run of the British Channel 5 series *The Battle of Britain: Three Days that Saved the Nation* (2020). *Der Spiegel* also touches on the Battle of Britain in an article entitled 'Fotoprojekt zur Luftschlacht um England: Als London brannte' (9 May 2016).

6. See P. Cronauer, *Luftschlacht um England – Unternehmen Adler: Wendepunkt der Geschichte* (Munich: GeraMond, 2019); J. Wehner, *»Technik können Sie von der Taktik nicht trennen«: Die Jagdflieger der Wehrmacht* (Frankfurt a. M; New York: Campus, 2022); H. P. Eisenbach and C. Dauselt, *Der Einsatz deutscher Sturzkampfflugzeuge gegen Polen, Frankreich und England 1939 und 1940: Eine Studie zur 'Grazer Sturzkampfgruppe' I./76 und I./3* (Aachen: Helios, 2019); F. Bell, *Britische Feindaufklärung im Zweiten Weltkrieg: Stellenwert und Wirkung der 'Human Intelligence' in der britischen Kriegsführung 1939–1945* (Paderborn: Ferdinand Schöningh, 2016).
7. A. Walker, *From Battle of Britain Airman to PoW Escapee: The Story of Ian Walker RAF* (Barnsley: Pen & Sword, 2017), 54.
8. Interview with Johannes Steinhoff, part of the history documentary by D. Hoffman, 'How Hitler Lost the War' (Varied Directions Production, 1989). [YouTube] Available online: www.youtube.com/watch?v=G3Qb4E-TdNo [Accessed 28/06/2021].
9. G. Rall, *Mein Flugbuch: Erinnerungen 1938–2004.* Edited by K. Braatz. (Moosburg: NeunundzwanzigSechs, 2004), 17.
10. '*Vorwort aus dem Vortrag vor dem Arbeitskreis Sächsischer Militärgeschichte am 04. März 1995 in Dresden*', 4 March 1995. Hauptstaatsarchiv, Dresden.
11. J. Steinhoff et al. (eds.), *Voices From the Third Reich: An Oral History* (Boston; Cambridge, MA: De Capo Press, 1994), 3.
12. J. Holland, 'Foreword', *Duel Under The Stars: The Memoir of a Luftwaffe Night Pilot in World War II* (Barnsley: Greenhill Books, 2020), 7–15; 9.
13. MF 10112/1 – 'First Hand Accounts of the Battle of Britain, collected by Alexander McKee. 2. Accounts by Luftwaffe personnel. General der Flieger a.D. Paul Deichmann, Chief of Staff, Luftflotte 2. Studiengruppe Luftwaffe bei der Führungsakademie der Bundeswehr – 7 August, 1959.' Royal Air Force Museum, London.
14. 3.2002.0904. 'Martin Meier an seine Ehefrau am 30.08.1940', France, 30 August 1940. This letter has been made publicly available on the MfK website, hence the naming of the Luftwaffe serviceman in full. Meier had served in the Luftwaffe

since August 1939 and was attached to a number of Luftwaffe air signals units, including *Luft-Nachrichten-Regiment 35*, before being reassigned to the German army's 14. *Panzerdivision* later in the war. Museum für Kommunikation, Berlin.

1. Corps of Vengeance

1. Report of Interrogation No. 5779; Report on Interrogation of: P/W Kessler, Ulrich, General d. Flieger (Maj. Gen.), *Kampfgeschwader 1* (Hindenburg), Capt. Halle, 20 September 1945. Donovan Nuremberg Trials Collection, Cornell University.
2. 010766. 'Interrogation of General Galland, famous fighter pilot of the German air force, and the birth, life and death of the German day fighter arm (1945)', 5. RAF Museum, London.
3. W. Deist, *The Wehrmacht and German Rearmament* (Basingstoke: Macmillan, 1981), 58.
4. 010766. Interrogation of General Galland.
5. Steinhoff, et al., *Voices from The Third Reich*, 7.
6. Ibid.
7. M. Alpert, *Franco and the Condor Legion: The Spanish Civil War in the Air* (London; New York: Bloomsbury Academic, 2019), 33.
8. H. J. Rieckhoff, *Trumpf oder Bluff? 12 Jahre Deutsche Luftwaffe* (Geneva: Inter Avia, 1945), 112.
9. 010766. Interrogation of General Galland, 6.
10. A. Saunders, *RAF Fighters vs Luftwaffe Bombers: Battle of Britain* (Osprey, 2020) [Kindle Edition], 51.
11. W. Lehweß-Litzmann, *Absturz ins Leben* (Querfurt: Dingsda, 1994), 111–12.
12. Saunders, *RAF Fighters vs. Luftwaffe Bombers*, 67.
13. J. S. Corum, *The Luftwaffe: Creating the Operational Air War, 1918–40* (Lawrence, KS: University Press of Kansas, 1997), 199.
14. N. Townson, *The Penguin History of Modern Spain: 1898 to the Present* (London: Penguin), 1924.
15. 'The Guernica Massacre', *Dundee Courier*, Thursday 29 April 1937, 6. The British Newspaper Archive.
16. Ibid.
17. 94/26/1, 'Private Papers of Major D. von Eichel-Schreiber', Documents. 2685. 12 January 1993. Imperial War Museum, London.
18. Rall, *Mein Flugbuch*, 24.
19. Ibid.
20. Ibid.
21. P. Kaplan, *Fighter Aces of the RAF in the Battle of Britain* (Barnsley: Pen & Sword,

2008) [Kindle Edition], 229–30.
22. Ibid.
23. AIR 2/2761. Visit by German Air Force Mission to England. Opened 28-1-37. 'Watch on Nazi Air Visitors by our Air Correspondent.' The National Archives, Kew.
24. Ibid.
25. Kaplan, *Fighter Aces of the RAF*, 231.
26. AIR 2/2761. No. 983. (45/215/37) British Embassy, Berlin. 3rd November, 1937. Visit by German Air Force Mission to England. Opened 28-1-37.
27. F. Bono, 'Why did Nazi Germany bomb four Spanish villages?', *El País* (4 May 2018). Available online: www.english.elpais.com/elpais/2018/05/03/inenglish/1525347336_693939.html [Accessed 10/10/2023].
28. As part of the swarm – known officially to the Luftwaffe as the *Schwarmgefechtslinie* ('swarm battle line') or *Schwarmwinkel* ('swarm angle') – the *Schwarmführer* ('swarm leader') was positioned slightly out in front, supported by a *Rottenflieger* ('wingman') on one side. On his other side was the *Rottenführer* ('pair leader'), who had his own wingman on his other side, sticking further out from the main three. See Wehner, *»Technik können Sie von der Taktik nicht trennen«*, 148–9.
29. Bob, 'Lecture as dialogue at Air Command and Staff College in Maxwell', Imperial War Museum, London.
30. Ibid.
31. S. Schüler-Springorum, *Krieg und Fliegen: Die Legion Condor im Spanischen Bürgerkrieg* (Paderborn: Ferdinand Schöningh, 2010), 19.
32. J. Corum, *Wolfram von Richthofen* (Lawrence: University Press of Kansas, 2008), 149.
33. S. W. Mitcham, *Eagles of the Third Reich: Leaders of the Luftwaffe in the Second World War*. 1988. (Manchester: Crecy Publishing, 2010), 47.
34. 15229. H. Göring, 'Speeches made by Goering and Hitler welcoming *Legion Condor* home from victories with Nationalist Army in Spain 6/6/1939.' Imperial War Museum, London.
35. Ibid.
36. S. L. Segal, *Mathematicians under the Nazis* (Princeton; Oxford: Princeton University Press, 2003), 284.
37. M. Wildt, *Hitler's Volksgemeinschaft and the Dynamics of Racial Exclusion: Violence against Jews in Provincial Germany*,

1919–1939. Translated from German by Bernard Heise. (New York; Oxford: Berghahn, 2014), 132.
38. Ibid.
39. 520–251. *Der Soldatenfreund (Ausgabe C) Jahrbuch für die Luftwaffe mit Kalendarium für 1939*, 19. Jahrgang. (Hanover: Adolf Sponholz Verlag, 1938), 1. The Imperial War Museum, London.
40. CAB 104/32. Foreign Affairs – Germany. The German Air Force, July 1939. 'From H.M. Representative at Berlin, No. 31 of 10th Jan. German Air Force, 21st Feb 1939. George Ogilvie-Forbes to Viscount Halifax. German Air Force: History. Part II. – From 1933, page 2.' The National Archives, Kew.
41. Ibid.
42. C. Epstein, *Nazi Germany: Confronting the Myths* (Chichester: Wiley Blackwell, 2015), 117.
43. 27.9.1938, 2.Kp./Fliegertechn. Schule, Berlin-Adlershof. Sondersammlungen, *Die Bibliothek für Zeitgeschichte*, Stuttgart. All names from this collection of Luftwaffe *Feldpost* have been redacted due to German privacy laws.
44. X003-8879. 'Private and Confidential Report on Visit to Germany. October 11th to 14th, 1938', by A. H. R. Fedden. RAF Museum, London.
45. Report of Interrogation No. 5779, P/W Kessler, Ulrich, General d. Flieger, 20 September 1945.
46. Ibid.
47. K. Scheffel, I./ *Sturzkampfgeschwader 77* Unit Diary (1940), 12. Kindly provided by Andy Saunders.
48. Ass.Arzt, I./Flak-Rgt. 508, Eger, 8 October 1938. *Bibliothek für Zeitgeschichte*, Stuttgart.
49. Steinhoff et al., *Voices From The Third Reich*, 106.
50. Ass.Arzt, I./Flak-Rgt. 508, Eger, 8 October 1938.
51. Bob, 'Lecture as dialogue at Air Command and Staff College in Maxwell', Imperial War Museum, London.
52. K. Scheffel, I./ *Sturzkampfgeschwader 77* Unit Diary (1941).
53. Ibid.

2. The Line Was Dead

1. Transcript for NMT 2: Milch Case. Official Transcript of the American Military Tribunal in the matter of the United States of America against Erhard Milch, defendant, sitting at Nuernberg, Germany, on 12 March 1947, *USA v. Erhard Milch*, HLS Nuremberg Trials Project, Harvard Law School.

2. Ibid.
3. Ibid.
4. Ibid.
5. Ibid.
6. Transcript for NMT 2: Milch Case.
7. Ibid.
8. Ibid.
9. P. W. Stahl, *Kampfflieger zwischen Eismeer und Sahara – In meinem Fall: Ju 88* (Stuttgart: Motorbuch, 1990), 7.
10. Steinhoff et al., *Voices from the Third Reich: An Oral History*, 65.
11. 10935. Neumann, Eduard Anton (Oral History). 14 October 1989 [German]. Imperial War Museum, London.
12. Steinhoff et al., *Voices from the Third Reich: An Oral History*, 65.
13. K. Bruns, 'Rechenschaftsbericht', in X004–8438. 'Generaloberst Bruno Loerzer geb. 22. Jan 1891 in Berlin.' The RAF Museum, London.
14. Ibid.
15. P. W. Hozzell, *Conversations with a Stuka Pilot* (Verdun, 2014) [Kindle Edition], 23–4.
16. Sold. Flg.Horst Finsterwalde, 20 August 1939. [Letter] Bibliothek für Zeitgeschichte, Stuttgart.
17. R. D. Müller, *Hitler's Wehrmacht, 1935–1945* (Lexington: The University Press of Kentucky, 2016) [Kindle Edition], loc. 460.
18. See J-D. G.G. Lepage, *Aircraft of the Luftwaffe, 1935–1945: An Illustrated Guide* (Jefferson, NC: McFarland & Company, 2009), 17–20.
19. R. L. Blanco, *The Luftwaffe in World War II: The Rise and Decline of the German Air Force* (New York: Julian Messner, 1987), 26–7.
20. Mitcham, *Eagles of the Third Reich*, 67.
21. Ibid.
22. Private Papers of Major H-E Bob. Imperial War Museum, London.
23. Blanco, *The Luftwaffe in World War II*, 27.
24. Ibid.
25. Sold. Lw.Bau-Kp.3/XII, Hüttenfeld-Seehof, 29 September 1939. Bibliothek für Zeitgeschichte, Stuttgart.
26. Flg. 3.Kp./Flg.Erg.Rgt.12, Fliegerhorst Stendal, 28 September 1939. Bibliothek für Zeitgeschichte, Stuttgart.
27. Ibid.
28. J. Steinhoff, *Messerschmitts Over Sicily: Diary of a Luftwaffe Fighter Commander* (Stackpole Books, 2023) [Kindle Edition], 54–7.
29. Steinhoff et al., *Voices from the Third Reich*, 80.
30. Steinhoff, *Messerschmitts Over Sicily*, 54–7.
31. 36 743. 7.Flugm.Kp./Lg.Nachr.Rgt.4, 11 October 1939. Bibliothek für Zeitgeschichte, Stuttgart.

32. H. Berg, *Sie nannten uns Helden: Der Stuka-Flieger Hans Deibl* (Kral: Berndorf, 2021), 18. It should be mentioned that the Stuka pilot Hans Deibl collaborated with the author Helmut Berg to create this memoir-like book. As such, although it is heavily constructed from their in-depth conversations and notes together, with Deibl still alive when it was published, Berg nevertheless retains some editorial control of the memoir.
33. M. Wildt, '"Eine neue Ordnung der ethnographischen Verhältnisse": Hitlers Reichstagsrede vom 6. Oktober 1939', *Zeithistorische Forschungen*, 129–37.
34. A. Hitler, 'Rede im Reichstag (6. Oktober 1939)', Zwangsmigrationen in Europa 1938–48: Die NS-Bevölkerungs- und Vernichtungspolitik für Osteuropa, *Friedrich Ebert Stiftung*. [Speech]
35. RG 242, Foreign Records Seized Collection. Records of Headquarters, German Air Force High Command (OKL) T-321, Roll 6. Chef des Ausbildungswesen, L.In.13 – Oberkommando der Luftwaffe. Az. 41 L 10 L.In.13/5d Nr. 2566/39, Berlin, 8. Oktober 1939. 'Bericht über die Wirkung der deutschen Luftangriffe gegen Krakau und Radom in luftschutztechnischer Hinsicht'. National Archives, College Park, Maryland (USA).
36. Ibid.
37. Ibid.
38. T321, Roll 12. Organisatorische Erfahrungen 1939/41–2. Polen 1939/40. National Archives, College Park, Maryland (USA).
39. L 26 877 = Mun.Ausg.St.d.Lw. 17/XII, Sold. 9 October 1939. [Letter] Bibliothek für Zeitgeschichte, Stuttgart.
40. 09.10.1939. L 08 504 = Gefreiter, Stab.Res. Flak-Abt.492. [Fieldpost] Sondersammlung, Bibliothek für Zeitgeschichte, Stuttgart.
41. 'The *Führer*'s speech to the Commanders in Chief', 22 August 1939. Translation of Document No.798-PS, Office of US Chief of Counsel. Donovan Nuremberg Trials Collection, Cornell University.
42. Rieckhoff, *Trumpf oder Bluff*, 93–4.
43. 28.9.1939. 3.Kp./Flg.Erg. Rgt.12 Fliegerhorst Stendal. [Fieldpost] Sondersammlung, Bibliothek für Zeitgeschichte, Stuttgart.

3. How Insane Is This War!

1. MF 10112/1 – 'First Hand Accounts of the Battle of Britain, Collected by Alexander McKee. 2.

Accounts by Luftwaffe personnel. Willibald Klein.' RAF Museum, London.
2. Documents Ger Misc 19 (147). 'Printout from a Letter from Johan Rausmayer, Langenlebarn 32, Deutsches Reich, dated December 1, 1939.' Penfriend Letter of a Young Austrian Airman, 1939. Imperial War Museum, London.
3. Hitler, 'Rede im Reichstag (6. Oktober 1939)', *Friedrich Ebert Stiftung*. [Speech]
4. Rieckhoff, *Trumpf oder Bluff*, 116.
5. D. C. Dildy, *Battle of Britain 1940: The Luftwaffe's 'Eagle Attack' (Air Campaign)* (Osprey Publishing, 2018) [Kindle Edition], 7.
6. Ibid.
7. T. Mason, 'British Air Power', in J. A. Olsen (ed.), *Global Air Power* (Washington, D.C.: Potomac Books, 2011), 7–63.
8. S. W. Mitcham, *The Men of Barbarossa: Commanders of the German Invasion of Russia, 1941* (Havertown, PA; Newbury: Casemate, 2009), 33–4.
9. R. Forcyzk, *We March Against England: Operation Sea Lion, 1940–41* (Osprey, 2016) [Kindle Edition], 43.
10. D. Dempster and D. Wood, *The Narrow Margin: The Battle of Britain and the Rise of Air Power, 1930–1940*. 1961. (Barnsley: Pen & Sword, 2010), 41.
11. S. Cox, 'A Comparative Analysis of RAF and Luftwaffe Intelligence in the Battle of Britain, 1940', *RAF Air and Space Power Review*, Vol. 3, Iss. 3 (2000), 35–55.
12. 'Directive No. 1 for the Conduct of the War', Supreme Commander Of The Armed Forces. Berlin, 31 August 1939. Lillian Goldman Law Library, *The Avalon Project*, Yale Law School. Available online: www.avalon.law.yale.edu/imt/wardir1.asp [Accessed 10/02/2024].
13. Ibid.
14. Rieckhoff, *Trumpf oder Bluff*, 116.
15. *Führer* Directive No. 9, 'Principles for the Conduct of the War against the Enemy's Economy', 29 November 1939, Documents on German Foreign Policy 1918–1945, Series D, vol. VIII (1954), 463–4.
16. W. Hubatsch, *Hitlers Weisungen für die Kriegführung, 1939–1945: Dokumente des Oberkommandos der Wehrmacht* (Bernard & Graefe, 1983), 40.
17. GB Exhibit 482, 'Allgemeines Stichwortverzeichnis zum Kr.T.B. der Skl C VIII',

10 October 1939. IMT Nuremberg Archives, H-4003, International Court of Justice. Taube Archive of the International Military Tribunal (IMT) at Nuremberg, 1945–46, Stanford University.
18. D. Caldwell, *JG 26: Top Guns of the Luftwaffe*. 1991. (Frontline Books, 2013) [Kindle Edition], 36.
19. R. Holmes, *The Battle of Heligoland Bight 1939: The Royal Air Force and the Luftwaffe's Baptism of Fire* (Grub Street Publishing, 2009) [Kindle Edition], loc. 1102.
20. S. Cox, 'Reaping the Whirlwind: Bomber Command's War', *Air Power Review*, Vol. 21, No. 1 (Apr 2018), 160–76; 161.
21. P. Zorner, *Nächte im Bomberstrom: Erinnerungen 1920–1950*. Edited by K. Braatz. (Moosburg: NeunundzwanzigSechs Verlag, 2007), 64.
22. Ibid.
23. Berlin-Lichterfelde, 11 January 1940. Bibliothek für Zeitgeschichte, Stuttgart.
24. 'Über den Shetland-Inseln. Besuch bei den England-Aufklärern. PK.' in W. Zuerl (ed.), *Das sind unsere Flieger: Erlebnisse und Heldentaten unserer Flieger im Kampf gegen England* (Munich: Curt Pechstein, 1941), 49.
25. Ibid.
26. L 35 204 = 6.Bttr./Res.Fstg. Flak-Abt.351, Linkenheim, Sold., 15 October 1939. [Letter] Bibliothek für Zeitgeschichte, Stuttgart.
27. 18 969 = 3.Stffl./J.G.52, Lt., 14 October 1939. [Letter] Bibliothek für Zeitgeschichte, Stuttgart.
28. L 10 807 = II.Gru./K.G. 55, Uffz, 19 November 1939. Bibliothek für Zeitgeschichte, Stuttgart.
29. L 09 214 = 13. Ln.Flugsich. Kp./Ln.Rgt.3, O'Gefr., 24 November 1939. Bibliothek für Zeitgeschichte, Stuttgart.
30. 3.2002.0817. '... An seine Eltern am 14.2.1939'. Museum für Kommunikation, Berlin.
31. Berg, *Sie nannten uns Helden*, 18.
32. Ibid.
33. Ibid.
34. D. Andrew, 'Strategic Culture in the Luftwaffe – Did it Exist in World War II and Has it Transitioned into the Air Force?', *Defence Studies*, 4:3 (2004), 361–86; 367–8.
35. 1998.A.0044. 'German Air Force collection'. W. F. Sheeley, Mc. Aero-Medical Research Section, Office of the Surgeon Headquarters, US Air Forces In Europe. Translation of the 'The Personnel Aptitude Testing Committee Of The German Air Force (Das

Personaleignungsprüfwesen der Luftwaffe)' – a report By S. Gerathewohl, c. 1945. David and Fela Shapell Family Collections, Conservation and Research Center, United States Holocaust Memorial Museum, Maryland (USA).
36. Berg, *Sie nannten uns Helden*, 23–4.
37. Ibid.
38. 1998.A.0044. 'German Air Force collection'. W. F. Sheeley, Mc. Aero-Medical Research Section.
39. Ibid.
40. Ibid.
41. J. S. Corum, 'Stärken und Schwächen der Luftwaffe: Führungsqualitäten und Führung im Zweiten Weltkrieg', in R. D. Müller and H. E. Volkmann (eds.), *Die Wehrmacht: Mythos und Realität* (Munich: Oldenbourg, 2012), 283–306.
42. J. Trigg, *The Air War Through German Eyes: How the Luftwaffe Lost the Skies over the Reich* (Amberley Publishing, 2024) [Kindle Edition], 23.
43. 3.2009.1998. 'Dohl an seine Frau und Töchter am 29.02.1940', 29 February 1940 [Feldpost] Museum für Kommunikation, Berlin. This letter is publicly available on the MfK website, hence the decision to name the letter's writer.
44. L 10 807 = II.Gru.K.G.55, Uffz., 19 February 1940. [Letter] Bibliothek für Zeitgeschichte, Stuttgart.
45. 21.2.1940. Hann. Münden [Letter] Sondersammlung, Bibliothek für, Stuttgart.
46. Innsbruck, 13 January 1940. [Letter] Bibliothek für Zeitgeschichte, Stuttgart.
47. Innsbruck-Mühlau, 17 March 1940. [Letter] Bibliothek für Zeitgeschichte, Stuttgart.
48. Premnitz, 20 March 1940. [Letter] Bibliothek für Zeitgeschichte, Stuttgart.
49. Hemmerde/Westf., 29 March 1940. [Letter] Bibliothek für Zeitgeschichte, Stuttgart.
50. R16267-A. 'Directive No. 6 for the Conduct of War', *Führer Directives and Other Top Level Directives of the German Armed Forces, 1939–1941*, Washington D.C. (1948)
51. J. Noakes and G. Pridham (eds.), *Nazism 1919–1945 Volume 3: Foreign Policy, War and Racial Extermination: A Documentary Reader* (Liverpool University Press, 2001), 156–7.
52. Ibid.
53. Mitcham, *Eagles of the Third Reich*, 81.
54. 3.3.1940. L 30 669 = Fl.H.Kdtr. (E) Kaltenkirchen [Fieldpost] Sondersammlung, Bibliothek für Zeitgeschichte (Württembergische Landesbibliothek), Stuttgart.

4. A Lobster Dinner

1. B. Carruthers, *Voices from the Luftwaffe* (Barnsley: Pen & Sword, 2012), 23.
2. Blanco, *The Luftwaffe in World War II*, 7.
3. A. R. A. Claasen, *Hitler's Northern War: The Luftwaffe's Ill-Fated Campaign, 1940–1945* (Lawrence, Kansas: University Press of Kansas, 2001), xviii.
4. Claasen, *Hitler's Northern War*, 41; 64.
5. Ibid.
6. Blanco, *The Luftwaffe in World War II*, 33.
7. Claasen, *Hitler's Northern War*, 52.
8. Blanco, *The Luftwaffe in World War II*, 7.
9. Ibid.
10. W. Johnen, *Duel Under The Stars: The Memoir of a Luftwaffe Night Pilot in World War II*. Translated from German by Greenhill Books. (Barnsley: Greenhill Books, 2020), 17–18.
11. Bruns, 'Rechenschaftsbericht', in X004-8438. The Royal Air Force Museum, London.
12. Ibid.
13. Lehweß-Litzmann, *Absturz ins Leben*, 139–40.
14. M. A. Boden, 'Operation Weserübung: Inter-service Cooperation and Use of Combined Arms Led to German Success in Norway', *Armor*, Vol. 110, Iss. 6 (Fort Knox, KY: US Army Armor Center, November–December 2001), 29–32; 30.
15. Mitcham, *Eagles of the Third Reich*, 77.
16. Ibid.
17. J. Cappelen, 'Foreword: Germany's Crimes Against Norway', Donovan Nuremberg Trials Collection, Cornell University.
18. Claasen, *Hitler's Northern War*, 79
19. Mitcham, *Eagles of the Third Reich*, 77
20. 'Germany's Crimes Against Norway', Donovan Nuremberg Trials Collection, Cornell University.
21. Ibid.
22. Claasen, *Hitler's Northern War*, 78
23. A. Byford, 'False Start: The Enduring Air Power Lessons of the Royal Air Force's Campaign in Norway, April–June 1940', *Air Power Review*, Vol. 13, Iss. 3 (April 2010), 119–42.
24. K. Tyezka, 'Auf Feindflug abgeschossen!', in W. Gericke, *Fallschirmjäger hier und da* (Berlin: Schützen, 1941), 139–41.
25. Ibid.
26. Lehweß-Litzmann, *Absturz ins Leben*, 143–4.
27. Ibid.
28. Ibid.
29. Ibid.

30. Lehweß-Litzmann, *Absturz ins Leben*, 141.
31. Ibid.
32. N. Hewitt, 'How the battle for Norway in 1940 saved Britain', *BBC History Extra*, 27 January 2023. Available online: www.historyextra.com/period/second-world-war/the-battle-for-norway-1940-the-forgotten-battle-of-britain/ [Accessed 10/04/2024].
33. S. Bungay, *The Most Dangerous Enemy* (London: Aurum, 2000), 105.
34. Ibid.
35. Lehweß-Litzmann, *Absturz ins Leben*, 141–4.
36. Byford, 'False Start', 119–42.
37. MF 10112/1 – 'First Hand Accounts of the Battle of Britain, Collected by Alexander McKee. 2. Accounts by Luftwaffe personnel. Willibald Klein.'
38. AIR 40/3071, S.R.A. 515: 'A 458 – Leutnant & A 430 – Oberleutnant', M.I.9.H, 11 September 1940. The National Archives, Kew.
39. Mitcham, *Eagles of the Third Reich*, 77–8.
40. MF 10112/1 – 'First Hand Accounts of the Battle of Britain, Collected by Alexander McKee. 2. Accounts by Luftwaffe personnel. Willibald Klein.'
41. X004-8438. Bruns, 'Rechenschaftsbericht'.
42. Lehweß-Litzmann, *Absturz ins Leben*, 141.
43. Hozzell, *Conversations with a Stuka Pilot*, 32.
44. Lehweß-Litzmann, *Absturz ins Leben*, 148.
45. R. Noppen, *Holland 1940: The Luftwaffe's first setback in the West* (Bloomsbury, 2021) [Kindle Edition], 4.
46. Ibid.
47. Ibid.
48. A. Hitler, 'Conference: 23rd November, 1939,' in 'Judgement: The Invasion of Belgium, The Netherlands and Luxemburg'. Lillian Goldman Law Library, *The Avalon Project*, Yale Law School. Available online: www.avalon.law.yale.edu/imt/judbelg.asp#:~:text=Breach%20of%20the%20neutrality%20of,then%20England%20and%20France%20will [Accessed 03/02/2024].
49. BArch RL 10/75. *Kampfgeschwader Hindenburg 1*, I A Nr. 124/40 gkdos. N.f.Kdr., 'Geschwaderbefehl Nr. 5: "Fall Rot" Neufassung', 8.4.40. Bundesarchiv, Freiburg-im-Breisgau.

5. Poor Poilu!
1. Noppen, *Holland 1940*, 31.
2. Mitcham, *Eagles of the Third Reich*, 79.

3. Blanco, *The Luftwaffe in World War II*, 34.
4. Bungay, *The Most Dangerous Enemy*, 105.
5. Blanco, *The Luftwaffe in World War II*, 37.
6. Ibid.
7. K. L. Schulz, 'So nahmen wir Waalhaven!', in W. Gericke, *Fallschirmjäger hier und da* (Berlin: Schützen, 1941), 102–5.
8. Blanco, *The Luftwaffe in World War II*, 37.
9. Ibid.
10. A. Sangster, *Field-Marshal Kesselring: Great Commander or War Criminal?* (Newcastle: Cambridge Scholars, 2015), 56.
11. 'Rotterdam - Königin der Maas - Opfer des Luftkrieges, aufgenommen von einem Schweizer nach dem 14. Mai 1940', *Zürcher Illustrierte* No. 50, Vol. 16. XVI. Jahrgang. 13 December 1940. E-Periodica, Eidgenössische Technische Hochschule, Zürich.
12. Schulz, 'So nahmen wir Waalhaven!', 102–5.
13. Stahl, *Kampfflieger zwischen Eismeer und Sahara*, 32.
14. Ibid.
15. Schulz, 'So nahmen wir Waalhaven!', 102–5.
16. Ibid.
17. C. J. Nolan, *The Concise Encyclopedia of World War II: Volume I* (Santa Barbara, CA; Denver, CO; Oxford: Greenwood, 2010), 150.
18. Ibid.
19. Blanco, *The Luftwaffe in World War II*, 39.
20. Ibid.
21. Mitcham, *Eagles of the Third Reich*, 79.
22. T. C. Imlay, 'A Reassessment of Anglo–French Strategy during the Phony War, 1939–1940', *The English Historical Review*, Vol. 119, No. 481 (Apr. 2004), pp. 333–72. Available online: www.jstor.org/stable/3490233 [Accessed 08/10/2020].
23. Mitcham, *Eagles of the Third Reich*, 79.
24. Ibid.
25. 00 093 = 2.Schwadr./Aufkl. Abt.263, 263.Inf.Div, 19 May 1940. [Letter] Bibliothek für Zeitgeschichte, Stuttgart.
26. K. H. Frieser, 'Die deutschen Blitzkriege: Operativer Triumph – strategische Tragödie', in R. D. Müller and H. E. Volkmann (eds.), *Die Wehrmacht: Mythos und Realität* (Munich: Oldenbourg, 2012), 182–96.
27. 10 203 = 2.Kp./Inf.Div. Nachr.Abt.267, 267.Inf.Div. 22.5.1940 [Letter] Bibliothek für Zeitgeschichte, Stuttgart.
28. H. Raffal, *Air Power and the Evacuation of Dunkirk: The RAF and Luftwaffe during Operation Dynamo, 26 May – 4 June 1940* (London: Bloomsbury, 2021) [Kindle Edition], 215.
29. Ibid.

30. Ibid.
31. Ibid.
32. T. Holmes, (2020), *Spitfire vs Bf 109: Battle of Britain* (Oxford; New York: Osprey, 2007), 20.
33. Bob, 'Lecture as dialogue at Air Command and Staff College in Maxwell', Imperial War Museum, London.
34. Ibid.
35. A. Galland, *The First and the Last*. 1954. (Reading Essentials: 2014) [Kindle Edition], 8.
36. 'Wir fliegen den Tod nach Flandern. PK.', in W. Zuerl (ed.) *Das sind unsere Flieger: Erlebnisse und Heldentaten unserer Flieger im Kampf gegen England* (Munich: Curt Pechstein, 1941).
37. Ibid.
38. H. Mahlke, *Memoirs of a Stuka Pilot*, 1993. Translated from German by J. Weal. (London: Frontline Books, 2013), 100.
39. The *SS-Verfügungstruppe* ('Combat Support Force', SS-VT) was founded in 1934 and, after serving as an armed SS wing for Hitler and the Nazi Party, was placed at the disposal of the German *Heer* in the Second World War. They were later incorporated into the *Waffen-SS* in April 1940.
40. 07 874 = Fu.Sturm/SS-Nachr. Tr. SS-Div.V.T. 30.5.1940. [Letter] Bibliothek für Zeitgeschichte, Stuttgart.
41. 31 501 = 5.Kp./Inf.Rgt.85, 10.Inf.Div., 5 June 1940. [Letter] Bibliothek für Zeitgeschichte, Stuttgart.
42. 11 025 B = 10.Bttr./Art. Rgt.5, 5.Inf.Div., 7 June 1940. [Letter] Bibliothek für Zeitgeschichte, Stuttgart.
43. Ibid.
44. L 10 807 = Feldwebel, II.Gru./K.G.55, 30 May 1940. [Letter] Sondersammlung, Bibliothek für Zeitgeschichte, Stuttgart.
45. H. Kohl, *Volltreffer!: Flieger zertrümmern ein Weltreich* (Reutlingen: Enßlin & Laiblin, 1941), 57.
46. L 09 214, 10.6.1940 = Obergefreiter, 13.Kp./Ln.Rgt.3 [Fieldpost] Bibliothek für Zeitgeschichte, Stuttgart.
47. Ibid.
48. B. Bond and M. Taylor, *The Battle for France & Flanders: Sixty Years On* (Barnsley: Pen & Sword, 2001), 60.
49. S. C. Tucker, *The Second World War* (London: Macmillan, 2003), 62.
50. L 30 458 = 2.Bttr./le.Flak-Abt.776, Wm, 15 June 1940 [Fieldpost] Bibliothek für Zeitgeschichte, Stuttgart.
51. L 30 458 = 2.Bttr./le.Flak-Abt.776, Wm, 15 June 1940.
52. Hamburg, 4.6.1940. [Letter] Bibliothek für Zeitgeschichte, Stuttgart.
53. Ibid.

54. Konstanz, 7.6.1940. [Letter] Bibliothek für Zeitgeschichte, Stuttgart.
55. Tübingen, 4.6.1940. [Letter] Bibliothek für Zeitgeschichte, Stuttgart.
56. Göttingen, 5.6. 1940. [Letter] Bibliothek für Zeitgeschichte, Stuttgart.
57. Hann. Münden, 17.6.1940. [Letter] Bibliothek für Zeitgeschichte, Stuttgart.

6. A Future Without Albion

1. Steinhoff et al., *Voices from the Third Reich*, 75.
2. Konstanz, 7 June 1940. [Letter] Bibliothek für Zeitgeschichte, Stuttgart.
3. Mahlke, *Memoirs of a Stuka Pilot*, 112.
4. L 06 561 = 3.Kp./Lw.Bau-Btl.4/VII. 28.6.1940. [Letter] Bibliothek für Zeitgeschichte, Stuttgart.
5. General der Flieger a.D. Johannes Fink, Kommodore KG 2 (Do 17), The RAF Museum, London.
6. L 10 807 = II.Gru./K.G.55, 29.6.1940. [Letter] Bibliothek für Zeitgeschichte, Stuttgart.
7. J. Müller-Marein, *Hölle über Frankreich: Unsere Luftgeschwader im Angriff*. (Berlin: Steiniger, 1940), 31.
8. H. Rökker, *Ausbildung und Einsatz eines Nachtjägers im II. Weltkrieg: Erinnerungen aus dem Kriegstagesbuch* (Zweibrücken: VDM, 2006), 14.
9. L 34 215 = Trsp.Kol.d.Lw.9/VII, 18.6.1940. [Letter] Bibliothek für Zeitgeschichte, Stuttgart.
10. Ibid.
11. Ibid.
12. R. Overy, *The Battle of Britain* (London: Penguin, 2000), 54.
13. H. Kürbs, *Die deutsche Luftwaffe: Ein Bildwerk* (Berlin: Junker & Dünnhaupt, 1936), 8–9.
14. Rökker, *Ausbildung und Einsatz eines Nachtjägers im II. Weltkrieg*, 21.
15. Ibid.
16. Berg, *Sie nannten uns Helden*, 30.
17. Ibid.
18. Ibid.
19. J. Kaufmann, *An Eagle's Odyssey: My Decade As a Pilot in Hitler's Luftwaffe*. 1989. Translated from German by J. Weal. (Barnsley: Greenhill Books, 2019), 70.
20. W. Murray, *Military Adaptation in War: With Fear of Change* (Cambridge: Cambridge University Press, 2014), 164.
21. Ibid.
22. Zorner, *Nächte im Bomberstrom*, 75.
23. Ibid.
24. S. Carlsen and M. Meyer, *Die Flugzeugführer-Ausbildung der Deutschen Luftwaffe, 1935–45: Von der Grundausbildung bis zur Blindflugschule, Band I*

25. (Zweibrücken: Nickel VDM, 1998), 3–5.
25. Zorner, *Nächte im Bomberstrom*, 76.
26. 1998.A.0044. 'The Treatment of Shock from Prolonged Exposure to Cold, Especially in Water', reported by Leo Alexander, Major, M.C., USA – CIOS Black List Item 24, Medical – Combined Intelligence Objectives Sub-Committee G-2 Division, SHAEF, APO 413 – 'German Air Force collection'. The Rubenstein Institute, United States Holocaust Memorial Museum, Bowie, Maryland.
27. H. W. Schmuhl, 'Hirnforschung und Krankenmord: Das Kaiser-Wilhelm-Institut für Hirnforschung 1937–1945', *Vierteljahrshefte für Zeitgeschichte*, Jahrgang 50, Heft 4 (2002), 559–609.
28. E. Malik et al., 'Franz Büchners Hypoxieforschung aus der Kriegszeit und der Nobelpreis', *Die Pathologie*, Iss. 44, (June 2022), 63–69; 69.
29. Ibid.
30. 1998.A.0044, 'The Treatment of Shock from Prolonged Exposure to Cold, Especially in Water', German Air Force collection, The Rubenstein Institute.
31. Ibid.
32. Mahlke, *Memoirs of a Stuka Pilot*, 114.
33. W. Churchill, 'Their Finest Hour', House of Commons, 18 June 1940 [Speech]. The International Churchill Society. Available online: www.winstonchurchill.org/resources/speeches/1940-the-finest-hour/their-finest-hour/ [Accessed 10 August 2023].
34. MF 10112/1. 'First Hand Accounts of the Battle of Britain, Collected by Alexander McKee. 2. Accounts by Luftwaffe personnel. General der Flieger a.D. Paul Deichmann, Chief of Staff, Luftflotte 2.' The RAF Museum, London.
35. L 39 829 = Nachsch. Kp.d.Lw.10/XI, 8 July 1940. [Letter] Bibliothek für Zeitgeschichte, Stuttgart.
36. 'Britische Luftpiraten', *Fehrbelliner Zeitung*, 28 June 1940. ZEFYS, Staatsbibliothek zu Berlin.
37. Müller-Marein, *Hölle über Frankreich*, 165.
38. T. Sysling, *A Boy From Amsterdam: An Autobiography* (Arvada: Lulu, 2012), 44.
39. G. Campion, *The Battle of Britain, 1945–1965: The Air Ministry and the Few* (London: Palgrave Macmillan, 2015), 13.
40. L 39 829 = Nachsch. Kp.d.Lw.10/XI Schwerin. 24.6.1940. [Letter] Bibliothek für Zeitgeschichte, Stuttgart.
41. D. Tillmann, 'Kriegsgesellschaft, Kriegsalltag und Kriegserleben

an der Heimatfront', in D. Tillmann and J. Rosenplänter (eds.), *Luftkrieg und 'Heimatfront': Kriegserleben in der NS-Gesellschaft in Kiel 1939–1945* (Kiel: Solivagus, 2020), 58–85; 65.
42. Ibid.
43. S. Scherreiks, '"Draussen war aber nur ein Toben, Krachen, Bersten und Donnern." Biographische Erinnerungen an die Luftangriffe auf Kiel', in D. Tillmann and J. Rosenplänter (eds.), *Luftkrieg und "Heimatfront": Kriegserleben in der NS-Gesellschaft in Kiel 1939–1945* (Kiel: Solivagus, 2020), 89–99; 89.
44. Ibid.
45. M. Rackwitz, '"Luftschutz ist Not!": Luftschutz, Flugabwehr und Bunkerbau in Kiel', in D. Tillmann and J. Rosenplänter (eds.), *Luftkrieg und 'Heimatfront': Kriegserleben in der NS-Gesellschaft in Kiel 1939–1945* (Kiel: Solivagus, 2020), 19–41; 312.
46. Tillmann, 'Kriegsgesellschaft, Kriegsalltag und Kriegserleben an der Heimatfront', 65.
47. M. Wiggam, 'The Blackout and the Idea of Community in Britain and Germany', in C. Baldoli et al., *Bombing, States and Peoples in Western Europe 1940–1945* (London: Bloomsbury, 2011), 43–58.
48. L 34 215 = Trsp.Kol.d.Lw.9/VII. 28.6.1940. [Letter] Bibliothek für Zeitgeschichte, Stuttgart.
49. Steinhoff et al., *Voices from the Third Reich*, 78.
50. 'Lt., Fallschirmschule Wittstock/Desse', 24 June 1940. [Letter] Bibliothek für Zeitgeschichte, Stuttgart.
51. 'Guernsey and Jersey mark eighty years since harbour bombings', *BBC News*, 28 June 2020. Available online: www.bbc.co.uk/news/world-europe-jersey-53134881 [Accessed 13/05/2024].
52. P. Sanders, *The British Channel Islands Under Occupation* (Jersey Heritage Trust, 2005), xx.
53. L 29 160 = Luftflotten-Kdo 6, 20 August 1940. [Letter] Bibliothek für Zeitgeschichte, Stuttgart; originally Generallandesarchiv Karlsruhe, 465 D 1312.
54. W. Fowler, *The Last Raid: The Commandos, Channel Islands and Final Nazi Raid* (Stroud: The History Press, 2016), 18.
55. Rall, *Mein Flugbuch*, 49.
56. Rieckhoff, *Trumpf oder Bluff?*, 111.
57. Rall, *Mein Flugbuch*, 49.
58. Ibid.

7. Flags Up! Hearts Up!

1. Rall, *Mein Flugbuch*, 50.
2. 'Der größte Feldzug aller Zeiten', *Baruther Anzeiger*,

3–4 July 1940. ZEFYS, Staatsbibliothek zu Berlin.
3. G. Weßler, 'Nun wollen wir Albion stürmen!', *Baruther Anzeiger*, 3–4 July 1940. ZEFYS, Staatsbibliothek zu Berlin.
4. Ibid.
5. 'Furcht und Verwirrung in London', *Briesetal-Bote*, 2 July 1940. ZEFYS, Staatsbibliothek zu Berlin.
6. MF 10112/1 – 'First Hand Accounts of the Battle of Britain. Willibald Klein.'
7. MF 10112/1 – 'First Hand Accounts of the Battle of Britain, Collected by Alexander McKee. 2. Accounts by Luftwaffe personnel. Adolf Galland.'
8. M. Arthur, *Last Of The Few: The Battle of Britain in the Words of the Pilots Who Won It* (London: Virgin, 2010), 273.
9. 'Part I: Ace of the Luftwaffe', *Adolf Galland: Legacy* [Documentary]. Virginia Bader Fine Arts & AeroCinema, 2010. Available online: www.youtube.com/watch?v=YnAVhtnHQM&t=1538s&ab_channel=LuftwaffeFighterAces [Accessed 20/02/2024].
10. Overy, *The Battle of Britain*, 18.
11. Dildy, *Battle of Britain 1940*, 12.
12. Bungay, *The Most Dangerous Enemy*, 146.
13. Overy, *The Battle of Britain*, 65.
14. Dildy, *Battle of Britain 1940*, 13.
15. K09/204. 'Operation 'Sea-Lion' (invasion of Britain: translations of twelve Top-Secret Directives for the above operation', Air Ministry, Air Historical Branch. The Imperial War Museum, London.
16. Ibid.
17. 07 874 = Fu.Sturm/SS-Nachr.Tr. SS-Div. V.T., Uffz., Biarritz, 2 July 1940. [Letter] Bibliothek für Zeitgeschichte, Stuttgart.
18. A. Tooze, *The Wages of Destruction: The Making and Breaking of the Nazi Economy* (London: Penguin, 2007), 382.
19. Ibid.
20. Overy, *The Battle of Britain*, 62.
21. A. Saunders, *Dowding's Despatch: The 1941 Battle of Britain Narrative Examined and Explained* (London: Grub Street, 2021), 30.
22. C. Bergström, *The Battle of Britain: An Epic Conflict Revisited* (Oxford: Casemate, 2015), 79.
23. Ibid.
24. 'Bomben auf britische Flugplätze', *Baruther Anzeiger*, 1–2 July 1940. ZEFYS, Staatsbibliothek zu Berlin.
25. D. Dildy and P. F. Crickmore, *To Defeat the Few: The Luftwaffe's Campaign to Destroy RAF Fighter Command, August–September 1940* (Osprey, 2020) [Kindle Edition], 251.

26. Saunders, *Stuka Attack!*, 18.
27. M. Spick, *Luftwaffe Fighter Aces: The Jagdflieger and their Combat Tactics and Techniques* (New York: Ivy Books, 1996), 57.
28. MF 10112/1 – 'First Hand Accounts of the Battle of Britain, Collected by Alexander McKee. 2. Accounts by Luftwaffe personnel. General der Flieger a.D. Johannes Fink, Kommodore KG 2 (Do 17), Arras, Kanal Kampfführer, July. General d.Fl.a.D. Johannes Fink, 1st & 2nd August, 1959.' The RAF Museum, London.
29. Bungay, *The Most Dangerous Enemy*, 146.
30. 'Ritterkreuz für den Führer eines Kampfgeschwaders', *Baruther Anzeiger*, 1–2 July 1940. ZEFYS, Staatsbibliothek zu Berlin.
31. Flg.H.Kp. Quedlinburg, 4 July 1940. [Letter] Bibliothek für Zeitgeschichte, Stuttgart.
32. Ibid.
33. K. Scheffel, I./Sturzkampfgeschwader 77 Unit Diary (1940), 23–30. Kindly provided by Peter Eisenbach. All Scheffel quotations are from his unit diary.
34. 'Harte Schläge gegen England', *Baruther Anzeiger*, 10–11 July 1940. ZEFYS, Staatsbibliothek zu Berlin.
35. Ibid.
36. 'Bei Alarm in den Luftschutzkeller', *Baruther Anzeiger*, 2 July 1940. ZEFYS, Staatsbibliothek zu Berlin.
37. Wesseln b. Hildesheim, 6 July 1940. [Letter] Bibliothek für Zeitgeschichte, Stuttgart.
38. Paderborn, 7 July 1940. [Letter] Bibliothek für Zeitgeschichte, Stuttgart.
39. Uelzen, 7 July 1940. [Letter] Bibliothek für Zeitgeschichte, Stuttgart.
40. Ibid.
41. Overy, *The Battle of Britain*, 57.
42. Ibid.
43. Wesseln b. Hildesheim, 14 July 1940. [Letter] Bibliothek für Zeitgeschichte, Stuttgart.
44. Flg., Wach-Kdo. Tessenow/ Mecklenburg, 14 July 1940. [Letter] Bibliothek für Zeitgeschichte, Stuttgart.
45. L 09 244 = 6. Bttr./Flak-Rgt.49, 13 July 1940. [Letter] Bibliothek für Zeitgeschichte, Stuttgart.
46. L 38 219 = III. Grup./K.G.2. Ass.Arzt. 6 July 1940. [Letter] Bibliothek für Zeitgeschichte, Stuttgart.
47. Ibid.

8. The Overture

1. 'Die Isolierung Englands bringt das Ende,' *Deutsche Luftwacht: Luftwelt*, Year 7, Number 14. Berlin, 15 July 1940. RAF Museum, London.
2. J. I. Israel, *Diasporas within a Diaspora: Jews, Crypto-Jews*

and the World of Maritime Empires (Boston; Leiden; Paderborn: Brill, 2002), 449–87.
3. C. Lannon, Gambling and/on the Exchange: The Victorian Novel and the Legitimization of the Stock Market [PhD Dissertation] Boston, MA: Boston College (December 2009), 10.
4. 'Die Isolierung Englands bringt das Ende', Deutsche Luftwacht: Luftwelt, 15 July 1940.
5. A. Hitler, 'Destiny Will Decide Who Is Right', Reichstag (Berlin), 6 October 1939 [Speech], in Vital Speeches of the Day, Vol. VI, pp. 2–12. Ibiblio, University of North Carolina at Chapel Hill.
6. W. Zuerl (ed.), Das sind unsere Flieger: Erlebnisse und Heldentaten unserer Flieger im Kampf gegen England (Munich: Curt Pechstein, 1941), 122.
7. Overy, The Battle of Britain, 55–6.
8. Bungay, The Most Dangerous Enemy, 122.
9. 'First Deliberations Regarding a Landing in England' ('Sea-Lion'), 12 July 1940. Imperial War Museum, London.
10. Ibid.
11. 'Die Isolierung Englands bringt das Ende', Deutsche Luftwacht: Luftwelt, 15 July 1940.
12. Dildy, Battle of Britain 1940, 13.
13. K. Bartz, Als der Himmel brannte: Der Weg der Deutschen Luftwaffe (Hanover: Adolf Sponholtz, 1955), 39.
14. Overy, The Battle of Britain, 56.
15. Dildy, Battle of Britain 1940, 83.
16. Overy, The Battle of Britain, 27.
17. A. Hitler, 'My Last Appeal to Great Britain', The Reichstag in Berlin [Speech], 19 July 1940, in Vital Speeches of the Day, Vol. VI, pp. 617–25. (Ibiblio, University of North Carolina at Chapel Hill/the Center for the Public Domain).
18. Ibid.
19. Ibid.
20. Overy, The Battle of Britain, 64.
21. Johnen, Duel Under The Stars, 20.
22. Zorner, Nächte im Bomberstrom, 63.
23. S. Born, 'Ein Stück Geschichte: Deutsche Nachnamen', Deutscher Akademischer Austauschdienst (DAAD), 7 June 2017. Available online: https://www.alumniportal-deutschland.org/de/magazin/deutschland/deutsche-nachnamen/ [Accessed 27/12/2024].
24. K. Mackowiak, 'Meier', Kann Spuren von Latein enthalten: Kleines Lexikon deutscher Wörter lateinischer Herkunft (Munich: C. H. Beck, 2023), 97.
25. H. W. Guggenheimer and E. H. Guggenheimer, Jewish Family Names and their Origins:

An Etymological Dictionary (Brooklyn, N.Y.: Ktav Publishing House, 1992), ixv.
26. Ibid.
27. Ibid.
28. Görlitz, 19 July 1940. [Letter] Bibliothek für Zeitgeschichte, Stuttgart.
29. 'Deutschlands Ueberlegenheit in der Luft', *Baruther Anzeiger*, 17–18 July 1940. ZEFYS, Staatsbibliothek zu Berlin.
30. 'England bleibt unbelehrbar', *Baruther Anzeiger*, 22–23 July 1940. ZEFYS, Staatsbibliothek zu Berlin.
31. Ibid.
32. 'Bedrohtes England', *Zürcher Illustrierte*, Issue 29, 19 July 1940. (E-Periodica, ETH Zürich).
33. 'Stahlhelme für die privaten Schießklubs – Zuchthausurteile am laufenden Band', *Baruther Anzeiger*, 12 July 1940. ZEFYS, Staatsbibliothek zu Berlin.
34. Ibid.
35. Konstanz, 23 July 1940. [Letter] Bibliothek für Zeitgeschichte, Stuttgart.
36. Tübingen, 23 July 1940. [Letter] Bibliothek für Zeitgeschichte, Stuttgart.
37. 'England hat gewählt', *Baruther Anzeiger*, 24–25 July 1940. ZEFYS, Staatsbibliothek zu Berlin.
38. Rall, *Mein Flugbuch*, 50. All of the Rall quotations in this section are from this source.
39. L 01 018 = 2. Bttr./Flak-Rgt.231, 22 July 1940. [Letter] Bibliothek für Zeitgeschichte, Stuttgart.
40. Ibid.
41. Ibid.
42. Dildy, *Battle of Britain 1940*, 43.
43. L 21 005 = 1. Flgh.Betr.Kp.Stuka-Geschw.2, 20 July 1940. [Letter] Bibliothek für Zeitgeschichte, Stuttgart.
44. Dildy, *To Defeat the Few*, 250.
45. Dildy, *Battle of Britain 1940*, 43.
46. Ibid.
47. L 13 353 = Flg. H. Kdtr. (E) 16/XI, 20 July 1940. [Letter] Bibliothek für Zeitgeschichte, Stuttgart.
48. Ibid.
49. Glashütten, 23 July 1940. [Letter] Bibliothek für Zeitgeschichte, Stuttgart.
50. Ibid.
51. Holland, *The Battle of Britain*, 437.
52. Bungay, *The Most Dangerous Enemy*, 111.
53. G. Hewitt, *The Royal Navy & the Defence of Great Britain April–October 1940* (Barnsley: Pen & Sword, 2008) [Kindle Edition], 58.
54. BArch RW 4/589, Oberkommando der Wehrmacht, '2. Ausfertigung: 'Ob.dl. hat 30.7.40 12.20 Uhr folgendes Fernschreiben erhalten', Führerhauptquartier, 30

July 1940. Bundesarchiv, Freiburg-im-Breisgau.
55. R. Hough and D. Richards, *Battle of Britain* (Barnsley: Pen & Sword, 1990), 137.
56. Rall, *Mein Flugbuch*, 54–5.

9. A Very Bad Piece of Work

1. L 38 219 = Ass. Arzt, III. Gru./K.G.2, 13 August 1940. [Letter] Bibliothek für Zeitgeschichte, Stuttgart.
2. Ibid.
3. L 38 219 = Ass. Arzt, III. Gru./K.G.2, 13 August 1940. [Letter] Bibliothek für Zeitgeschichte, Stuttgart.
4. L 33 946, Uffz., Ldsschtz.Zug d.Lw. 16/III – Oranienburg, 1 August 1940. [Letter] Bibliothek für Zeitgeschichte, Stuttgart.
5. Sold., Wachkommando Tessenow/Mecklenburg, 13 August 1940. [Letter] Bibliothek für Zeitgeschichte, Stuttgart.
6. BArch RW 4/590. A. Jodl, 'Beurteilung der Lage, wie sich nach den Auffassungen von Heer und Kriegsmarine für eine Landung in England ergibt', Berlin, 13 August 1940. Bundesarchiv, Freiburg-im-Breisgau.
7. Ibid.
8. BArch RW 4/590. Oberkommando der Wehrmacht, 'WFA/Abt. L. I. Nr. 33189/40 g.K.Chefs, Betr: Seelöwe, 4. Ausfertigung', F.H.Qu., 1 August 1940. Bundesarchiv, Freiburg-im-Breisgau.
9. MF 10112/1 – 'First Hand Accounts of the Battle of Britain. General der Flieger a.D. Paul Deichmann.'
10. Mitcham, *Eagles of the Third Reich*, 100.
11. Rieckhoff, *Trumpf oder Bluff*, 218–19.
12. Bishop, *Battle of Britain*, 120.
13. Ibid.
14. J. Arquilla, *Why The Axis Lost* (Jefferson, NC: McFarland & Company, 2020), 50.
15. BArch RW 4/590. A. Jodl, 'Beurteilung der Lage, wie sich nach den Auffassungen von Heer und Kriegsmarine für eine Landung in England ergibt', Berlin, 13 August 1940. Bundesarchiv, Freiburg-im-Breisgau.
16. Cox, 'A Comparative Analysis of RAF and Luftwaffe Intelligence in the Battle of Britain, 1940', 35–55.
17. Ibid.
18. Ibid.
19. Nr. 334 von 3.8.1940, 'Überblick über britische Abwehrmassnahmen auf der Erde'. Tägliche Lageberichte West des Führungsstabes der Luftwaffe. Deutsch–Russisches Projekt zur Digitalisierung Deutscher Dokumente in Archiven der Russischen Föderation.

This digitalized archive is an online collaborative research collection between the German academic research organization Max Weber Stiftung and the Russian GARF, RGASPI, RGVA, and CAMO archives.
20. Ibid.
21. Documents. 11945. K. G. Raynor (originally, 'From the Other Side'), in Private Papers of K. G. Raynor, Box No: 02/19/1. The Imperial War Museum, London.
22. Spick, *Luftwaffe Fighter Aces*, 51.
23. Ibid.
24. J. Starkey, *The RAF's Cross-Channel Offensive: Circuses, Ramrods, Rhubarbs & Rodeos, 1941–1942* (Yorkshire; Philadelphia: Air World, 2022), 49.
25. G. Bailey, 'The Narrow Margin of Criticality: The Question of the Supply of 100-Octane Fuel in the Battle of Britain', *The English Historical Review*, Vol. 123, No. 501 (Oxford University Press, April 2008), 394–411.
26. R. Hough and D. Richards, *The Battle of Britain: The Jubilee History*. 1990. (Barnsley: Pen & Sword, 2007), 387.
27. Bungay, *The Most Dangerous Enemy*, 266.
28. MF 10112/1 – 'First Hand Accounts of the Battle of Britain, Collected by Alexander McKee. 2. Accounts by Luftwaffe personnel. Adolf Galland.'
29. MF 10112/1 – 'First Hand Accounts of the Battle of Britain, Collected by Alexander McKee. Brigadegeneral Johannes Steinhoff, 2nd Staffel, JG 52 (Me 109), Peuplingues, Calais.'
30. 'Heritage: Hawker Hurricane', BAE Systems. Available online: www.baesystems.com/en-uk/heritage/hurricane#:~:text=The%20Hawker%20Hurricane%20was%20a,in%20the%20Battle%20of%20Britain [Accessed 02/06/2024].
31. Overy, *The Battle of Britain*, 36.
32. R. Overy, *The Battle* (London: Penguin, 2000), 85.
33. A. Stewart, *They Flew Hurricanes* (Barnsley: Pen & Sword, 2005), 8.
34. Galland, *The First and The Last*, 6–7.
35. A. Deere, *Nine Lives*. 1959. (Manchester: Crécy, 2019), 83–8.
36. Rall, *Mein Flugbuch*, 54.
37. Mahlke, *Memoirs of a Stuka Pilot*, 121.
38. Ibid.
39. Dildy and Crickmore, *To Defeat the Few*, 392–3.
40. Blanco, *The Luftwaffe in World War II*, 66.
41. Dildy, *Battle of Britain 1940*, 87.

42. Akte 99. Der Oberbefehlshaber der Luftwaffe, Führungsstab Ic, 'Lagebericht Nr. 342', H.Qu., 13 August 1940. Tägliche Lageberichte West des Führungsstabes der Luftwaffe. Deutsch–Russisches Projekt zur Digitalisierung Deutscher Dokumente in Archiven der Russischen Föderation.
43. L 36 412 = 4.Flgh.Betr.Kp./K.G.51, 12 August 1940. [Letter] Bibliothek für Zeitgeschichte, Stuttgart.
44. O. Bechtle, BArch N 502/26, 'Der Einsatz der Luftwaffe gegen England'[Lecture] Berlin, 2 February 1944. Bundesarchiv, Freiburg-im-Breisgau. Bechtle is often identified as a Luftwaffe intelligence officer attached to *Kampfgeschwader* 2, but he started off as a flight instructor and then became a Dornier Do 17 bomber pilot with KG 2 during the war.
45. Akte 99. 'Wetterablauf am 13. August 1940', Anlage 1 zum Lagebericht, Nr. 343 vom 14.8.1940. Tägliche Lageberichte West des Führungsstabes der Luftwaffe. Deutsch–Russisches Projekt zur Digitalisierung Deutscher Dokumente in Archiven der Russischen Föderation.
46. Dildy and Crickmore, *To Defeat the Few*, 347.
47. MF 10112/1 – 'First Hand Accounts of the Battle of Britain, Collected by Alexander McKee. 2. Accounts by Luftwaffe personnel. General der Flieger a.D. Johannes Fink, Kommodore KG 2 (Do 17), Arras, Kanal Kampfführer, July. General d.Fl.a.D. Johannes Fink, 1st & 2nd August, 1959.' The RAF Museum, London.
48. Ibid.
49. J. Willis, *Churchill's Few: The Battle of Britain*. 1985. (Mensch Publishing, 2023) [Kindle Edition], 115.
50. T. Heath, *In Furious Skies: Flying with Hitler's Luftwaffe in the Second World War* (Barnsley: Pen & Sword, 2022), 106.
51. Dildy, *Battle of Britain 1940*, 98.
52. Holland, *The Battle of Britain*, 454.
53. 'Neue Luftangriffe auf Englands Süd- und Südostküste', *Baruther Anzeiger*, 14–15 August 1940. ZEFYS, Staatsbibliothek zu Berlin.
54. Dildy and Crickmore, *To Defeat the Few*, 356.
55. 'Der pausenlose Würgegriff', *Baruther Anzeiger*, 14–15 August 1940. ZEFYS, Staatsbibliothek zu Berlin.
56. Ibid.
57. Bechtle, BArch N 502/26, 'Der Einsatz der Luftwaffe

gegen England', Bundesarchiv, Freiburg-im-Breisgau.
58. MF 10112/1 – 'First Hand Accounts of the Battle of Britain, Collected by Alexander McKee. 2. Accounts by Luftwaffe personnel. General der Flieger a.D. Johannes Fink, Kommodore KG 2 (Do 17), Arras, Kanal Kampfführer, July. General d.Fl.a.D. Johannes Fink, 1st & 2nd August, 1959.' RAF Museum, London.

10. The Hour of Judgement

1. 'Die Stunde des Gerichts nahe!', *Rheinsberger Zeitung*, 16 August 1940. ZEFYS, Staatsbibliothek zu Berlin.
2. A. Claasen, *Dogfight: The Battle of Britain* (Titirangi: Exisle, 2012), 108.
3. D. Caldwell, *JG 26 Luftwaffe Fighter Wing War Diary, Volume I: 1939–42* (Mechanicsburg, PA: Stackpole, 2012), 57.
4. Bergström, *The Battle of Britain*, 130.
5. Mitcham, *Eagles of the Third Reich*, 95
6. Dildy, *Battle of Britain 1940*, 130.
7. Bungay, *The Most Dangerous Enemy*, 231.
8. K. Scheffel, I./Sturzkampfgeschwader 77 Unit Diary (1940), 39–40.
9. Ibid.
10. J. Ward, *Hitler's Stuka Squadrons: The Ju 87 at War, 1936–1945* (St. Paul, MN: MBI, 2004), 109.
11. Hozzell, *Conversations With A Stuka Pilot*, 44.
12. Saunders, *Stuka Attack!*, 164–5.
13. T. Newdick, *The World's Most Powerful Military Aircraft* (New York: Rosen Publishing, 2017), 24.
14. Saunders, *Stuka Attack!*, 164–5.
15. Bechtle, BArch N 502/26, 'Der Einsatz der Luftwaffe gegen England', Bundesarchiv, Freiburg-im-Breisgau.
16. Ibid.
17. Bergström, *The Battle of Britain*, 132.
18. Ibid.
19. Galland, *The First and The Last*, 11.
20. P. G. Eriksson, *Alarmstart: The German Fighter Pilot's Experience in the Second World War: Northwestern Europe – from the Battle of Britain to the Battle of Germany* (Amberley Publishing, 2017) [Kindle Edition], 55.
21. Rieckhoff, *Trumpf oder Bluff*, 221–2.
22. W. Baumbach, *Zu Spät? Aufstieg und Untergang der deutschen Luftwaffe. 1949.* (Stuttgart: Motorbuch, 1978), 102.
23. General der Fl. A.D. Paul Deichmann, Studiengruppe

Luftwaffe bei der Führungsakademie der Bundeswehr – 7 August 1959.
24. Eriksson, *Alarmstart*, 107.
25. Cox, 'A Comparative Analysis of RAF and Luftwaffe Intelligence in the Battle of Britain, 1940', 35–55.
26. Wehner, »*Technik können Sie von der Taktik nicht trennen*«, 91–2.
27. Dildy and Crickmore, *To Defeat the Few*, 419.
28. Dempster and Wood, *The Narrow Margin*, 170.
29. P. Addison and J. Crang, *The Burning Blue: A New History of the Battle of Britain* (Faber & Faber, 2010) [Kindle Edition], 99.
30. Saunders, *Dowding's Despatch*, 59.
31. G. Simpson, *A History of the Battle of Britain Fighter Association: Commemorating the Few* (Barnsley: Pen & Sword), 135.
32. Saunders, *Dowding's Despatch*, 60.
33. K. Ford, *The Bruneval Raid: Operation Biting 1942* (Osprey, 2006), 10.
34. Bergström, *The Battle of Britain*, 84–5.
35. P. J. Hugill, *Global Communications since 1844: Geopolitics and Technology* (Baltimore; London: Johns Hopkins University Press, 1999), 199.
36. J. R. Lindsay, *Information Technology and Military Power* (Cornell University Press, 2020), 82.
37. Hugill, *Global Communications since 1844*, 194.
38. Dildy, *Battle of Britain 1940*, 94.
39. Galland, *The First and the Last*, 17–18.
40. Bob, 'Lecture as dialogue at Air Command and Staff College in Maxwell', Imperial War Museum, London.
41. Ibid.
42. MISC 3404 244/2. D. L. Armitage, 'Battle of Britain', as part of 'Questionnaires, 1 – 100 for Richard Hough and Denis Richards, Golden Jubilee History of the Battle of Britain', c. 1988. Imperial War Museum, London.
43. MISC 244/2. F. D. Hughes, 'Answer to Questionnaire', in 'Questionnaires 101–229', for Richard Hough and Denis Richards, Golden Jubilee History of the Battle of Britain', 25 July 1988. Imperial War Museum, London.
44. MISC 3404 244/1. H. V. Cossons, 'Into the Battle of Britain: Letter to Historians of Golden Jubilee History of the Battle of Britain', 8 April 1988 [Letter] Imperial War Museum, London.
45. Ibid.
46. Overy, *The Battle of Britain*, 67.
47. Bechtle, BArch N 502/26, 'Der Einsatz der Luftwaffe gegen England', Bundesarchiv, Freiburg-im-Breisgau.

48. A. B. Downes, *Targeting Civilians in War* (Ithaca, N.Y.: Cornell University Press, 2010), 148.
49. Cox, 'A Comparative Analysis of RAF and Luftwaffe Intelligence in the Battle of Britain, 1940', 35–55.
50. Bergström, *The Battle of Britain*, 83.
51. Ibid.
52. Saunders, *RAF Fighters vs Luftwaffe Bombers*, 146.
53. Bungay, *The Most Dangerous Enemy*, 94–5.
54. Ibid.
55. Rall, *Mein Flugbuch*, 59.
56. AIR 40/3071, S.R.A. 523, 'A 430 – Oberleutnant and A 458 – Oberleutnant', M.I.9.H., 12 September 1940. The National Archives, Kew.
57. Rall, *Mein Flugbuch*, 60.
58. Andrew, 'Strategic Culture in the Luftwaffe', 361–86.
59. Galland, *The First and the Last*, 22.
60. Ibid.
61. Bergström, *Battle of Britain*, 136.
62. Corum, 'Stärken und Schwächen der Luftwaffe', 283–306.
63. Bungay, *The Most Dangerous Enemy*, 120.
64. Dildy, *Battle of Britain 1940*, 80.
65. Andrew, 'Strategic Culture in the Luftwaffe', 361–86.
66. Ibid.
67. Reel 2828, Frame 42. 'Untersuchungen und Mitteilung Nr. 645: Statistische Zusammenfassung der Unfälle im Flugbetrieb von September 1939 bis Dezember 1940 nach flugmedizinischen Gesichtspunkten', Deutsche Versuchsanstalt für Luftfahrt E.V., Berlin-Adlershof (1941). Steven F. Udvar-Hazy Center, Smithsonian National Air and Space Museum, Chantilly (Virginia).
68. Ibid.
69. Ibid.
70. T321, Roll 39 (2881) Luftgaukommando Westfrankreich, 'Tagesbefehl Nr. 48, Betr. Kraftfahrunfälle', Hauptquartier, den 22.8.40. *Gruppe* II, Nr. (1044)/40. National Archives, College Park, Maryland (USA).
71. AIR 40/3071. 'S.R.A. 569, 'A 494 – Oberleutnant (Bomber – Observer) & A 490 – Leutnant (Bomber – Pilot)', M.I.9.H, 19 September 1940. The National Archives, Kew.
72. BArch RL 7-3/712. Anlage zu Gat. 1043/40, 'Ständiger Gruppenbefehl Nr. 1', III./K. Gesch.Gen.Wever 4, Ia/Ib, Gefechtsstand, den 14.8.1940, 1. Bundesarchiv, Freiburg-im-Breisgau.
73. Stahl, *Kampfflieger zwischen Eismeer und Sahara*, 31.
74. Bungay, *The Most Dangerous Enemy*, 48.
75. Ibid.
76. Saunders, *RAF Fighters vs Luftwaffe Bombers*, 7–8.

77. Dildy, *Battle of Britain 1940*, 33.
78. Ibid.
79. C. Goss, *Dornier Do 17 Units of World War 2* (Osprey, 2019) [eBook], 12.
80. Bungay, *The Most Dangerous Enemy*, 41.
81. Steinhoff et al., *Voices from the Third Reich*, 81.

11. Don't Talk so Loud!

1. 'Farm "At Home" for Nazi Airmen', *Newcastle Journal*, Friday 16 August 1940. The British Newspaper Archive (British Library).
2. Steinhoff et al., *Voices From The Third Reich*, 78–9.
3. Ibid.
4. Ibid.
5. 14/21/1. Hadley, 'Service with a Smile: A Record of Four Years', Imperial War Museum, London.
6. J. L. Hodson, 'Captured Airmen Vary Vastly', *Aberdeen Press and Journal*, Saturday 28 September 1940. The British Newspaper Archive (British Library)
7. IWM 10092. W. D. David, 'Oral History' [Interview], Imperial War Museum (18 January 1988).
8. A. G. Ethridge, 'Letter to Mr Denis Richard & Richard Hough', 4 April 1988.
9. H. Fry, *The Walls Have Ears: The Greatest Intelligence Operation of World War II* (New Haven; London: Yale University Press, 2017) [Kindle Edition], 21.
10. WO 208. 'Prisoners of War Section – Subseries', The National Archives, Kew. Available at: www.discovery.nationalarchives.gov.uk/details/r/C77595 [Accessed 3 January 2024].
11. S. Neitzel and H. Welzer, *Soldaten – On Fighting, Killing and Dying: The Secret Second World War Tapes of German POWs* (Simon & Schuster, 2012) [Kindle Edition], loc. 6596.
12. Ibid.
13. Fry, *The Walls Have Ears*, 44.
14. Bell, *Britische Feindaufklärung im Zweiten Weltkrieg*, 165.
15. Ibid.
16. Ibid.
17. AIR 40/2429. S. D. Felkin, 'Further Report on the He. 115, S4 + IH of 1/506, brought down in the Sea 40 miles off Grimsby, on 8/12/1940', Report No. 942/1940, 15 December 1940. Reports by Prisoners of War, Air Ministry/Royal Air Force. The National Archives, Kew.
18. Dildy, *Battle of Britain 1940*, 36.
19. Ibid.
20. Ibid.
21. AIR 40/3071, S.R.A. 515: 'A 458 – Leutnant & A 430 – Oberleutnant', M.I.9.H., 11 September 1940. The National Archives, Kew.

22. AIR 30/3071, 'Oberleutnants', M.I.9.H, 11 September 1940. The National Archives, Kew.
23. Ibid.
24. Ibid.
25. AIR 40/3071, 'Oberleutnant', M.I.9.H, 11 September 1940. The National Archives, Kew.
26. Neitzel & Welzer, *Soldaten*, 839.
27. AIR 40/2429. 'Extracts from Losses, Morale etc. for November', A.I.1. (k) Report No. 953/1940. The National Archives, Kew.
28. AIR 30/3071. 'S.R.A. No. 555: Ob.Lt. A 460 (F), Ob.Lt. A436 (F) & Ob.Lt. A.433 (F)', M.I.9.H., 18 September 1940. The National Archives, Kew.
29. AIR 30/3071. 'Hauptmann – Bomber Observer & Leutnant – Bomber Pilot', September 1940. The National Archives, Kew.
30. AIR 40/3071. 'S.R.A. No. 605, 'A 519 – Oberleutnant (Fighter Pilot), A 508 – Oberleutnant (Bomber Pilot on Reconnaissance), & A 509 – Leutnant (Fighter Pilot)'. M.I.9.H, 23 September 1940. The National Archives, Kew.
31. AIR 40/3071, S.R.A. 506: 'A 450 – Hauptmann and A 449 – Oberleutnant', M.I.9.H, 11 September 1940. The National Archives, Kew.
32. AIR 40/3071. 'S.R.A. 579 – A 493 – Feldwebel (Fighter Pilot) & A 488 – Feldwebel (Long Range Escort Fighter Pilot), M.I.9.H, 21 September 1940. The National Archives, Kew.
33. Ibid.
34. Fry, *The Walls Have Ears*, 55–6.
35. Ibid.
36. Fry, *The Walls Have Ears*, 55–6.
37. AIR 40/3071. 'S.R.A. No. 713, 'A 504 – Major (Bomber Gruppenkommandeur), A 545 – Hauptmann (Bomber Observer), & A 520 – Oberleutnant (Fighter Pilot)', 8 October 1940. The National Archives, Kew.
38. 'Inside Germany', Winter 1940. The National Archives, Kew.
39. AIR 40/3071. 'S.R.A. 550: A 430 – Oberleutnant – Fighter Pilot & A 480 – Oberleutnant – Bomber Pilot', 19 September 1940. The National Archives, Kew.
40. AIR 40/3071. S.R.A. 582: A 430 – Oberleutnant (Fighter Pilot) & A 489 – Oberleutnant (Bomber – Pilot), M.I.9.H, 21 September 1940. The National Archives, Kew.
41. 'Inside Germany', Winter 1940. The National Archives, Kew.
42. AIR 40/2429. S. D. Felkin, A.I.1 Report No. 939/1940, 'General Report', 28 November 1940. The National Archives, Kew.
43. Ibid.
44. AIR 30/3071. 'S.R.A. 535: A 539 – A 430 – Oberleutnant

and A 480 – Oberleutnant', M.I.9.H, 14 September 1940. The National Archives, Kew.
45. AIR 40/3071, S.R.A. 514: 'A 450 – Hauptmann & A 449 – Oberleutnant', M.I.9.H., 11 September 1940. The National Archives, Kew.
46. AIR 40/3071, S.R.A. 520, 'A 422 – Oberleutnant, A 460 – Oberleutnant, and A 340 – Oberleutnant', M.I.9.H, 18 September 1940. The National Archives, Kew.
47. AIR 40/3071, S.R.A. 548, 'A 474 – Unteroffizier (Bomber) and A 456 – Feldwebel (Bomber)', M.I.9.H, 17 September 1940. The National Archives, Kew.

12. Like a Thunderbolt He Falls

1. 'Die Luftoffensive gegen England hat begonnen!', *Deutsche Luftwacht: Luftwelt*, Year 7, No. 17, Berlin, 1 September 1940. RAF Museum, London.
2. '505 Britenflugzeuge in einer Woche – Gestern allein 106 englische Maschinen vernichtet', *Briesetal-Bote*, 16 August 1940. ZEFYS, Staatsbibliothek zu Berlin.
3. Ibid.
4. G. Campion, *The Good Fight: Battle of Britain Propaganda and the Few* (New York: Palgrave Macmillan, 2009), 55.
5. Wehner, »*Technik können Sie von der Taktik nicht trennen*«, 217.
6. Ibid.
7. Ibid.
8. Ibid.
9. Ibid.
10. BArch N 667/7. Die Vergrößerung der deutschen Fallschirmtruppe im Sommer 1940, 6. Bundesarchiv, Freiburg-im-Breisgau.
11. Corum, *Creating the Operational Air War*, 247.
12. Ibid.
13. Ibid.
14. Bestand 500, Findbuch 12488, Akte 1133 (1). 'Unterlagen der Ia-Abteilung des Infanterieregiments 135: Merkblätter für das Unternehmen "Seelöwe", Weisungen für die Zusammenarbeit mit der Luftwaffe, Ausbildungshinweise, Bedarfsmeldungen für den Seetransport, Feindlageberichte des XXXX II. Armeekorps, Befehle der 45. Infanteriedivision für das Unternehmen ua (1)'. 9 August 1940–12 April 1941. Deutsch–Russisches Projekt zur Digitalisierung Deutscher Dokumente in Archiven der Russischen Föderation.
15. 'Von Oberst Model, Der Chef des Generalstabes',

Armeeoberkommando 16, Ia/Koluft, Br.B.Nr. 1369/40 g.Kdos. A.H.Qu. 26 August 1940. Deutsch–Russisches Projekt zur Digitalisierung Deutscher Dokumente in Archiven der Russischen Föderation.

[16] Akte 556. A.H.Qu., den 25.8.1940. Kommandeur der Luftwaffe beim A.O.K.16, 'Studie über den Einsatz der Heeresfliegerverbände und Flakartillerie im Fall "Seelöwe"'. Deutsch–Russisches Projekt zur Digitalisierung Deutscher Dokumente in Archiven der Russischen Föderation.

[17] Kommandeur der Luftwaffe beim A.O.K. 16 -Ia- A.H.Qu., den 30.8.1940 – 220 Ausfertigungen – 'Richtlinien für Verbreitung und Ausbildung bei den Aufklärungsstaffeln der 16. Armee für Fall "Seelöwe"'. Deutsch–Russisches Projekt zur Digitalisierung Deutscher Dokumente in Archiven der Russischen Föderation.

[18] Unterlagen der Ia-Abteilung des Generalkommandos des XIII. Armeekorps: Zeittafel für "Seelöwe", Deutsch–Russisches Projekt zur Digitalisierung Deutscher Dokumente in Archiven der Russischen Föderation.

[19] J. Klatte, *Kampfflieger in der Luftschlacht um England. Ein ungewöhnlicher Werdegang: Persönliche Berichte aus den Jahren 1930–1941* (Berlin: Cardamina, 2021), 186–7.

[20] 'Deutschland beherrscht die Luft', *Briestal-Bote*, 2 September 1940. ZEFYS, Staatsbibliothek zu Berlin.

[21] Bergström, *Battle of Britain*, 178.

[22] Cox, 'A Comparative Analysis of RAF and Luftwaffe Intelligence in the Battle of Britain, 1940', 35–55.

[23] K. Deumling, *41 Sekunden bis zum Einschlag: Als Bomberpilot im Kampfgeschwader 100 Wiking mit der geheimen Fernlenkbombe Fritz X* (Springe: H.E.K. Creativ, 2008), 20–1.

[24] T 321, Roll 32. 'Richtlinien für die Waffenausbildung der Flugzeugführer für Kampfverbände', 24 August 1940. National Archives, College Park, Maryland (USA).

[25] Ibid.

[26] T 321, Roll 32. 'Richtlinien für die Waffenausbildung von Bordfunkern', 24 August 1940. National Archives, College Park, Maryland (USA).

[27] T 321, Roll 32. 'Richtlinien für die Ausbildung von Fliegerbordschützen für Kampfverbände', 24 August 1940. National Archives, College Park, Maryland (USA).

28. Bergström, *The Battle of Britain*, 105.
29. T 321, Roll 32. 'Richtlinien für die Waffenausbildung von Jagdflugzeugführern', 24 August 1940. National Archives, College Park, Maryland (USA).
30. T 321, Roll 32. 'Richtlinien für die Waffenausbildung von Jagdflugzeugführern', 24 August 1940.
31. Galland, *The First and the Last*, 20.
32. Akte 100. 'Sonstiges: Gefangenenaussagen', in 'Lagebericht Nr. 359', Der Oberbefehlshaber der Luftwaffe, Führungsstab Ic, Nr. 16300/40 g. H.Qu, 30 August 1940. Deutsch–Russisches Projekt zur Digitalisierung Deutscher Dokumente in Archiven der Russischen Föderation.
33. Ibid.
34. Baumbach, *Zu Spät?*, 114–15.
35. Ibid.
36. U. Steinhilper, *Spitfire on my Tail* (Bromley: Independent Books, 1990), 283.
37. Ibid.
38. MF 10112/1 – 'First Hand Accounts of the Battle of Britain, Collected by Alexander McKee. 2. Accounts by Luftwaffe personnel. Adolf Galland.' The RAF Museum, London.
39. MF 10112/1 – 'First Hand Accounts of the Battle of Britain, Collected by Alexander McKee. 2. Accounts by Luftwaffe personnel. Oberst Carl Viek, – 31st July, 1959 (interview). Chief of Staff, Jafü 2, Wissant, Oberstleutnant.' [Microfilm]. The RAF Museum, London.
40. 1998.A.0044. 'German Air Force collection: Flying Personnel Research Committee. Neuropsychiatric Organizations in the German Air Force (Report CIOS Trip No. 277). Wing Commander Denis Williams', June 1945. David and Fela Shapell Family Collections, Conservation and Research Center, United States Holocaust Memorial Museum, Maryland (USA).
41. MF 10112/1 – 'First Hand Accounts of the Battle of Britain, Oberst Carl Viek – 31st July, 1959 (interview).
42. Caldwell, *JG 26 Luftwaffe Fighter Wing War Diary*, 70.
43. 1998.A.0044. 'German Air Force collection'. Headquarters, Air Disarmament Command (Prov), United States Strategic Air Forces in Europe. Office of the Surgeon, AAF Station 549. App.7a 16 December 1944. Medical Intelligence Bulletin, Number 17. 'Psychiatry and Psychology in the Luftwaffe.' United States Holocaust Memorial Museum, Maryland (USA).

44. 1998.A.0044. 'German Air Force collection: Flying Personnel Research Committee. Neuropsychiatric Organizations in the German Air Force (Report CIOS Trip No. 277). Wing Commander Denis Williams', June 1945.
45. Ibid.
46. Ibid.
47. Ibid.
48. 1998.A.0044. Headquarters, Air Disarmament Command (Prov), United States Strategic Air Forces in Europe. Office of the Surgeon, AAF Station 549. App.7a 16 December 1944. Medical Intelligence Bulletin, Number 17. 'Psychiatry and Psychology in the Luftwaffe.'
49. 1998.A.0044. 'German Air Force collection: Flying Personnel Research Committee. Neuropsychiatric Organizations in the German Air Force (Report CIOS Trip No. 277). Wing Commander Denis Williams', June 1945.
50. Ibid.
51. S. S. Gartner, 'Wartime Strategic Assessment: Concepts and Challenges', in L. J. Blanken et al. (eds.). *Assessing War: The Challenge of Measuring Success and Failure* (Washington, D.C: Georgetown University Press, 2015), 30–48; 38.
52. C. Everitt and M. Middlebrook, *The Bomber Command War Diaries: An Operational Reference Book, 1939–1945* (Barnsley: Pen & Sword, 2014), 76–7.
53. Berlin-Zehlendorf, 26 August 1940 [Letter] Bibliothek für Zeitgeschichte, Stuttgart.
54. T. Heath, *Hitler's Girls: Doves Amongst Eagles* (Barnsley: Pen & Sword, 2017), 95.
55. Ibid.
56. Emmendingen, 26 August 1940. [Letter] Bibliothek für Zeitgeschichte, Stuttgart.
57. Berlin-Wilmersdorf, 28 August 1940. [Letter] Bibliothek für Zeitgeschichte, Stuttgart.
58. N98.6., 'Luftwaffenhelfer; Erinnerungen'. Bibliothek für Zeitgeschichte, Stuttgart.
59. Göttingen, 1 September 1940. [Letter] Bibliothek für Zeitgeschichte, Stuttgart.
60. Berlin-Zehlendorf, 31 August 1940. [Letter] Bibliothek für Zeitgeschichte, Stuttgart.
61. Berlin-Wilmersdorf, 1 September 1940. [Letter] Bibliothek für Zeitgeschichte, Stuttgart.
62. 3.2002.0904. Martin Meier an seine Ehefrau am 30.08.1940, France, 30 August 1940 [Feldpost] Museum für Kommunikation, Berlin. This letter is available publicly on the MfK website, hence the writer being named.
63. Ibid.
64. 04 086 = R.A.D.-Abt.8/173, 2 September 1940 [Letter]

13. You Dummkopf!

1. Heath, *Hitler's Girls*, 104.
2. Ibid.
3. A. Hitler, 'Nr. 93: Führerrede vom 4. September 1940 zur Eröffnung des Kriegswinterhilfswerks 1940/41 in Berlin' [Speech] 4 September 1940.
4. S. F. Kellerhoff, 'Wir werden ihre Städte ausradieren!', *Die Welt*, 4 September 2020. Available online: www.welt.de/geschichte/zweiter-weltkrieg/article215020180/Hitlers-Luftkrieg-Wir-werden-ihre-Staedte-ausradieren.html [Accessed 02/01/2024].
5. R. Overy, 'From "Uralbomber" to "Amerikabomber": The Luftwaffe and strategic bombing', *Journal of Strategic Studies*, Vol. 1, No.2 (1978), 154–78. Available online: www.tandfonline.com/doi/pdf/10.1080/014023978084 36996?needAccess=true&in stName=University+of+Hull [Accessed 01/02/2022].
6. Documents.10205, Box No: Misc 74 (1107), 'Operation Sea Lion Directives: 'Unternehmen "Seelöwe" – (Operation Sealion) – Invasion of U.K., 1940'. Keitel/Jodl, 'O.K.W.: 'Sea-Lion', Fuehrer's Headquarters, 3.9.1940. The Imperial War Museum, London.
7. Ibid.
8. Wotrum/Pommern, 5 September 1940 [Letter] Bibliothek für Zeitgeschichte, Stuttgart.
9. T. Newdick (ed.), *Air Combat From World War I to the Present Day* (London: Amber Books, 2019) [eBook], 52.
10. Dempster and Wood, *The Narrow Margin*, 218.
11. Blanco, *The Luftwaffe in World War II*, 71.
12. Hamburg, 9 September 1940. [Letter] Bibliothek für Zeitgeschichte, Stuttgart.
13. Bungay, *The Most Dangerous Enemy*, 123.
14. Dildy, *Battle of Britain 1940*, 7.
15. L 13 353 = Flg.H.Kdtr. (E) 16/XI, Sold., 8 September 1940. [Letter] Bibliothek für Zeitgeschichte, Stuttgart.
16. L 36 412 = 4.Flgh.Betr.Kp./ K.G. 51, Sold., 9 September 1940. [Letter] Bibliothek für Zeitgeschichte, Stuttgart.
17. Bob, 'Lecture as dialogue at Air Command and Staff College in Maxwell', Imperial War Museum, London.
18. AIR 40/2429. S. D. Felkin, A.I.1 Report No. 911/1940, 'Gas, Anti-Gas, Smoke and Invasion', 19 November 1940. The National Archives, Kew.
19. Ibid.
20. Documents.10205, Box No: Misc 74 (1107), 'Operation Sea

Lion Directives: 'Unternehmen "Seelöwe" – (Operation Sealion) – Invasion of U.K., 1940'. Keitel/Jodl, 'O.K.W.: 'Sea-Lion', Fuehrer's Headquarters, 3.9.1940. The Imperial War Museum, London.

21. Heeresgruppenkommando A, in Nr. 1457/40, Koluft, H. Qu., 9.9.1940. 'Vernebelung bei "Seelöwe"', Unterlagen des Koluft beim AOK 16: taktische Weisungen und Verfügungen des OKH und des OKM, Weisungen der Heeresgruppe A für den Einsatz der Luftwaffe und das Unternehmen "Seelöwe", Übersicht der unterstellten Verbände, Befehle für die Verneblung des Landungsraumes, Anordnungen für die nachrichtentechnische Sicherstellung, Funkpläne, Merkblätter u.a. (1). Deutsch–Russisches Projekt zur Digitalisierung Deutscher Dokumente in Archiven der Russischen Föderation.

22. Heeresgruppenkommando A, in Nr. 1457/40, Koluft, H. Qu.,'Verwendung künstlichen Nebels während einer Landung', Unterlagen des Koluft beim AOK 16: taktische Weisungen und Verfügungen des OKH und des OKM, Weisungen der Heeresgruppe A für den Einsatz der Luftwaffe und das Unternehmen "Seelöwe", Übersicht der unterstellten Verbände, Befehle für die Verneblung des Landungsraumes, Anordnungen für die nachrichtentechnische Sicherstellung, Funkpläne, Merkblätter u.a. (1). Deutsch–Russisches Projekt zur Digitalisierung Deutscher Dokumente in Archiven der Russischen Föderation.

23. Ibid.

24. Eisenbach and Dauselt, *Der Einsatz deutscher Sturzkampfflugzeuge gegen Polen, Frankreich und England 1939 und 1940*, 194.

25. Ibid.

26. A.O.K. 16, Planspiel VII. A.LK. 'Aufgetretene Fragen von allgemeinem Interesse während des Planspieles des VII.A.K.', 7 September 1940. Unterlagen der Ia-Abteilung des Generalkommandos des VII. Armeekorps: Weisungen und Befehle vorgesetzter Dienststellen zu "Seelöwe". Deutsch–Russisches Projekt zur Digitalisierung Deutscher Dokumente in Archiven der Russischen Föderation.

27. Oberkommando des Heeres, Gen St d H.Abt.Fremde Heere West, H.Q., OKH, den 7.9.1940. 'Merkblatt: Englische Guerilla – Kriegsführung.' Unterlagen des Koluft beim AOK 16.

Deutsch-Russisches Projekt zur Digitalisierung Deutscher Dokumente in Archiven der Russischen Föderation.

28. Ibid.

29. Unterlagen der Ia-Abteilung der Heeresgruppe A: Schriftwechsel mit dem OKH, der Kriegsmarine, der Luftwaffe und dem AOK 9 sowie 16 zum Unternehmen "Seelöwe", Ausbildungshinweise für die Landung in England, Befehle und Weisungen für "Seelöwe". Deutsch-Russisches Projekt zur Digitalisierung Deutscher Dokumente in Archiven der Russischen Föderation.

30. 'Anlage 1 zu Gen.Kdo. XIII.A.K. Ia, Nr. 840/40: Feindnachrichtenblatt (zugleich Beurteilung der Feindlage)', 12 September 1940. Deutsch-Russisches Projekt zur Digitalisierung Deutscher Dokumente in Archiven der Russischen Föderation.

31. 'Kommandeur der Luftwaffe beim A.O.K. 16 – Ia – Aufklärungsaufträge für die 1.(F)/22 (Ergänzungen zu den "Besonderen Anordnungen für die Luftaufklärung am X-Tag")', 15 September 1940. Deutsch-Russisches Projekt zur Digitalisierung Deutscher Dokumente in Archiven der Russischen Föderation.

32. 45. Division, Infantry Regiment 135, Div.St.Qu., den 8.9.40. 'Betr: Ausbildung für das Unternehmen "Seelöwe"'. Deutsch-Russisches Projekt zur Digitalisierung Deutscher Dokumente in Archiven der Russischen Föderation.

33. Oberkommando der Kriegsmarine, B.Nr.1233/40 g.Kdos.2.Abt.Skl. Berlin, den 11. September 1940. 'Führungsmaßnahmen für den Erkennungsdienst zwischen Kriegsmarine und Luftwaffe beim Unternehmen "Seelöwe"'. Deutsch-Russisches Projekt zur Digitalisierung Deutscher Dokumente in Archiven der Russischen Föderation.

34. Stahl, *Kampfflieger zwischen Eismeer und Sahara*, 108–9.

35. Ibid.

36. Ibid.

37. Baumbach, *Zu Spät?*, 117.

38. Ibid.

39. Bechtle, BArch N 502/26, 'Der Einsatz der Luftwaffe gegen England'.

40. Galland, *The First and the Last*, 24.

41. Dildy and Crickmore, *To Defeat the Few*, 491.

42. Cox, 'A Comparative Analysis of RAF and Luftwaffe Intelligence in the Battle of Britain, 1940', 425–43.

43. Bungay, *The Most Dangerous Enemy*, 301.

44. P. Kaplan, *The Few: Fight for the Skies* (Barnsley: Pen & Sword, 2014), 70.
45. MISC 244/2. M. C. Maxwell, 'Letter to Richard Hough & Denis Richards', in 'Questionnaires 101–229', for Richard Hough & Denis Richards, Golden Jubilee History of the Battle of Britain', c. 1988. Imperial War Museum, London.
46. Mitcham, *Eagles of the Third Reich*, 100.
47. Ibid.
48. L 36 412 = 4.Flgh.Betr.Kp./K.G. 51, Sold., 13 September 1940 [Letter] Bibliothek für Zeitgeschichte, Stuttgart.
49. Keitel and Jodl, 'O.K.W. Extract', Berlin, 14.9.1940.
50. Bergström, *The Battle of Britain*, 220.
51. BArch N 667/1, K. Student, 'Erinnerungen eines Soldaten', 203. Bundesarchiv, Freiburg-im-Breisgau.
52. MF 10112/1 – 'First Hand Accounts of the Battle of Britain, Otto Wolfgang Bechtle.' The RAF Museum, London.
53. 'Pausenlose Angriffe auf London: Britische Jagdabwehr versagte', *Briesetal Bote*, 17 September 1940.
54. Dildy and Crickmore, *To Defeat the Few*, 551.
55. Ibid.
56. Document No. 1842-PS, 'Partial Translation: Notes on the discussion of the Reich Foreign Minister von Ribbentrop with the Duce in the presence of Count Ciano as well as the ambassadors von Mackensen and Alfieri, in Rome on the 19th September 1940.' In *Nazi Conspiracy and Aggression Volume IV*, Office of the United States Chief Counsel for Prosecution of Axis Criminality (Washington, DC: United States Government Printing Office, 1946). The Avalon Project, Lillian Goldman Law Library, Yale Law School.
57. Ibid.
58. Student, '*Erinnerungen eines Soldaten*', 206.

14. Great Dark Bloodstains

1. Forczyk, *We March Against England*, 265.
2. T971, Roll 3. 4376/112. The Von Rohden Collection on the Role of the German Air Force (RG 242 Foreign Records Seized Collection). H. Rauch, 'Report on the operation against England, which had been planned in the year 1940', by Maj. Rauch. Code Name: "Sea Lion". (Bericht über das im Jahre 1940 geplant gewesene Englandunternehmen von Maj Rauch – Deckname: "Seeloewe"). Prisoner of War

Camp (OKL), Berchtesgaden, 30 July 1945.
3. Akte 610. Luftgaustab z.b.V.300, Stabsquartier, den 27.11.40. Deutsch-Russisches Projekt zur Digitalisierung Deutscher Dokumente in Archiven der Russischen Föderation.
4. S. Lazar, *Sparing Civilians* (Oxford: Oxford University Press, 2015), 21.
5. W. D. Rubenstein, *A History of the Jews in the English-Speaking World: Great Britain* (Basingstoke: Macmillan, 1996), 224.
6. R. van Trombley, 'Wormhoudt Massacre (1940)', in A. Mikaberidze (ed.), *Atrocities, Massacres, and War Crimes: An Encyclopaedia* (Santa Barbara, CA: ABC-Clio, 2013), 716.
7. 'Unterlagen der Ic-Abteilung des Generalkommandos des XIII. Armeekorps: Schriftverkehr zum Einsatz des Sonderverbandes Lehrregiment "Brandenburg" bei "Seeloewe", Verbesserungsvorschläge für das Unternehmen, Gliederungsübersichten, Weisung des Korps für das Unternehmen, Aufklärungsweisungen für die Luftwaffe ua (1)', 24 August 1940–17 February 1941.
8. Ibid.
9. 'Lehrregiment Brandenburg, Hptm. V. Hippel: Nebenabdruck an A.O.K 16, Ic/A.O., 17. Division, 35. Division.' Generalkommando XIII, A.K., O.U., 13 September 1940.
10. Ibid.
11. N. Ohler, *Blitzed: Drugs in Nazi Germany* (London: Penguin, 2015) 1.
12. Luftgaustab z.b.V.300, Stabsquartier, den 27.11.40.
13. BArch RW 4/592, Oberkommando der Wehrmacht, 'WFSt/Abt. L (I) Nr. 33 318/40 g.K. Chefs., 6. Ausfertigung.' F.H.Qu., 12 October 1940. Bundesarchiv, Freiburg-im-Breisgau.
14. Ibid.
15. BArch RW 4/590. Oberkommando der Wehrmacht, WFA/Abt. L (I) Nr. 35 190/40 g.K.Chefs. 'Betr: Richtlinien für die Feindtäuschung (England).' F.H.Qu. 7 August 1940; BArch RW 4/592. Oberkommando der Wehrmacht, 'WFSt/Abt. L (I) Nr. 33 331/40 g.K. Chefs: Richtlinien für die Feindtäuschung (Seelöwe)', F.H.Qu., 22 October 1940. Bundesarchiv, Freiburg-im-Breisgau.
16. Ibid.
17. BArch RW 4/592, Oberkommando der Wehrmacht, 'WFSt/Abt. L (I) Nr. 33 318/40 g.K. Chefs., 6.

Ausfertigung.' F.H.Qu., 12 October 1940.
18. BArch RW 4/590. Oberkommando der Wehrmacht, 'WFSt/Abt. L (I) Nr. 33 331/40 g.K. Chefs: Richtlinien für die Feindtäuschung (Seelöwe)', F.H.Qu., 22 October 1940. Bundesarchiv, Freiburg-im-Breisgau.
19. BArch RW 4/592, Nr. 33 331/40, 'Richtlinien für die Feindtäuschung (Seelöwe)', F.H.Qu., 22 October 1940. Bundesarchiv, Freiburg-im-Breisgau.
20. Overy, *The Battle of Britain*, 94.
21. BArch RW 4/592, Oberkommando der Wehrmacht, 'WFSt/Abt. L (I) Nr. 33 318/40 g.K. Chefs., 6. Ausfertigung.' F.H.Qu., 12 October 1940.
22. Galland, *The First and the Last*, 13–19; 22–30.
23. Baumbach, *Zu Spät?*, 121.
24. Galland, *The First and the Last*, 28.
25. Dildy, *Battle of Britain 1940*, 100–1.
26. Ibid.
27. Bob, 'Lecture as dialogue at Air Command and Staff College in Maxwell', Imperial War Museum, London.
28. C. Goss, 'What was the Impact of the Luftwaffe's "Tip And Run" Bombing Attacks, March 1942 – June 1943?', *Air Power Review*, Vol. 4, Iss. 4. (2001), 92–117; 97.
29. Ibid.
30. Ibid.
31. Bergström, *The Battle of Britain*, 688–9.
32. Galland, *The First and the Last*, 29.
33. Caldwell, *JG 26 Luftwaffe Fighter Wing War Diary*, 75–6.
34. Ibid.
35. Eriksson, *Alarmstart*, 163.
36. Ibid.
37. Galland, *The First and the Last*, 29.
38. Goss, 'What was the Impact of the Luftwaffe's "Tip And Run" Bombing Attacks', 97.
39. Galland, *The First and the Last*, 29.
40. Ibid.
41. Ibid.
42. Eriksson, *Alarmstart*, 111.
43. Ibid.
44. Ibid.
45. Reel 2828, Frame 42. 'Untersuchungen und Mitteilung Nr. 645: Statistische Zusammenfassung der Unfälle im Flugbetrieb von September 1939 bis Dezember 1940 nach flugmedizinischen Gesichtspunkten', *Deutsche Versuchsanstalt für Luftfahrt E.V.*, Berlin-Adlershof (1941).
46. H. Knoke, *I Flew for the Führer: The Memoirs of a Luftwaffe Fighter Pilot*. 1991. (Barnsley: Greenhill, 2020), 17–19.
47. Ibid.

48. 1998.A.0044. 'German Air Force collection'. W. F. Sheeley, Mc. Aero-Medical Research Section.
49. AIR 40/3071, 'A 450 – Hauptmann & A 449 – Oberleutnant', M.I.9.h., 12 September 1940. The National Archives, Kew.
50. Ibid.
51. Saunders, *RAF Fighters vs Luftwaffe Bombers*, 20.
52. BArch RL 7-3/823, 'Fernschreiben an Flieger-Korps I, IV, V, Seenotdienst, Jafü 3, Aufkl.Gr.123, Luftgau VII/XII/XIII, Westfrankreich. Tagesbefehl', 18 October 1940. Bundesarchiv, Freiburg-im-Breisgau.
53. Ibid.
54. Eriksson, *Alarmstart*, 111.
55. Saunders, *RAF Fighters vs Luftwaffe Bombers*, 20.
56. Galland, *The First and the Last*, 29.
57. Eriksson, *Alarmstart*, 112.
58. P. Kaplan, *Fighter Aces of the Luftwaffe in World War II* (Barnsley: Pen & Sword, 2006), 75.
59. J. T. Correll, 'How the Luftwaffe Lost the Battle of Britain', *Air & Space Forces Magazine*, 1 August 2008. Available online: www.airandspaceforces.com/article/0808battle/ [Accessed 20 May 2024].
60. Ibid. 1,023 aircraft were lost from Fighter Command, 376 from Bomber Command, and 148 from Coastal Command.

15. Dante's *Inferno*

1. 'Einmotorenflug mit Do 17 Z. Auszüge aus Gefechtsberichten der K.Gr. 606. Anlage 5 zum Lagebericht Nr. 405 vom 15.10.1940.' 1.) Flugzeugmuster – Do 17Z, Taktische Zeichen: 7T + EL. Besatzung – Lt. Z. S. Havemann, Beobachter u. Kdt. Feldw. Selig, Flugzeugführer, Gefr. Faehrmann, Bordfunker, Hgefr. Dorfschmidt, Bordmechaniker. Deutsch-Russisches Projekt zur Digitalisierung Deutscher Dokumente in Archiven der Russischen Föderation.
2. MF 10112/1 – 'First Hand Accounts of the Battle of Britain, Collected by Alexander McKee. 2. Accounts by Luftwaffe personnel. General der Flieger a.D. Johannes Fink, Kommodore KG 2 (Do 17), Arras, Kanal Kampfführer, July. General d.Fl.a.D. Johannes Fink, 1st & 2nd August, 1959.' The RAF Museum, London.
3. MF 10112/1 – 'First Hand Accounts of the Battle of Britain, Collected by Alexander McKee. 2.

Accounts by Luftwaffe personnel.' Hauptmann a.D. Rudolf Lamberty, F/F to S/K, Staffel 9, KG 76 (Do. 17) Cormeilles-en-Vexin. Herr Rudolf Lamberty – 30th July, 1959. KG/76 (LF2: Kesselring). Arthur, *Last Of The Few*, 270.
4. MF 10112/1 – 'First Hand Accounts of the Battle of Britain, Collected by Alexander McKee. 2. Accounts by Luftwaffe personnel. General der Flieger a.D. Johannes Fink, Kommodore KG 2 (Do 17), Arras, Kanal Kampfführer, July. General d.Fl.a.D. Johannes Fink, 1st & 2nd August, 1959.' The RAF Museum, London.
5. AIR 40/2429. A.I.1 (k) Report No. 953/1940, 'Losses, Morale, etc. for November 1940'. The National Archives, Kew.
6. Transcript for NMT 1: Medical Case: Official Transcript of the American Military Tribunal in the matter of the United States of America, against Karl Brandt, et al., defendants, sitting at Nurnberg, Germany', 28 April 1947 (Harvard Law School Library Nuremberg Trials Project).
7. Anlage 25 zu Luftgaustab z.b.V. 300, Abt. Ia/Qu, Nr. 110/40 g. Kdos., v. 27.11.40, 'Merkblatt über erste Hilfe bei Bergung aus Seenot', 27 November 1940. Unterlagen des Marineverbindungsoffiziers beim AOK 16: Lageberichte der Seekriegsleitung, Befehle des Marinebefehlshabers Kanalküste, Besprechungsnotizen, Denkschrift der Luftwaffe zum Seetransport für "Seelöwe", Organisationsübersichten für das Unternehmen, Merkblätter ua (1). Deutsch–Russisches Projekt zur Digitalisierung Deutscher Dokumente in Archiven der Russischen Föderation.
8. Ibid.
9. A. Saunders, *Bf 109 E: Battle of Britain* (Oxford: Osprey, 2024), 38.
10. Anlage 25 zu Luftgaustab z.b.V. 300, Abt. Ia/Qu, Nr. 110/40 g. Kdos., v. 27.11.40, 'Merkblatt über erste Hilfe bei Bergung aus Seenot', 27 November 1940.
11. Ibid.
12. 'Transcript for NMT 1: Medical Case: Official Transcript of the American Military Tribunal in the matter of the United States of America, against Karl Brandt, et al, defendants, sitting at Nurnberg, Germany, 24 February 1947 (Harvard Law School Library Nuremberg Trials Project).

13. The Rubenstein Institute, 1998.A.0044 – 'German Air Force collection'. 'D (Luft) 1208, Supplied for every Buoy. Instructions for the Use of the "Rescue Buoy of the Inspector General of Equipment". Published by L C 8 in conjunction with L I N 16. Translation of German Handbook, "Rettungsboje Generalluftflugzeugmeister", September 1940, Berlin'. The German Minister of Aviation and Commander of the Air Force, Technical Office, Section LC 8. Berlin, 26 September 1940. David and Fela Shapell Family Collections, Conservation and Research Center, United States Holocaust Memorial Museum, Maryland (USA).
14. Ibid.
15. Ibid.
16. Anlage 25 zu Luftgaustab z.b.V. 300, Abt. Ia/Qu, Nr. 110/40 g. Kdos., v. 27.11.40, 'Merkblatt über erste Hilfe bei Bergung aus Seenot', 27 November 1940.
17. AIR 40/2429. A.I.1 (k) Report No. 953/1940, 'Losses, Morale, etc. for November – Extract on Losses'; Report No. 942/1940. S. D. Felkin, 'Further Report on the He. 115, S4 + IH of 1/506, brought down in the Sea 40 miles off Grimsby, on 8/12/1940', 15 December 1940. The National Archives, Kew.
18. AIR 40/2429. E. S. Toll, 'Report by Interrogation Officer on Three Dead German Airmen Brought into Grimsby on 6th November 1940', Interrogation Officer, attached Royal Air Force Station, Digby, Lincs.
19. 'Inquests on Airmen', *Irish Independent*, Monday 28 October 1940. The British Newspaper Archive (British Library).
20. MF 10112/1 – 'First Hand Accounts of the Battle of Britain, Otto Wolfgang Bechtle'. The RAF Museum, London.
21. The Rubenstein Institute, 1998.A.0044 – 'German Air Force collection'. Headquarters, United States Strategic Air Forces in Europe, Office of Asst. Chief of Staff, A-2. AAF Sta. 379, APO 633, US Army, 16 August 1945. 'Air Staff Post Hostilities Intelligence Requirements on German Air Force', Aviation Medicine, Section IV. United States Holocaust Memorial Museum.
22. Ibid.
23. Ibid.
24. Ibid.
25. Ibid.
26. Akte 101. Nr. 20240/40, Der Oberbefehlshaber der

Luftwaffe, 'Lagebericht Nr. 416: Gefangenenaussage über die Zustände in London', H. Qu., 26 October 1940. Deutsch–Russisches Projekt zur Digitalisierung Deutscher Dokumente in Archiven der Russischen Föderation.

27. Ibid.
28. Akte 101. Nr. 19610/40, Der Oberbefehlshaber der Luftwaffe, Nr. 19610/40, 'Lagebericht Nr. 407: Einzelnachrichten über London', H.Qu., 17 October 1940. Deutsch–Russisches Projekt zur Digitalisierung Deutscher Dokumente in Archiven der Russischen Föderation.
29. Ibid.
30. Ibid.
31. Akte 101. Nr. 20510/40, Der Oberbefehlshaber der Luftwaffe, 'Lagebericht Nr. 421: Sonstiges – Angriffswirkungen auf London', H.Qu., 31 October 1940. Deutsch–Russisches Projekt zur Digitalisierung Deutscher Dokumente in Archiven der Russischen Föderation.
32. Akte 101. Nr. 19650/40, Der Oberbefehlshaber der Luftwaffe, 'Lagebericht Nr. 408: Angriffswirkung auf G.B., insbesondere auf London', H. Qu., 18 October 1940. Deutsch–Russisches Projekt zur Digitalisierung Deutscher Dokumente in Archiven der Russischen Föderation.
33. Akte 101. Nr. 19860/40, Der Oberbefehlshaber der Luftwaffe, 'Lagebericht Nr. 411: Sonstiges', H.Qu., 21 October 1940. Deutsch–Russisches Projekt zur Digitalisierung Deutscher Dokumente in Archiven der Russischen Föderation.
34. S. Wynn, *Churchill's Flawed Decisions: Errors in Office of the Greatest Briton* (Barnsley; Philadelphia: Pen & Sword Military, 2020), 69.
35. Dildy, *Battle of Britain 1940*, 113
36. Bergström, *The Battle of Britain*, 148–9.
37. J. Hale, *The Blitz 1940–41: The Luftwaffe's biggest strategic bombing campaign* (New York: Bloomsbury USA, 2023), 16.
38. M. I. Handel, *War, Strategy and Intelligence* (Abingdon: Frank Cass, 1989), 182.
39. Hale, *The Blitz 1940–41*, 28–30.
40. Rieckhoff, *Trumpf oder Bluff*, 225.
41. B. Johnson, *The Secret War* (Barnsley: Pen & Sword, 2004), 44.
42. Bergström, *The Battle of Britain*, 441–2.
43. Ibid.
44. Hale, *The Blitz 1940–41*, 18.
45. Ibid.
46. Generalmajor a.D. Paul Weitkus, CO II Gruppe, KG

2 (Do.17), St.Leger, near Arras. 14th/15th November, Coventry. The RAF Museum, London.
47. Stahl, *Kampfflieger zwischen Eismeer und Sahara*, 67–8.
48. Ibid.
49. Klatte, *Kampfflieger in der Luftschlacht um England*, 195.
50. M. Smith, *Britain and 1940: History, Myth and Popular Memory* (London; New York: Routledge, 2000), 84.
51. Baumbach, *Zu Spät?*, 117.
52. T-321, Roll 54. 'Tagesbefehl des Reichsmarschalls am 21.11.1940', 21 November 1940. National Archives, College Park, Maryland (USA).
53. MF 10112/1 – 'First Hand Accounts of the Battle of Britain, Otto Wolfgang Bechtle.' The RAF Museum, London.
54. Stahl, *Kampfflieger zwischen Eismeer und Sahara*. 144.

16. Flying, Sleeping, Eating, Waiting

1. BArch RW 4/592, Oberkommando der Wehrmacht, WFSt/Abt. L (I) Nr. 00811/40 g.Kdos – Betr: 'Luftangriffe im Kanalgebiet', F.H.Qu., 2 October 1940. Bundesarchiv, Freiburg im Breisgau.
2. N98.6. 'Luftwaffenhelfer; Erinnerungen'. Bibliothek für Zeitgeschichte, Stuttgart.
3. Ibid.
4. Darmstadt, 11 October 1940. [Letter] Bibliothek für Zeitgeschichte, Stuttgart.
5. S. D. Felkin, "Inside Germany"/'Further Report on Me. 109 of 2/J.G.26 brought down at Udimore, Kent, 28.11.40: Previous Report No.941/1940'.
6. L 38545, L.G.P.A. Hamburg (Flakscheinwerfer-Division), Sold., 4 December 1940. [Letter] Museum für Kommunikation, Berlin.
7. Ibid.
8. Private Papers of K. G. Raynor, Box No: 02/19/1. The Imperial War Museum, London.
9. Mitcham, *Eagles of the Third Reich*, 101.
10. Baumbach, *Zu Spät?*, 117.
11. Stahl, *Kampfflieger zwischen Eismeer und Sahara*, 93.
12. W. Mork, 'Werner Mork: Gedanken über den Krieg', *Lebendiges Museum Online*. Deutsches Historisches Museum, Berlin. Available online: www.dhm.de/lemo/zeitzeugen/werner-mork-gedanken-%C3%BCber-den-krieg [Accessed 10/01/2024].
13. Ibid.
14. Ibid.

15. Der Chef der Luftflotte 3 und Befehlshaber West Sperrle, 'Luftflottenkommando 3, Nr. 46/41, H.Qu., am 20.3.41.', Heeres Gruppe A Ia Nr. 257/41 g. Kdos. Chefsache, 20 March 1941.Unterlagen der Ia-Abteilung der Heeresgruppe A: Schriftwechsel mit dem OKH, der Kriegsmarine, der Luftwaffe und dem AOK 9 sowie 16 zum Unternehmen "Seelöwe"'. Deutsch–Russisches Projekt zur Digitalisierung Deutscher Dokumente in Archiven der Russischen Föderation.
16. Ibid.
17. Tillmann, 'Kriegsgesellschaft, Kriegsalltag und Kriegserleben an der Heimatfront', 69.
18. Ibid.
19. Ibid.
20. Ibid.
21. L 29 646 = Rgts.Stab/Flak-Rgt.22, Sold., 20.4.1941. [Letter] Museum für Kommunikation, Berlin.
22. L 09 244 = 6.Bttr./Flak-Rgt.49, Uffz., 10 May 1941. [Letter] Bibliothek für Zeitgeschichte, Stuttgart.
23. L 22 143 = Parkschiff der Lw. "Bukarest", Uffz., 26 May 1941. [Letter] Bibliothek für Zeitgeschichte, Stuttgart.
24. Stahl, Kampfflieger zwischen Eismeer und Sahara, 143–44.
25. Ibid.
26. E. R. Hooten, Eagle in Flames: The Fall of the Luftwaffe (London: Weidenfeld & Nicolson, 1997), 33.
27. Ibid.
28. D. Stahel, Operation Barbarossa and Germany's Defeat in the East (Cambridge: Cambridge University Press, 2010), 124.
29. Lehweß-Litzmann, Absturz ins Leben, 156–7.
30. Ibid.
31. L 09 244 = 6.Bttr./Flak-Rgt.46, Bremen, Uffz., 1 June 1941. [Letter] Bibliothek für Zeitgeschichte, Stuttgart.
32. K. Smith, Conflict over Convoys: Anglo–American Logistics Diplomacy in the Second World War (Cambridge: Cambridge University Press, 1996), 271.
33. A. Axelrod, The Real History of World War II: A New Look at the Past (New York: Stirling, 2008), 80.
34. Bell, Britische Feindaufklärung im Zweiten Weltkrieg, 124–5.
35. Axelrod, The Real History of World War II, 80.
36. J. Shields, 'The Battle of Britain: A Not So Narrow Margin', Air Power Review, Vol. 18, Iss. 2 (April 2015), 182–97.
37. Bungay, The Most Dangerous Enemy, 368.
38. Ibid.
39. Auswärtiges Amt, 'No. 532 – Führer's Directive: Directive No. 21: Operation Barbarossa', Führer's

Headquarters, 18 December 1940, in *Documents on German Foreign Policy, 1918–1945* (Washington, D.C.: United States Government Printing Office, 1960), 899.
40. Ibid.
41. Lehweß-Litzmann, *Absturz ins Leben*, 157–8.
42. 'Das britisch-bolschewistische Komplott', *Sorauer Tageblatt*, 23 June 1941. ZEFYS, Staatsbibliothek zu Berlin.
43. L 33 281 = *Unteroffizier*, Fliegerhorst Lyon, 23 June 1941. [Letter] Sondersammlung, Bibliothek für Zeitgeschichte, Stuttgart.
44. Private Papers of K. G. Raynor, Box No: 02/19/1. The Imperial War Museum, London.
45. Lehweß-Litzmann, *Absturz ins Leben*, 157–8.
46. Bell, *Britische Feindaufklärung im Zweiten Weltkrieg*, 124–5.
47. Ibid.
48. Scriba, 'Die "Luftschlacht um England"', *Deutsches Historisches Museum*.
49. Correll, 'How the Luftwaffe Lost the Battle of Britain', *Air & Space Forces Magazine*, 1 August 2008.
50. D. Uziel, *Arming the Luftwaffe: The German Aviation Industry in World War II* (Jefferson: McFarland, 2011), 13.
51. Blanco, *The Luftwaffe in World War II*, 98.
52. Uziel, *Arming the Luftwaffe*, 13.
53. Ibid.
54. 'Bomben fallen auf London', *Sorauer Tageblatt*, 24 January 1944. ZEFYS, Staatsbibliothek zu Berlin.
55. Bechtle, BArch N 502/26, 'Der Einsatz der Luftwaffe gegen England', Bundesarchiv, Freiburg-im-Breisgau.
56. Ibid.
57. Ibid.
58. Ibid.
59. Ibid.
60. Bechtle, BArch N 502/26, 'Der Einsatz der Luftwaffe gegen England', Bundesarchiv, Freiburg-im-Breisgau.

17. The Other Faculty

1. T971, Roll 3. 4376/112. The Von Rohden Collection on the Role of the German Air Force (RG 242 Foreign Records Seized Collection). H. Rauch, 'Report on the operation against England, which had been planned in the year 1940, by Maj. Rauch. Code name: "Sea Lion".'
2. Documents. 10205, Box No: Misc 74 (1107), 'Operation Sea Lion Directives: 'Unternehmen "Seelöwe"'– (Operation Sealion) – Invasion of U.K., 1940', The Imperial War Museum, London.
3. C. Goss, 'The Design and Evolution of the Junkers

Ju 88', *Key Aero*, 18 February 2022. Available online: www.key.aero/article/design-and-evolution-junkers-ju-88 [Accessed 3 January 2024]

4. N-9618. 'Interrogation of Reich Marshal Hermann Goering by General Carl Spaatz at Ritter Schule, Augsburg, Germany on May 10, 1945.' World War II Operational Collection, Ike Skelton Combined Arms Research Library Digital Library. Available online: https://cgsc.contentdm.oclc.org/digital/collection/p4013coll8/id/4149/ [Accessed 12/02/2024].

5. Baumbach, *Zu Spät?*, 59.

6. N-9618. 'Interrogation of Reich Marshal Hermann Goering', May 10, 1945.

7. "Detailed Interrogation Report / Keitel's and Kesselring's Replys to Questions Cabled From Washington / SECRET", 21 April 1945. Cornell University Law Library Donovan Nuremberg Trials Collection. Available online: www.reader.library.cornell.edu/docviewer/digital?id=nur01613#page/5/mode/1up [Accessed 20/01/2024]

8. Student, 'Erinnerungen eines Soldaten', 206.

9. 'Hermann Göring', *Holocaust Encyclopedia (United States Holocaust Memorial Museum)*. Available online: www.encyclopedia.ushmm.org/content/en/article/hermann-goering [Accessed 07/05/2021].

10. 'Defense Exhibit GOERING No. 7', H-1674, IMT Nuremberg Archives, International Court of Justice. Stanford Digital Repository, Stanford University. Available Online: www.purl.stanford.edu/st843th6939 [Accessed 20/01/2024]

11. Ibid.

12. A. Kesselring, *Soldat bis zum letzten Tag*. 1953. (Schnellbach: Siegfried Bublies, 2000), 88.

13. Ibid.

14. Student, 'Erinnerungen eines Soldaten', 197–8.

15. Baumbach, *Zu Spät?*, 113.

16. W. A. Fletcher, 'Foreword', in R. Raiber (ed.) *Field Marshal Albert Kesselring, Via Rasella, and the Ginny Mission* (Newark: University of Delaware Press, 2008), 9–11.

17. Steinhoff, *Messerschmitts Over Sicily*, loc. 3213.

18. S. P. MacKenzie, *The Battle of Britain on Screen: 'The Few' in British Film and Television Drama* (Edinburgh University Press, 2007), 81.

19. Ibid.

20. Ibid.

21. Kaplan, *Fighter Aces of the Luftwaffe*, 14.

22. Ibid.

23. *The Air Force Law Review* (Maxwell Air Force Base: Office of the Judge Advocate General, United States Air Force, 1990), 108–9.
24. AIR 40/3071. 'S.R.A 675: A 557 – Gefreiter (Bomber Gunner) & A 564 – Gefreiter (Fighter-Pilot), M.I.9.H, 4 October 1940.
25. S. Cox, 'Battle of Britain: Despatch By Air Chief Marshal Sir Hugh C. T. Dowding, Copy No. 19', *Air Power Review*, Vol. 18, Iss. 2 (Apr 2015), 70–136.
26. Kaplan, *Fighter Aces of the Luftwaffe*, 16.
27. *The Air Force Law Review*, 108–9.
28. Mackenzie, *The Battle of Britain on Screen*, 82.
29. Steinhoff et. al, *Voices From The Third Reich*, 81.
30. Steinhilper, *Spitfire On My Tail*, 277.
31. Ibid.
32. W. S. Churchill, *Their Finest Hour: Volume II, The Second World War*, 1949. (London: Penguin, 1985), 284.
33. Bestand 500, Findbuch 12452, Akte 101. Lageberichte West des Führungsstabes der Luftwaffe (1c), 'Rettungen durch Seenotdienst', Anlage 5 zum Lagebericht Nr. 396 vom 5. Okt. 1940, 1–31 October 1940, 158. Deutsch-Russisches Projekt zur Digitalisierung Deutscher Dokumente in Archiven der Russischen Föderation.
34. 'General Report, A.I.1 (k) Report No. 939/1940.'
35. Steinhilper, *Spitfire On My Tail*, 277.
36. Holland, *The Battle of Britain*, 450.
37. Churchill, *Their Finest Hour: Volume II, The Second World War*, 284.
38. Article 17, 'Rules concerning the Control of Wireless Telegraphy in Time of War and Air Warfare.' Drafted by a Commission of Jurists at the Hague, December 1922–February 1923. *International Committee of the Red Cross*. Available online: https://ihl-databases.icrc.org/en/ihl-treaties/hague-rules-1923/article-17?activeTab= [Accessed 03/03/2024].
39. C. Antonopoulos, *Non-Participation in Armed Conflict: Continuity and Modern Challenges to the Law of Neutrality* (Cambridge: Cambridge University Press, 2022), 34.
40. See M. Paris, *Warrior Nation: Images of War in British Popular Culture, 1850 – 2000* (London: Reaktion, 2000).
41. *Battle of Britain* (1969) [Film] Directed by G. Hamilton. United Kingdom: Spitfire Productions.

42. Rökker, *Ausbildung und Einsatz eines Nachtjägers im II. Weltkrieg*, 23.

18. Better Liars than Flyers

1. Arthur, *Last Of The Few*, 132–3.
2. *The Air Force Law Review*, 110.
3. Hauptmann a.D. Rudolf Lamberty, F/F to S/K, Staffel 9, KG 76 (Do. 17) Cormeilles-en-Vexin. Herr Rudolf Lamberty – 30 July 1959. KG/76 (LF2: Kesselring).
4. Arthur, *Last Of The Few*, 270.
5. Ibid.
6. C. Peniston-Bird, 'Commemorating Invisible Men: Reserved Occupations in Bronze and Stone', in L. Robb and J. Pattinson (eds), *Men, Masculinities and Male Culture in the Second World War* (London: Palgrave Macmillan, 2018); 189–214.
7. Ibid.
8. MISC 3404 244/1. 'Last burst hit Spitfire', Friday 4 September 1987, p. 11. The Imperial War Museum, London.
9. 'It Was Two Years Ago – Do You Remember? 185 German Raiders Crashed In a Day – "Divine Intervention"', *Bradford Observer*, Tuesday 15 September 1942.
10. J. Mevius, 'Repräsentationen der Battle of Britain und die Konstruktion nationaler Identität', PhD Thesis, Ruprecht-Karls-Universität (29 May 2008), 157.
11. Nuremberg Trial Proceedings Vol. 9, Eightieth Day, Morning Session, Wednesday, 13 March 1946. The Avalon Project, Lillian Goldman Law Library, Yale Law School.
12. Ibid.
13. J. Steinhoff et al., *Voices from the Third Reich*, 352.
14. W. A. Urbanowicz, 'Letter to Richard Hough & Denis Richards', 10 May 1988, in 'Questionnaires 101–229', for the 'Golden Jubilee History of the Battle of Britain'. Imperial War Museum, London.
15. Ibid.
16. Neitzel and Welzer, *Soldaten*, loc. 863.
17. Ibid.
18. AIR 40/2429. Reports by Prisoners of War, Air Ministry/Royal Air Force. 'Inside Germany', Winter 1940.
19. G. de Syon, 'The Einsatzgruppen and the Issue of "Ordinary Men"', in J. C. Friedman (ed.), *The Routledge History of the Holocaust* (London; New York: Routledge, 2011) 148–55.
20. Neitzel and Welzer, *Soldaten*, loc. 2669.
21. Beckermann, *Jenseits des Krieges* (Vienna: Aichholzer Filmproduktion, 2002)

[Documentary: English Subtitles].
22. AIR 40/3071. S.R.A. No. 818: A 607 – Hauptmann (Bomber Observer), A 604 – Oberleutnant (Fighter Pilot) & A 601 – Leutnant (Bomber Pilot), 25–7 October 1940. The National Archives, Kew.
23. Transcript for NMT 2: Milch Case. Official Transcript of the American Military Tribunal in the matter of the United States of America against Erhard Milch, defendant, sitting at Nurnberg, Germany, on 3 January 1947, *USA v. Erhard Milch*, HLS Nuremberg Trials Project, Harvard Law School.
24. Ibid.
25. Ibid.
26. *Obergefreiter* = Ln.Kp.(mot.) Flughafenbereich Radom, 4 January 1941. [Letter] Bibliothek für Zeitgeschichte, Stuttgart.
27. L 34 215 = Trsp.Kol.d.Lw.9/VII, 23 April 1941. Bibliothek für Zeitgeschichte, Stuttgart.
28. Lehweß-Litzmann, *Absturz ins Leben*, 152.
29. Ibid.
30. 'Transcript for NMT 1: Medical Case: Official Transcript of the American Military Tribunal in the matter of the United States of America, against Karl Brandt, et al., defendants, sitting at Nurnberg, Germany', 10 December 1946 (Harvard Law School Library Nuremberg Trials Project).
31. Ibid.
32. N. Farron, 'Rascher and the "Russians": human experimentation on Soviet prisoners in Dachau – a new perspective', in P. Weindling (ed.), *From Clinic to Concentration Camp: Reassessing Nazi Medical and Racial Research, 1933–1945* (London; New York: Routledge, 2017), 257–71.
33. Ibid.

Conclusion

1. N 60/137 – Bd. 4: 'Ferdinand Schörner. Generalfeldmarschall in letzter Stunde', Bundesarchiv, Freiburg-im-Breisgau.
2. Forczyk, *We March Against England*, 13.
3. Corum, 'Stärken und Schwächen der Luftwaffe', 283–306.
4. Mevius, 'Repräsentationen der Battle of Britain und die Konstruktion nationaler Identität', 279.
5. Lehweß-Litzmann, *Absturz ins Leben*, 157–8.
6. Ibid.
7. Rieckhoff, *Trumpf oder Bluff?*, 287.

8. Bechtle, BArch N 502/26, 'Der Einsatz der Luftwaffe gegen England', Bundesarchiv, Freiburg-im-Breisgau.
9. Steinhoff et al., *Voices From The Third Reich*, 82.
10. Private Papers of K. G. Raynor, Box No: 02/19/1. The Imperial War Museum, London.
11. General der Fl. A. D. Paul Deichmann, *Studiengruppe Luftwaffe bei der Führungsakademie der Bundeswehr* – 7 August 1959.
12. V. Mikula, *Stuka: Tatsachenbericht*. 1965. (Munich: Wilhelm Heyne, 1981), 89.

Index

A
Africa (North), 287, 292
Anderson, Michael: *The Dam Busters*, 313
antisemitism/persecution of Jews, 130, 195–6, 245, 321–2, 324, 331
  in France, 328
  in Italy, 305–6
  'Jewish plutocrats' in Britain, 8, 124, 131–2, 194–5, 290–1
  in Poland, 40–1, 325–6, 327–8
  in the Sudetenland, 25–6
  see also Holocaust
appeasement, 24
Atlantic, 304–5, 336
  Battle of, 111
Austria/Austrians, 29, 43, 69, 336
  Anschluss, 22, 324–5
  native Nazis, 22, 324
  Stiftskaserne Flak Tower, Vienna, 51
  Zeltweg airfield, 97

B
the Balkans, 287, 291, 292, 293, 336
Baltic Sea, 59, 106
Bartels, Werner, 103, 185
Baumbach, Werner, 169, 211, 233, 251, 282, 304–5

Bechtle, Otto-Wolfgang, 157, 160, 167, 236, 269, 278, 294–6, 297, 337
Becker-Freyseng, Hermann, 331, 333
Belgium/Belgians, 70–1, 76–7, 91, 124, 124–5, 134–5, 138–9, 153, 163, 233, 304–5
  Army Air Force, 73, 76
  German invasion (*Fall Gelb*/Case Yellow), 57, 68, 73, 76–7
  Mechelen incident, 57, 71
  see also Low Countries
Bob, Hans-Ekkehard, 2, 20, 26, 36, 81, 109, 172–3, 226–7, 252
Britain
  Aberdeen Press and Journal, 187
  Admiralty, 138
  Air Ministry Secret Intelligence Service Branch, 274
  'Albion', 101–2, 106, 108
  armament industry, 125
  Army, 81, 110, 149, 174
    7th Devons, 186
    Anti-Aircraft Command, 133, 149, 231, 237, 277, 287
    radar, 147, 150, 156, 170, 171–2, 234, 252
      Chain Home (CH), 170, 171, 251

Britain (*continued*)
    Women's Auxiliary Air Force (WAAF), 171, 187–8
  commandos, 328
Baedeker Raids, 293
BBC, 186–7, 197, 198
Birmingham, 263, 277, 288
the Blitz, 275, 279–96, 304
  'Baby Blitz', 293–4
  beginning of, 224–5, 233, 234, 236, 259
  as a change of Luftwaffe tactic, 3, 3–4, 223, 224–5, 230
  declining effect, 257, 282, 283, 287, 290, 292, 330, 331
  failure of, 238
  faith in, 226, 231, 250
  at night, 263, 267
  outside London, 288–9
  over London, 250–1, 272–3, 288–9, 313, 332
  retaliation, 270
  see also under London; World War II: *Blitzkrieg*
Bradford Observer (newspaper), 320–1
Brighton, 224, 229
Bristol, 288, 293
  Aeroplane Company, 319
British Empire, 109–10, 123, 124, 126, 277, 304
  Commonwealth personnel, 124
British Expeditionary Force (BEF), 44, 79, 87, 107, 119
'British War', 157
Canterbury, 224, 229–30, 236
Cardiff, 288, 293
Civilian Repair Organisation (CRO), 176
Combined Intelligence Committee, British and American, 298

Combined Services Detailed Interrogation Centre (CSDUC UK), 189–90
Convoy Outbound Atlantic (OA-178), 112
Coventry, 273, 288
  Cathedral, 277
  Unternehmen Mondscheinsonate (Operation Moonlight Sonata), 275–7
Dover, 107, 135, 136, 137, 148, 154, 156, 229–30, 243
  White Cliffs, 101, 118–19, 135
Dundee Courier, 15
enemy aircrew inquests, 267–9
English Channel, 149, 249–50, 310
  action over the Channel, 83, 295, 335–6
  Allied shipping, 98, 107, 111, 112, 114–15, 134, 154
  challenges as a waterway, 105, 106, 204–5, 212, 228, 231, 264, 265–6, 268, 338
  Channel Islands, 104–5, 117, 125
  Kanalkampf (Channel Battle), 118, 128, 135, 140, 164, 250
    beginning, 111–17, 119, 121, 126, 137–8
    failure of, 143–4, 201, 259
    night-time combat, 129
    threat to Britain, 231
  proximity of England, 135, 138, 304–5
  psychological barrier, 125–6, 135, 136, 139, 147, 172
  Strait of Dover, 138
Glasgow, 287, 288
guerrilla warfare, 228–9
House of Commons, 132
Hull, 286, 287, 288, 293

Isle of Wight, 154, 170
Liverpool, 275, 287, 288
Local Defence Volunteers/Home
    Guard, 133, 150, 174
London, 132
  Bethnal Green, 215
  the Blitz, 224–6, 230–8, 250–1,
      288–9, 293, 302, 313, 314
      see also Britain: the Blitz
  Bond Street, 103
  Buckingham Palace, 236–7
  defence of, 185, 225
  docks, 185, 232, 236
  East Ham, 215
  'Hardest Night', 287, 322
  Hitler's view, 45, 109
  Hyde Park, 193
  Jagdbomber attacks, 252–4,
      257, 278, 280–1
  Oxford Street, 272
  Piccadilly Circus, 273
  propaganda regarding morale,
      270–3
  St Paul's Cathedral, 273
  vulnerability (or not) to
      Luftwaffe attack, 2, 44–5,
      108, 156, 199
Maidstone, 229–30, 236
Norwich, 210, 224
Observer Corps, 170–1, 252
Operation Steinbock 'Baby Blitz',
    293, 294
'pirate island', 2, 7, 87, 338,
    see also Royal Air Force:
    '*Luftpiraten*'
Plymouth, 148, 288
Portland, 112, 115, 148, 224, 238
Portsmouth, 148, 156, 258, 288
prisoners, 4, 183–99, 227, 263,
    267, 274, 281, 298, 310,
    311, 323, 325
Royal Navy, 45, 68, 69, 79, 125,
    132, 140, 149, 156, 163
  Fleet Air Arm, 133

  sea rescue service, 310
  Shetland Islands, 43, 48–9
  Southampton, 224, 288
  Territorial Army, 174
  War Cabinet, 6, 131, 310
    see also Royal Air Force
Bruns, Karl, 33, 61–2, 69–70
Büchner, Franz, 98

C
Chamberlain, Neville, 22–3,
    24, 44
Churchill, Winston, 130
  Hitler's view of, 44, 126–7,
      130–1, 223
  Luftwaffe view of, 195, 284
  propaganda against, 48–9, 131,
      132, 163, 206
  questionable decisions, 309–10,
      311, 312
  quoted, 99, 112
  retaliating to the Blitz, 215,
      301–2, 309–10
Czechoslovakia/Czechs, 24–6, 29,
    34, 41, 124
  Bohemia, 29
  Slovakia/Slovakian, 29, 124
  Sudetenland, 22–3, 24, 25–6,
      27, 29
Czech, William, 66, 67

D
Dahm, Franz, 1–3
Dalwigk zu Lichtenfels,
    Friedrich-Karl von, 114,
    115, 116, 117
Deere, Alan, 153–4
Deibl, Hans, 38–9, 50–1, 52–3,
    96, 270
Deichmann, Paul, 6, 99, 104,
    147, 149, 169, 225, 295,
    338, 339
Denmark, 59, 61, 68, 249, 338
  Aalborg, 60, 61, 62, 66

Denmark (*continued*)
Copenhagen, 60, 61
German invasion (*Unternehmen 'Weserübung Süd'*/ Operation Weser Exercise South), 59–62, 69–70, 79
Hawker Furies, 60
Deumling, Klaus, 206–7
Diehl, Hermann, 47–8
Dowding, Hugh, 111–12, 170, 171–2, 213, 300, 307–8, 320
Dutch
see Netherlands

E
Ehrlich, Lothar, 136, 137
England
see Britain

F
Falck, Wolfgang, 16–17, 47, 48, 168
Felmy, Helmuth, 44, 45
Fink, Johannes, 93–4, 113, 157–8, 160, 262, 263
First World War
see World War I
France/French, 101, 163, 322
Armée de l'air (air force), 73, 77, 78, 86
Bloch-152, 77
Morane-Saulnier MS-406, 77
army, 15, 44, 56, 68, 77–8, 85
Battle of France, 84, 98, 99, 264, 338
Beauvais, 233
Brest, 249, 262, 263
Brittany, 46, 148
Calais, 118–19, 134, 135, 307, 318
Cherbourg, 98, 165, 166, 263, 285
Dunkirk, 79, 84, 86, 87, 174, 175, 186, 226, 295
air battle, 79, 80, 81–3, 91–2, 153–4, 185
Luftwaffe misses chance to eliminate BEF, 81, 87, 109, 118
Operation Dynamo, 79, 84
propaganda value to Germany, 107–8
Fall of, 83, 106, 110, 226
German invasion of France (*Fall Rot*/Case Red)/Battle of France, 59–60, 68, 83–7, 92–5, 96–7, 97–9, 105, 114, 127, 138, 192, 193
German occupation, 118, 135, 143, 144, 172, 194, 198, 218, 282, 284, 328
Germany's war with France and/or Britain anticipated/doubted, 24, 29, 30, 41–2, 42, 44, 45, 56–7, 59, 71–2, 87, 121
Hazebrouck, 82–3
Lyon, 291
Maginot Line, 71, 78
Maltot, 165, 166
northern France and Brittany, 46, 56, 71, 73, 78, 107, 245, 262
Paris, 83, 85, 94, 180
Champs Elysées, 103
St Omer, 247, 283, 284
Théville, 114, 117, 154
western France, 179, 285
Wormhoudt Massacre, 245
see also Low Countries: German invasion
Franco, Francisco, 13, 20, 241

G
Galland, Adolf, 94, 103, 176, 185, 206, 207, 209, 250–1
and *Battle of Britain* film, 306–7, 308

INDEX

on British fighter planes and
technology, 117–18, 152,
153, 171–2
on Dunkirk, 81, 109
on the English Channel, 212
on *Fall Gelb*/Case Yellow
invasion of the Low
Countries, 76–7
on Luftwaffe leadership, 168,
253, 254. 256–7, 258
on the Luftwaffe's rebirth, 12
over Poland, 252
in Spain, 13–14, 252
talking to Hitler, 109
Geneva Convention, 188, 310, 312
Gerathewohl, Siegfried, 52, 256
Germany
Berghof, Bavarian Alps, 30
Berlin, 31–2, 34, 37, 48, 95–6,
100, 235
Aviation Technical School, 23
Bonn, 37
Bremen, 288
Döberitz, Brandenburg, 16–17
Dresden, 215
Eisernes Kreuz II (Iron Cross
2nd Class), 191, 326
Freya radar, 47–8, 171
Gestapo, 222
Graf Zeppelin (airship), 171
Hamburg, 57, 62, 86, 215, 225,
280, 286
Heer (army), 249, 283, 284
16th Army, 244
Army Group A, 228, 244,
284–5, 297
Army Higher Command
AOK-9, 229
AOK-16, 203–5, 228,
229–30, 244
Artillery Regiment-3, 83
Artillery Regiment
Hohmann, 243
Condor Legion (*Legion Condor*),

13–14, 16, 20, 21, 39,
256–7
cooperation with Luftwaffe,
94, 244, 245
Fall of France, 83
General Command XIII, 229
Infantry Regiment-85, 83
Infantry Regiment-134
45th Division, 230
Infantry Regiment-135, 203
invasion of Denmark
(*Unternehmen 'Weserübung
Süd'*/Operation Weser
Exercise South), 60
invasion of Norway
(*Unternehmen
'Weserübung Nord'*/
Operation Weser Exercise
North), 69
invasion of Poland (*Fall Weiss*/
Case White), 35, 44
invasion of the Low Countries
and northern France
(*Fall Gelb*/Case Yellow),
71, 77, 78
Nachrichtentruppe
(Signal Corps), 283
Panzerdivision (tank divisions),
35, 77–8, 78
participation in Spanish Civil
War, 19
planning the invasion of Britain,
110, 123, 126, 132–3,
146–9, 203, 227–8, 229–30
reconnaissance units, 78,
204, 205
XIII. Army Corps
Lehrregiment
Brandenburg z,b,V. 800
(commando units), 246
Heligoland, 106
Battle of Heligoland Bight, 47
Jever, 47, 105
Kiel, 102, 103, 118, 285–6

Germany (*continued*)
Kinderlandverschickung (KLV/
Evacuation of Children),
280
Kriegsmarine (German navy)
coastal protection boat, 67–8
enmity with Luftwaffe, 46–7
Maritime Warfare Command,
46–7
presence (and failure) in the
Baltic, 59, 60, 62, 64,
69, 125
radar, 47–8
role in (and unreadiness for)
the assault on Britain, 110,
126, 140, 147, 148, 149,
228, 230, 244, 284
U-Booten (submarines), 47, 146
Lufthansa, 11, 62
Mannheim, 41, 120
Mecklenburg, 119, 145, 224
Militär & Geschichte
(magazine), 3
Munich, 98, 273
Beer Hall Putsch, 45
peace talks (crisis), 22–3
National Socialism/Nazism
extermination of 'asocials', 245,
322, 331
press/propaganda, 2, 48–9,
195, 206, 233, 272, 284,
289–90, 343
Baruther Anzeiger
(newspaper), 107–8, 112,
113, 114, 117–18, 131–2,
133, 134, 158, 159, 160
Briesetal-Bote (newspaper),
108, 202, 205–6, 236–7
Der Stürmer (newspaper), 197
Deutsche Luftwacht
(magazine), 123, 124–5,
126, 201
Fehrbelliner Zeitung
(newspaper), 100

Hölle über Frankreich (Hell
over France/propaganda
book), 94, 100–1
Rheinsberger Zeitung
(newspaper), 163
Sorauer Tageblatt
(newspaper), 290–1, 293–4
Nuremberg
Rallies, 197
Trials, 63, 301, 305, 321,
327, 331
Oberkommando der Wehrmacht
(OKW/High Command
of the Military), 249–50,
279–80, 338
briefed by Hitler on Poland, 30–1
'Guidelines for the Deception
of the Enemy', 248–9
Jodl, Alfred (*Generalleutnant*/
Chief of Operations), 110,
146, 223, 227, 235
Keitel, Wilhelm (*Generaloberst*/
Head), 110, 151, 223, 227,
235, 248
report on prospects for
Adlertag, 146–7, 149
stance on the proposed
invasion of Britain, 203,
223, 224, 227
Rhineland, 21
Ritterkreuz (Knight's Cross), 207
Ruhr, 56, 128, 128–9, 129
Schutzstaffel (SS), 22, 84, 102, 111,
244–5, 321–2, 324, 328, 330
*1st SS Panzerdivision
Leibstandarte SS Adolf
Hitler*, 245
Black Book, 330
Combat Support Force, 111
Einsatzgruppen
(Task Forces), 245
Operation Tannenberg, 245
Waffen-SS, 325
Westwall (Siegfried Line), 49

Würzburg radar, 171
Goebbels, Joseph, 118, 186–7
Göring, Hermann, 31, 45, 59, 75, 112, 127, 129–30, 164, 169, 197, 304
  Blitz as new Luftwaffe tactic, 3, 224–6, 230, 231, 238, 277–8, 286–7
  cordiality with RAF/London, 16–17, 18
  deflecting blame, 299–302
  'Documents Pertaining to England's Sole Guilt', 301–2
  failure to empathize with pilots, 106, 110, 170, 177, 178, 234, 257–8
  intervention in Spanish Civil War, 12–13, 20–1
  lack of respect from fliers, 176, 177, 253, 338
  Luftwaffe as 'corps of vengeance', 5, 12
  Luftwaffe surreptitiously reborn, 11–12, 20–1, 27
  as *Reichsmarschall* of the Luftwaffe, 6, 37–8, 42, 84, 96, 127, 144, 339
  undiplomatic/inappropriate remarks, 17, 23–4, 215, 221–2
Great Britain
see Britain
Great War
see World War I

H
Hague Rules of Air Warfare, 312
Hamilton, Guy: *Battle of Britain*, 306–7, 308, 311, 312–15
  'Battle of the Air' sequence, 313–14
Henderson, Nevile, 18, 23
Henze, Karl, 115, 116, 117, 164–5
Hiekel, Kurt, 66–7

Himmler, Heinrich, 22, 331–2, 333
Hitler, Adolf, 103, 130, 195
  'Direction No. 1 for the Conduct of the War', 45
  'Directive No. 16', 126
  'Directive No. 21', 290
  frustrated by and then backing the Blitz, 3, 223
  as the Führer
  ambivalence towards Britain, 24, 29–30, 109–10, 241, 301–2, 330, 335–6, 338
  announces invasion of Poland, 32, 33
  apparent determination to engage with Britain, 140–1, 394
  blowing his own trumpet, 29
  empire-building as *Lebensraum* (living space), 21, 22–3, 26, 27, 34, 71–2, 290
  encourages vengeful sentiments against Britain, 87
  '*Führer*'s Hammer'/eagles (Luftwaffe's inspiration), 2–3, 7, 51
  and Göring, 12, 23, 31, 257, 258, 279–80, 299–303
  havering over British invasion, 41, 45–6, 81, 113, 146–7, 237, 243, 244, 246, 248
  induces Wehrmacht to breach international neutrality, 70–1
  inspiring "god-like trust"/ claiming divine support, 23, 27, 85–6, 95, 131, 197
  launches *Blitzkrieg*, 35
  meeting Mussolini, 56
  relishes collapse of Poland, 39
  rewards Luftwaffe commanders, 127
  tries to play Churchill, 44
  turns his attention to the Soviet Union, 290

Hitler, Adolf (*continued*)
  'Instructions for the Conduct of the Air and Sea War against England', 304
  invasion of France and the Low Countries, 56–7
  invasion of Poland, 30–1, 39
  invasion of the Soviet Union, 290
  'Last Appeal to Reason', 126–8, 130–1, 132, 133, 134
  preparing for war with Britain, 125–8, 140, 222–3
  'Principles for the Conduct of the War against the Enemy's Economy', 46
  recognizes Franco's Spanish government, 13
  'The War against England', 110–11
Holland
  see Netherlands
Holocaust, 301
Houwald, Wolf-Heinrich von, 136, 137
Hozzell, Paul-Werner, 33–4, 70, 167

I
Imperial Airways, 11
Italy, 123
  Rome, 237

J
Jeschonnek, Hans, 44, 71, 125, 126, 178
  Studieplan (Study Plan), 44–5

K
Kesselring, Albert, 127, 225, 234, 242, 244, 297, 300–1, 302–4, 305–6
  on *Adlertag* (Eagle Day), 157, 160
  on concentration camps, 321
  parroting Göring, 253

Kessler, Ulrich, 11–12, 23–4
  in London, 24
Klein, Willibald, 43, 69, 108–9
Knappe, Walter, 91–2, 185–6
Knoke, Heinz, 255, 305
Kutsche, Theodor, 66, 67

L
Lamberty, Rudolf, 263, 318
League of Nations, 21
Lehweß-Litzmann, Walter, 330, 336, 337
  on the attack on the Soviet Union, 290, 291–2, 293
  letter from his pilot brother, 14
  on the Norway invasion, 62, 68, 69, 70, 328, 329
  raids on London, 287–8
Low Countries/Benelux Countries
  German invasion (*Fall Gelb*/Case Yellow), 56–7, 59–60, 68, 70–7, 81–3, 93, 95, 103, 127, 138, 148, 223–4
Luftwaffe
  7. Air Division, 62, 74
  9. Air Division, 242–3
  abgeflogen (war neurosis/suicide rate), 213–14, 269–70, 338
  Adlerangriff (Eagle Attack), 174, 178, 181, 183, 190–1, 258
  Adlertag (Eagle Day/first day of Eagle Attack), 143–9, 157–60, 164, 249, 309
  and *Kanalkampf*, 137–8, 259
  loss rate, 164, 175, 238
  planning and anticipating, 141, 146, 147, 150, 156, 251
  second phase, 201–2, 203, 205–7, 209–10, 214, 218, 219
  switch to the Blitz, 224, 234, 253, 304

INDEX

aeromedical experiments, 331–3
Air Battle over the German Bight, 48
Air District Staff
  Command-5, 285
  Luftgaustab (logistics advance guard)
  England, 247, 339
  z.b.V-300, 244
airfield construction battalion, 93, 244
air intelligence service, 149, 150, 175, 234, 270–2
5th Department, 45, 149–50
Air Signals
  Regiment, 6, 50. 59, 74, 84–5, 150, 244
  3rd Department, 149–50
  Regiment-35, 218–19
anoxia (oxygen deprivation), 98, 219
armoured train unit, 145
Aufklärungsgruppe
  (Reconnaissance Group)
  F-123 Dornier Do-17, 104
  reconnaissance crews generally, 62–3, 243, 275, 309
bombers, 37, 73, 134, 138, 224, 259
Dornier Do-17, 14–15, 19, 73, 93, 157, 169, 174, 180, 181, 261–2, 278
Heinkel He-111, 19, 43, 64, 169, 180, 181, 227, 263, 299, 313, 319–20, 324
B-1, 14
bombing of Rotterdam, 74
invasion of Norway, 66, 69
invasion of the Low Countries, 73
numerous but unreliable, 14, 66–8

Junkers
Ju-52, 32
Ju-88 *Schnellbomber*
  (fast bomber), 169, 180–1, 185, 230, 232, 257, 263, 276, 283, 299
Central Weather Service Group, 157
Coastal Reconnaissance Units, 191
Condor Legion (*Legion Condor*), 13–14, 16, 20, 21, 39, 256–7
Construction, 243
  Company-3.XII, 36–7
culture of, 15
Erholungsheim
  (convalescent home), 213
Erprobungsgruppe-210
  (test wing), 251, 252
Fallschirmjäger (paratroopers), 6, 8, 103–4, 150, 203, 229, 302–3, 311–12
Belgian invasion, 76
Netherlands invasion, 74–6
Scandinavian invasion, 60, 62, 63, 64, 65
Flakkorps
Branch-776, 85
crews and batteries generally, 6, 54, 77–8, 86, 119, 120, 145, 228, 244, 281–2, 314
Heavy Flak Unit-492, 41
II (2nd), 243
Regiment-22, 286
Regiment-49
  6, Battery, 286, 288
Fliegerkorps (Air Corps)
II (2nd), 147, 242–3
Jafü-3, 285
VIII (8th), 98, 112, 242–3, 246
X (10th), 63, 64
'Führer's Hammer', 2–3
Guard Command, 119

Luftwaffe (*continued*)
  High Command, 256
  injuries, 254–6
  Italian Division, 242–3
  Jagdgeschwader (fighter wing),
      48, 76, 77–8, 134
    in the Blitz, 224, 232, 234
    C, 20, 169
    compared to British fighters,
        79–80, 151–2, 153, 176,
        177–8, 259
    compared to Stukas, 135, 225
    D, 20
    Daimler-Benz 601-A engine, 80
    E, 36, 153, 169
    E-1, 105
    E-3, 80
    E-4, 105
    Fighter Control/command,
        172, 323
    Gruppen generally, 253, 256
    I. *Gruppe* JG-21, 20
    III. *Gruppe* JG-52, 136, 137, 154
    Jagdbomber/'*Jabos*' (Bf-109
        fighter-bombers), 251–5,
        258–9
      Jaboangriff (Jabo attack),
          252, 257, 258
    Jagdflieger (fighter pilots), 4,
        206, 234, 252–3, 306
      Tag der Jagdflieger
          (Fighter Pilots' Day), 78
    JG-3, 177, 212
    JG-26
      2. *Staffel*, 281
    JG-51
      3. *Staffel*, 311
    JG-52, 253
    JG-53, 47
    JG-54, 49
    JG-132, 16
    Kanalkampf (Channel Battle),
        112, 135–8, 147, 173, 317
    Messerschmitt AG, 47
  Messerschmitt Bf-108/Me-108,
      57
  Messerschmitt Bf-109/Me-109,
      47, 63, 96–7, 116, 168,
      169, 234, 307, 311
  Messerschmitt Bf-110/Me-110
      Zerstörer (destroyer), 47,
      157, 224, 257
    in the Blitz, 224
    compared to Bf-109, 167–8,
        254
    compared to British fighters,
        80–1, 169–70, 176, 259
    Kanalkampf
        (Channel Battle), 129
    MG 17 machine-guns, 168
    MG FF cannons, 168
    over France, 79–80
    over Norway, 62–3
    pilots, 96
  over France, 77, 78, 80
  over Norway, 62–3
  over Poland, 36
  over Rotterdam, 75
  over Spain, 19
  pilots, 177–8, 185, 211, 313
  see also *Jagdbomber* above
  Kampfgeschwader
      (bomber wing), 96–7, 234
    I. *Gruppe*, 114, 165
    KG-1 'Hindenburg', 71, 227
    KG-2, 24, 93, 113, 143, 157,
        227, 269, 276
      9. *Staffel*, 278
      III. *Gruppe*, 120
    KG-4 'General Wever', 61,
        73, 180
    KG-26, 125
    KG-30 'Adler', 233, 242, 283
    KG-51, 156–7, 226, 235
    KG-53, 324
    KG-54, 66, 68, 205
    KG-55, 14, 84
      II. *Gruppe*, 50, 55, 94, 114

KG-76, 263
KG-77, 190
KG-100 'Wiking', 206
KG-606, 261, 263
Kanalkrankheit
   (Channel Sickness), 212
Koluft-16, 204
Luftflotte (Air Fleet) Command,
   104, 242, 243
Luftflotte-1, 35, 44
Luftflotte-2, 42, 46, 74, 110,
   125, 126, 229–30, 244,
   287, 297
Luftflotte-3, 110, 125. 126,
   244, 275, 284–5, 287
Luftflotte-4, 35
Luftflotte-5, 110, 125, 126,
   164, 284–5
Nachtjagdgeschwader
   (night-fighter wing), 128
night-fighters generally, 48,
   61, 94, 113, 129, 261–3,
   267–9, 287
NJG-1, 128
navigational systems
   Knickebein (Crooked Leg)
   42–48 MHz radio beam
   navigation and bombing
   system, 273–4
navigators, 91
X-Verfahren (X System), 274–6
pre- and early-war relations with
   RAF (and with England/
   Britain), 16–18, 23–4,
   29–30, 31, 43–4
prisoners, 209–10
propaganda, 49
recruitment and training, 50–5
Regiment-2, 50
Regiment-49, 120
Regiment-231, 134
Schwarm flying formation, 19–20
Schwarzer Donnerstag (Black
   Thursday), 163–4, 170

Sea Rescue Service, 106, 116,
   285, 310–11
   Arado Ar-196, 310
   Dornier Do-18, 310
   Dornier Do-24, 310
   Heinkel He-59, 309, 310, 311
Seeflieger (naval aviators),
   138, 205
Staffeln (squadrons)
   1. *Staffel*, 164–5
   8. *Staffel*, 155
   9. *Staffel*, 157
Station Command 16/XI,
   225–6
Sturzkampfgeschwader (dive-
   bomber wing), 25, 134
Ju-87 *Sturzkampfflugzeug*
   (Stuka dive-bomber), 37,
   78, 164, 187–8, 190,
   264, 285
   dive-bombing capacity, 18–19
   'Hardest Day', 164, 167
   over France, 77–8, 79, 82,
      83, 85
   over the Channel, 83, 107,
      112, 115–17, 135–8,
      154–5, 165–7
   pilots, 24, 33–4, 38–9, 50, 70,
      82, 93, 96, 114, 155–6,
      164–5, 313
   StG-1 III. *Gruppe*, 112
   testing, 26–7
Ju-88, 31, 42, 73
L/StG-77, 114
StG-2, 138
StG-77, 167
training, 95–8, 105, 207–9
transport ship *Bukarest*, 286
transport wing, 96–7
   column 9/VII, 94, 103
   Focke-Wulf-200, 62
   Junkers
      Ju-52, 26, 62, 63, 69, 77
      Ju-90, 62

Luftwaffe (*continued*)
   watchkeepers, 119, 145
   Zerstöreergeschwader
      (destroyer wing)
      Bf-110, 96, 168–9
      ZG-1, 168
      ZG-76, 47, 169
Luxembourg, 73

M
Mahlke, Helmut, 82, 93, 98–9, 154–6
Mikula, Valentin: *Stuka: Tatsachenbefricht* (Battle Report), 340
Milch, Erhard, 15–16, 17, 18, 29, 31, 127, 171, 178, 195, 233, 327
Mölders, Werner, 47, 94, 151–2, 206, 207, 256–7
Mork, Werner, 283, 284
Mussolini, Benito, 22–3, 54, 237

N
Netherlands/Holland/Dutch, 57, 71, 73–4, 101–2, 124–5, 148, 163, 233, 249, 303, 304–5
   Army Aviation Brigade (air force), 73–4
   Rotterdam, 74–6, 91, 100, 230, 338
   see also Low Countries
North Sea, 47, 49, 61, 223, 338
Norway, 59, 105, 124–5, 147, 163, 241–2, 249, 305, 328–30, 338
   Air Force, 68, 69
   Bodø, 63
   German invasion (*Unternehmen 'Weserübung Nord'/* Operation Weser Exercise North), 35, 46, 59–60, 62–72, 79, 125
   Narvik, 63, 64, 68, 69
      Battle of Narvik, 64–6, 69
      Operation Claymore, 328–9
   Oslo, 63, 64, 66, 68, 70
   Stavanger, 63, 64
   Trondheim, 63, 64, 66

O
Ostmann, Klaus, 155

P
Park, Keith, 234–5
Pervitin, 247
Poland/Polish, 41–2, 50, 79, 124, 159, 163, 193–4, 252, 322–3, 327, 331
   Air Force, 35–6, 40–1, 322
      PZL P.11 fighter, 36
   Danzig/Gdańsk, 29, 34
   German invasion (*Fall Weiss/* Case White), 27, 30–1, 33–7, 38–41, 44, 60, 61, 94, 103, 125, 323–4, 325–6
   Polish forces fighting with the Allies, 68
   Soviet invasion, 36
   Warsaw, 61, 100
      Ghetto Uprising, 321–2
Prisoners of War (PoWs), 245
   see also Britain: prisoners

R
Raeder, Erich, 125, 126
Rall, Günther, 5, 105, 106, 107, 134, 135–7, 141, 154, 177
Rascher, Sigmund, 331–2, 333
Rauch, Heinrich, 243–4, 297–8
Raynor, Kurt Gerhard, 151, 282, 291, 338
Rechtle, Otto-Wolfgang, 157
Red Cross/red crosses, 63, 308, 309–10, 311–12
Reuters, 236
Ribbentrop, Joachim von, 34, 237

INDEX

Richthofen, Manfred von, 17
Richthofen, Wolfram von,
    39–40, 112
Rieckhoff, Herbert, 12, 13, 42,
    44, 46, 106, 148, 168–9,
    275, 337
Rökker, Hans, 94, 95–6, 314
Romania, 111
Royal Air Force (RAF)
  airfields, 2, 156, 158, 234
  Balloon Command, 133, 163
  Bomber Command, 86, 130,
      279, 289, 342
    Armstrong Whitworth
        Whitley, 129
    Bristol Blenheims, 37, 78, 153,
        158, 259, 309
    Fairey Battles, 78
    "feeble", 54
    Handley Page Hampdens, 215
    raids over Germany, 6, 54, 55, 61,
        103, 117–18, 130–1, 145,
        215–18, 270, 281, 301–2
      Berlin, 215, 217, 218, 221–3,
          224–5, 280, 286, 302,
          313, 314
      Bremen, 286
      Emmendingen, 216
      Göttingen, 217
      Hamburg, 225, 280, 286
      Kiel, 102–3, 285, 286
      Mannheim, 120
      Munich, 273
      Ruhr Valley, 128
      Tessenow, Mecklenburg, 119
    'Striking Force', 128
    Vickers Wellington bombers,
        47, 48, 215
  Coastal Command, 158, 165,
      210, 279, 342
    Avro Ansons, 158
  'The Few', 342
  Fighter Command, 124, 135,
      148–9, 156, 158–9, 173–4,
      175–6, 213, 224, 323,
      337, 342
    Battle of Britain Day, 235–6
    Boulton Paul Defiant, 166, 174
    Browning machine-guns, 153–4
    comparative losses (and top-ups),
        111–12, 112–13, 158–9,
        163–4, 175–6, 202, 206,
        234, 235–6, 259, 289
    de Havilland/Rotol propellers, 152
    de Wilde incendiary bullets,
        153, 174
    Gloster Sea Gladiators, 66, 68
    'Greatest Day', 164
    Hawker Hurricane, 232, 234, 319
      compared with German planes,
          80–1, 168, 169, 253,
          295, 332
      Luftwaffe pilots' view, 151,
          152–3, 179, 295
      shot down, 112, 202, 209, 259,
          317–18
    and the Luftwaffe's Bf-110
        fighters, 79, 81, 202
    and the Luftwaffe's *Jagdflieger*/
        Bf-109 fighter bombers, 4,
        8, 19, 112, 152, 208, 234,
        254, 306, 313
    No. 11 Group, 234
    No. 13 Group, 164
    No. 43 Squadron, 317, 318
    No. 152 Squadron, 319, 320
    No. 213 Squadron, 187–8
    No. 249 Squadron, 308
    No. 257 Squadron, 209
    No. 264 Squadron, 174
    No. 266 Squadron, 173
    No. 617 Squadron, 313
    and radar, 18, 150, 170–1, 172
    Rolls-Royce Merlin engine, 80
    sea rescue service, 310
    Spitfires, 69, 112, 117, 158,
        163, 183, 185, 232,
        234, 319

Royal Air Force (*continued*)
  compared with German planes, 80–1, 168, 253, 295, 332
    fast attack, 115, 136
    Luftwaffe pilots' view, 151–4, 177, 283, 307, 309
    Mk-I, 152
    Mk-II, 332
    shot down, 137, 202, 209, 210, 259, 320
    'Spitfire Summer', 124, 153
    Supermarine Mk-Ia, 80
    'Vic' flying formation, 19
    'Finest Hour', 8, 321, 342, 343
    '*Luftpiraten*' (air pirates), 100, 223, 259
    see also under Luftwaffe

S
Saarland, 21
Salmond, Sir John, 11–12
Scandinavia
  see Denmark; Norway; Sweden
Scapa Flow, 55–6
Scheffel, Kurt, 24–5, 26, 27, 114, 115–17, 118–19, 164–7
Schmid, Josef 'Beppo', 45, 149, 176, 225
  Studie Blau (Study Blue), 45
Schörner, Ferdinand, 335–6
Schroedter, Rolf, 263–4, 264
Schulz, Karl Lothar, 74, 75, 76
Second World War
  see World War II
Simpson, John, 317–18
Soviet Union (USSR/Russia), 34, 36, 71, 287, 290, 291, 292, 300, 318, 331
  Battle of Stalingrad, 336
  Moscow Peace Treaty (pact with Finland), 55–6
  Operation Barbarossa, 287, 290–1, 291, 292

Red Army, 15
Spaatz, Carl A., 299–300
Spain/Spanish, 125, 252
  Biscay, 249
  Civil War, 12–13, 18, 20, 21, 27, 33, 136
  Nationalist
    government, 13, 20, 241
    military, 19
  Second Republic, 13
  Übung Rügen
    (Luftwaffe intervention), 13–14, 21, 39–40
  generally, 19–20, 35
  Guernica, 15
Sperrle, Hugo, 127, 234, 284–5
Stahl, Peter W., 31–2, 75, 180–1, 230–3, 276–7, 278, 283, 286–7
Steinhilper, Ulrich, 211–12, 309–10, 311
Steinhoff, Johannes, 5, 12, 37–8, 152, 176, 181, 194, 306, 308, 321–2, 334, 337–8
stock market, 123, 124, 290
Stoffregen, Erich, 230–1, 278, 286–7
Streib, Werner, 129
Student, Kurt, 62, 74, 236, 238, 301, 304
Sturmabteilung
  (SA/Storm Troopers), 324
Sweden, 59, 338
Stockholm, 132

T
Third Reich
  Air Ministry, 11, 14, 16, 23, 37, 96, 144, 175, 178, 299
  Reich Chancellery, 29
  Reichsarbeitsdienst
    (Reich Labour Service) Battalion-8-173, 219

Reichstag, 39, 126, 127, 132, 301, 314
'thousand-year Reich', 335
Trenchard, Hugh, 16–17
Trübenbach, Hanns, 253, 254, 258, 258–9
Tyezka, Karl, 64

U
Udet, Ernst, 178
Urbanowicz, Witold A., 322–3
US Strategic Air Forces (USSTAF), 299

V
Viek, Carl, 212–13
Vogel, Helena, 221–2, 224–5

W
Werra, Franz von, 195–6
World War I, 5, 16–17, 20, 85, 92–3, 94, 256–7, 290–1
Bewegungskrieg (mobile warfare), 35

Treaty of Versailles, 5, 39, 92–3
ban on German military aviation, 5, 11
as unfinished business, 32–5, 38, 40, 42, 178
World War II, 3, 6, 17, 19, 25, 183, 305
begins, 32, 41–2
Blitzkrieg (lightning war), 18, 27, 33, 35, 39, 60, 73, 92, 147, 148, 159, 337
Blumenkriege (flower wars), 27
post-war, 306, 333–4
Sitzkrieg (sitting war)/Phoney War/*drôle de guerre* (strange war), 44, 47, 48, 49, 50, 51, 54, 55, 56, 57, 77, 153

Z
Zorner, Paul, 48, 97–8, 129
Zürcher Illustrierte (Swiss magazine), 132–3